ECOLOGICAL IMBALANCE OF THE GANGA RIVER SYSTEM:
ITS IMPACT ON AQUACULTURE

Parmila Ray

2024

Daya Publications House®

A Division of

Astral International Pvt. Ltd.

New Delhi – 110 002

First Published, 1998

Reprinted, 2024

ISBN: 978-81-7035-195-5 (HB)

Published by : **Daya Publishing House®**
A Division of
Astral International Pvt. Ltd.
– ISO 9001:2015 Certified Company –
4736/23, Ansari Road, Darya Ganj
New Delhi-110 002
Ph. 011-43549197, 23278134
E-mail: info@astralint.com
Website: www.astralint.com

The Time has Come

When

Man Can no longer Continue

Using

The Land, Sea and Air

as his

TRASH BASKET

We must find ways to Cycle

WASTES

Into

ECONOMY

(National Academy of Sciences Report)

1966

ACKNOWLEDGEMENTS

While writing my previous book "Aquaculture in Sundarban Delta — its perspective", I was envisaged on several occasions on the importance and role of the Ganga river system that has contributed substantially towards formation of the Sundarban Delta. This has induced me to find out the present status and fate of this Ganga river system. It did not take a tangible form until I discussed my object in details with my friend and colleague Dr. P.R. Sen, Principal Investigator, CIFA (ICAR) who not only welcomed my ideas but also encouraged me to reveal valuable informations about this river system, as it might be a valuable contribution of the time and will be useful not only to limnologists but also to various concerned states for their developmental work. Dr. Sen has always been cooperative in periodic discussions and offering valuable suggestions. My greatest obligations has been incurred to him.

To Dr. Parmeshwaran Ayyar former Director, Central Inland Capture Fisheries Research Institute (ICAR), Barrackpore, who took interest in completing the document and generously helped me by providing various reports and documents for my use and which I gratefully acknowledge.

My special thanks are due to Dr. K.L. Sehgal former Director, National Research Centre on Cold Water Fisheries (ICAR) who felt moral obligations to see that this valuable document is published, to Mr. B.B. Ghosh, former Principal Investigator and to Dr. D.K. De, Principal Investigator, Central Inland Capture Fisheries Research Institute (ICAR), Barrackpore, for their valuable constructive criticism, to Mrs. Anjali De, Librarian CICFRI, Barrackpore for rendering various help and to Mr. Pramtosh Ghosh, Secretary, Fish Producers Association, West Bengal, for his encouragement.

A word of appreciation is also due to Director, Department of Fisheries, Government of West Bengal for providing the photograph of "Sukharia Sabuj Deep" with required information, Dr. R.L. Singh, author "India, A Regional Geography", and editors of different daily news papers like, 'The Statesman', 'Bartaman', 'Ananda Bazar Patrika', 'Aajkal' and the weekly magazine, 'Ananda Mela' for using their published data and photographs. Thanks are also due to my artist Shri Bhaskar Ghatak who took pains in drawing some valuable plates for the document. Finally, to my dear nephews, messers, Sujit Roy and Shouvik Roy whom I owe the greatest for their constant inspiration throughout the preparation of this manuscript.

PREFACE

It is at a time when every year during monsoon the daily newspapers cover the heading of prolonged devastating floods caused in different major rivers and their tributaries in the north and peninsular India causing inundation to large areas of the state. Simultaneously, the opposite picture like drought/late monsoon causing scarcity of water for normal use is also not uncommon. Added to it, high abstraction of water from the rivers or streams through perennial canals for use in agricultural purposes is also reported. Utilization of large volume of water by growing industries and cities who in turn dispose untreated complex waste water indiscriminately in the river or streams round the year, construction of dams and barrages as a measure of flood control, thereby allow very little flow downstream, defective fish pass which prevents fish migration, high silting in rivers or streams loose their water holding capacity and during summer the flow becomes alarmingly low and thus upsets the dilution ratio to a great extent. All such factors collectively is heading towards ecological imbalance day by day and turning from bad to worse, resulting in appreciable loss of biomass including prized fishes. This shifting of critical balance of the ecosystem and their impact on aquaculture as a whole have induced the author to trace out facts which upset the ecological balance of the Ganga River system at this moment that it is facing, inspite of its recuperative capacity.

The Ganga is the life line for few hundred millions of people. It is regarded as India's most important river mainly because a substantial part of the river and its tributaries pass through the hospitable terrains of the plain. After mendering a distance of about 2290 km. through the plains of Uttar Pradesh, Bihar and West Bengal, the river joins the Bay of Bengal. Its basin however drains through eight states like Himachal Pradesh, Punjab, Haryana, Union territory of Delhi, Uttar Pradesh, Rajasthan, Madhya Pradesh, Bihar and West Bengal.

The Ganga rises in the Garhwal Himalaya (30° 55' N : 79°7' E) under the name Bhagirathi about 4100 m above the sea level. It is below the confluence of Alakananda and Devaprayag that the United stream is generally known as river Ganga. Though not a clear cut physical unit, the middle Ganga plain (24°30' N - 27°50' N : 81° 47 '- 87°50' E) is a large physical area (144, 409 km^2). It covers easternly Uttar Pradesh and Bihar plains in their entirety, lying on either side of the Ganga and the Ghagra with Himalayan and peninsular ramparts on the north and south respectively. The lower Ganga region (21°25' - 26°50' N and 86°30' - 89°58' E) known as Bhagirathi embraces the area of about 64, 468 $Km.^2$

This river once held a copious biological wealth characterised by its rich faunistic diversity. Apart from being the original abode of the most prized carp species of the sub-continent viz. *Catla catla, Labeo rohita, Cirrhinas mrigala* and *Labeo calbasu,* the river sustained fisheries of large cat fishes, mahaseers, hilsa and other miscellaneous fishes. It was a major source of riverine spawn which meet the carp seed requirements of the culture sector to the tune of about 30 per cent.

With the growing population, the Ganga river basin representing more than a quarter of the total graphic area of the country has witnessed an expeditious pace of irrigation, industrialisation and urbanisation during the past few decades leaving trails of adverse impact on the riverine habitat and its fisheries. Apart from a steady decline of the fish yield there has been an alarming downswing in the population of prized major carps and hilsa fisheries in favour of cat fishes and carp minnows.

The output of major carp spawn has also registered a declining trend in the spawn collection centres all along the river course. Though over exploitation of fish and the changing socio-economic climate have played, a definite role in the decline of fish yield, the over-whelming influence of environmental stress together with the human activities in various forms on the habitat is largely responsible for the qualitative and quantitative deterioration of the Ganga fisheries.

Nature induced factors like cyclone, flood, geographical annihilation etc. together with increasing human settlements and activities play a tremendous role in altering the configuration of the river system. All these have resulted in raising of river bed appreciably by accumulation of recurring annual silt precipitation, reduction in water holding capacity, increase shape factor, slope of the bed and decrease velocity of water which had helped further silt accumulation in the absence of dredging and causing change in the course of the river, flood etc. Abstraction of water in appreciable volume for various uses like agriculture, industrial processing and urban development together with failure or late monsoon complicates the situation further.

Dams and barrages also create big sand bars above the barrage and thus reduce the flow of water downstream. All these are experienced in the Ganga river system in which biomass are surviving. Some examples will clarify the position how the river is fast heading towards ecological disaster.

The forest area in the Gangetic basin is under severe stress and strain due to high rate of deforestation under the present system of forest environment. This has resulted in severe silting in the river basin and the river is fast loosing its water holding capacity. The slope is also very much affected. For example, the slope varies from 1 in 67 (Hilly region of Rishikesh) to 1 in 4100 upto Allahabad (800 km.) From Allahabad to Farakka, it is 1 in 13,800 (1005 km.). From Farakka to Nawabganj (230 km.), it is 1 in 23,000 and further down to deltaic tidal plain (240 km.) it is 1 in 24,000. All these show that water velocity is also affected particularly during non-monsoon months. This is in sharp contrast with the earlier records around 1869 between Allahabad and Rajmahal when depth ranged 3-4.57 m during summer with water velocity 1.34 m/sec. against 3.12 - 3.57 m during other periods. Size and number of deep pools are now appreciably reduced.

Further, though not infrequently when the Ganga itself remains in spate and thwarts the incoming water from its numerous tributaries in the region, large and small, contributing substantially during monsoon when tributaries gather run offs from hilly terrain and large areas get flooded. Considerable silting is effected and the outfalls of nearby smaller tributaries get locked up. For instance, the Rapti, and the Ghagra in Uttar Pradesh, the Kosi and Son in Bihar and the Ajoy and the Damodar in West Bengal and the Ganga itself are the examples and facilitate shifting their courses resulting in remnants of their former beds in the form of ox-bow lakes, mender loops, dead arms, chars, tals and ruins of settlements etc. Such water areas (disconnected pools) both seasonal and perennial are known to be potentially rich source of fish supply. Except Bihar, neither Uttar Pradesh nor West Bengal has any clear cut information on the number of Ox-bow lakes, chars and mans of varying sizes.

In the estuarine stretch of Bhagirathi - Hooghly, the river bed at the source is about 7.3 metre higher than the main Ganga bed. This happened sometime between 1795 and 1805 due to sudden tectonic movement resulting to a dangerous geographical annihilation at Rajmahal and its neighbouring areas, the Ganga changed its course of flow shifting the major floods to the Padma. Further, diversion of the Damodar, down its present channel with its outfall below Uluberia not only sealed the fate of the Jamuna in West Bengal but also deprived the Hooghly river. The major flow it receives is only during monsoon when the water brings heavy of silt load. Accumulation of high rate of silt that takes place daily in the Hooghly estuary during monsoon is 17.25 lakh f^3 which works out to 20 crore f^3 in about four months only and out of which three fourth settle and that cannot move with the water, as the Damodar with a greater velocity (1ft/mile) joins the Hooghly (flowing 3"/mile) at right angle and arrest the flow of the Hooghly. The silt that settles are again brought back upstream alongwith the spring tide, but cannot be pushed to the sea due to low water velocity (Velocity of water, 7 ft/sec and above only can remove the deposited silt). Thus the estuary suffers from low water flow (except monsoon) and silting occurs almost round the year. The river is subjected to indiscriminate use of water for agriculture, industrial and urban consumption. In a balance sheet of annual consumption of freshwater and corresponding waste water released (from industries and big cities) show that average water consumption is around 80 per cent and the lion share is taken for irrigation through major canal projects in the Ganga basin. The waste water released from various industries and cities contain organic, inorganic, metallic and pesticidal materials. From one estimate, it is stated that seasonwise ratio of waste to river water per month clearly stated that the Ganga in Uttar Pradesh suffers throughout the year (dilution ratio 1:1) during monsoon and post-monsoon period, while in winter and summer the waste exceeds the flow; West Bengal suffers during post-monsoon (1: 8), winter (1 : 9) and summer (1 : 7), while Bihar suffers only during winter (1 : 7) and summer (1 : 5). Thus it is obvious that solution to pollution by dilution may not hold good in country like India where considerable share is drawn for agricultural purposes.

Although easily biodegradable organic wastes have tremendous fertility effect, their presence in excess (untreated one) has a considerable demand on the oxygen budget, although with tremendous resilience the river losses its recuperative capacity in some stretches. The others have appreciable toxic impact of metals, pesticides and other non-biodegradable agents on the fish physiology through their bio-accumulation in various organs of fish and other aquatic organisms including fish food. Nevertheless the problem is more acute in the estuary on account of to and fro movement of water containing toxicants due to tidal influence. All these will have direct influence on their mobility, survival, growth, maturation and reproduction and ultimately on fish production.

Intersection of many perennial streams and rivers by dams, barrages, weirs etc. restricts the normal movement and migration of fish in the absence of any scientific construction of fish pass. Farakka barrage is one of them.

Fishes congregate below the barrage during monsoon but cannot migrate upstream. Further, due to its presence large sand bars appear above the barrage and thus prevents, free flow of water downstream (Farakka and Durgapur barrages). Similar results are evident in the upstream barrages both in the main river and its tributaries. Moreover, during low flow period of drought, fishes take shelter in pools from where they are indiscriminantly fished leaving very few broad fishes for reproduction.

The resultant effect of such stress and strain certainly have some effect in the composition and production of fish and other bio-mass. Progressive changes in the fish composition and production have been noticed not only in the freshwater zone but also in the estuary. In the freshwater zone of the Ganga and the Jamuna, catfishes started dominating in place of major carps perhaps due to the absence of adequate large deep pools where they could safely harbour. Due to inadequate living space total production of carps have gone down. According to an estimate, in 1987 it has been noticed that in the middle and lower stretches of the Ganga, the average total yield has declined from 50.3 kg/ha/yr from pre 1961 to 22.0 kg/ha/yr in the post 1972. The average yield rate of major carp dropped from 13.3 kg/ha/yr to only 4.6 kg/ha/yr. Hilsa was abundant till 1972 at all the centres in the lower Ganga above Farakka. At Allahabad the fluctuation was between 7.90 and 32.50 t and at Buxar, 7.6 and 112.7 t during 1958-72. During post Farakka period (1973 and 1984) hilsa landing ranged from 0.13 to 12.2 t and nil to 71.9 t at the above mentioned centres. Further, observation showed record drop in landing of hilsa i.e. 19 t to 0.9 t from Allahabad and 36 t to 0.34 t at Buxar and 4 t to 0.82 t at Bhagalpur during pre and post 1972 periods.

Depth of water being low due to heavy silt precipitation (Physiographical changes), the qualitative and quantitative transfer in the effort pattern for fishing become inevitable. This is how low energy fishing devices under stressed condition have also changed. Thus the crafts and tackles involve still lower energy in operation with sufficient better output.

In the estuary, the highly industrialised zone suffers from fall in production compared to other zones during pre-Farakka period; while during post-Farakka period due to change in salinity on account of shifting of gradient zone and marine zone into gradient zone, the general habital of the biological population abruptly changed. Some of the euryhaline species show a decline trend in the upper and gradient zone although no significant changes were recorded in the total landings. Nevertheless, Farakka barrage has become an important point of indiscriminate fishing centre for operation of hilsa during monsoon at the barrage before their spawning. This also has resulted hilsa a threatened species in the upper zone.

It is a common phenomenon that each year during autumn the riverine catch consists of a sizable percentage of major carp-fingerlings although it lasts for a short duration. It is revealed that their main sources are creeks, shallow pools, rivulets, inundated fields etc. which are resulted from the receding of the flooded river during autumn months and change in the course of main river and tributaries like Ghagra, Son, Kosi etc. Rivers, during low water level, contribute substantial juvenile fishery. During 1952 from a stretch of 208 km (Mughalsarai to Patna) a total of 3,05,512 specimens of all three major carps *i.e.* mrigal, catla and rohu, weighing 12.7 t was estimated to have been captured in about a fortnight. In 1953, 3,28,408 specimens were captured in 12 days. Further in Kosi-dhar, mass killing of fingerlings and juvenile carps were estimated to be more than 3.7 t/day on some days. In such low lying areas, the practice of pen-culture appeared to be profitable.

In the estuary, however, adequate data on capture of such early juveniles are not available. However, wanton destruction of young ones of some commercially important fishes are captured in the upper stretch of Calcutta.

Available information reveal that tributaries like Son and Kosi meeting the Ganga in the middle and lower stretches yield large quantities of fish seed. According to a Government of India report published in 1966, which revealed an average of about 2010 million spawn from Bihar and 1200 million from West Bengal were produced. Production from West Bengal appears to have very much declined with an average of about 530 million during 1971-75. The best collection in recent past was recorded in 1973 (1375 million).

In the estuary, prior to 1950, the spawning ground of hilsa was restricted upto Palta which subsequently extended towards Medgachi (Freshwater zone). Later on in seventies and after, the area extended between Khusiogoli and Uchipur. From studies it reveals that breeding occurs round the year with a peak during post-monsoon period, and the second one occurs half way through the winter season.

The potamological aspects of the Ganga river system including the estuary in which fishes and other biological organisms are surviving under stressed condition are highlighted in the monograph. Although aquatic biomass are surviving, a progressive decline in fish population and fish food organism in presence of toxic materials in water and soil under low water level is evident both in freshwater and estuarine condition. Annual destruction of spawns and juvenile fishes in a large scale and how they can be protected, are discussed. Since the vast dredging in the entire river system is not possible due to economic reasons (except where it is unavoidable) alternative measures suggested may be experimented. The fish-pass or ladder may be provided scientifically so that fish may ascend and migrate freely round the year. This may augment the fish production and maintain a balance of fish population in the entire Gangetic system. Finally, the various aspects discussed in this monograph are as follows.

High rate of deforestation, recurring annual silt precipitation, high value of shape factor reduced water holding capacity of the river and its tributaries. Low water flow and its velocity, shifting of river course and its tributaries resulting in formation of ox-bow lakes, mender loops, dead arms, chors, tals and ruins of settlements *etc.* (disconnected seasonal and perennial pools), abstruction of water for agriculture, industrial and urban use, nature of waste water disposed, pollution load, their ratio with unpolluted water in different States, changes in physicochemical and biological parameters of water and soil, fishing gear, effort and fish production, progressive changes in the fish size their composition, fishery in the shallow inundated areas (ox-bow lakes, chars, mans *etc.*) status of juvenile fishery in the total catch, fish seed resources—its future, breeding grounds, impact of dam and barrages on fishery potential etc, and overall stressed environment and the aquatic life and finally conservation and management of the river system under existing condition etc.

The seemingly vast fishery potentials of the Ganga river system, the largest river system in the country called for a thorough scientific appraisal of the magnitude of its resources and evolving of suitable measures for rational exploitation of the same are highlighted. Such dissemination of information will be of immense help to researchers to acquaint them with complicated problems of the ecosystem and their interpretation and for the planners responsible for development of fisheries in the Ganga river system.

Once the silt load is adequately controlled, disconnected shallow pools (ox-bow lakes, mans, chars etc.) formed as a result of shifting of river course are scientifically utilized for pen-culture for producing juvenile (immatured fish) fishes which can be released at various centres in the Ganga river system (not in the present condition), the pen-culture in running water of the parennial canal water used for irrigation and finally utilize the treated waste water material as fertilizer, the river system will not only prove its potential resources by producing fish abundantly but also may once again come back as the life line for millions of people as was in the past.

—P. RAY

CONTENTS

	Page
Acknowledgements	v
Preface	vii
1. Introduction	1
2. Physiography and Hydrology	4
3. Tributaries and Their Role	17
4. Major Canal Projects in the Ganga Basin	22
5. Irrigation Projects and Abstration of Water	26
6. Deforestation Vis-a-Vis Soil Erosion	29
7. Sediment Load and Nature of Soil	30
8. Remnants of Riverbeds Due to Shifting of Its Course	31
9. Industrial Wastes — Their Pollution Load Disposed into the Ganga	34
10. Ecological Imbalance of the Ganga	42
11. Impact on Biota Around the Outfall and the Main River—Their Pollution Status	51
12. Limnological Studies	69
I. Physico-chemical Characteristics	
II. Biological Characteristics	
13. Fish and Fisheries	102
I. Commercial Catch	
II. Status of Juvenile Fishery in Commercial Stock	
III. Fishseed Resources	
IV. Prospects of Beel/Maan Fishery	
V. Fish Culture in Fluviatile waters	
VI. Fishing Gear and Their Mode of Operation	
14. Tributaries—Their Pollution Problems and Impact on Biomass	137
15. Overall Stressed Environment and Biotic Life	155
16. Environmental Changes Vis-a-Vis Riverine Ecology	172
17. The Ganga and the Bhagirathi - Hooghly—Their Past and Present	174
18. Conservation and Management of the River System	186
19. Suggestions for Future Investigation	192
20. A Word on "GAP"- A Scientific Study	194
Appendices	198
References	227
Author Index	239

1

Introduction

The Ganga originates from the confluence of Bhagirathi and Alakananda in the Himalaya. It is a snowfed perennial river. After traversing a distance of about 250 km. in the Himalayas it enters the plains at Haridwar and after mendering a distance of about 2290 km. through the plains of Uttar Pradesh, Bihar and West Bengal joins the Bay of Bengal (Fig. 1). Total drainage basin of the Ganga extends over a million sq. km. The water level fluctuates widely due to influence of periodic character of south-west and north-east monsoon. Added to it a number of tributaries contribute substantially to the Ganga in different states (Fig. 1). It is estimated that the surface flow of water is around 4,90,000 mm^3. The high precipitation brings enormous quantity of silt and alluvial soil alongwith the water. The river has tremendous potentiality and its water resources are being largely used for agriculture through irrigation, fish culture, power generation, industry, domestic consumption, navigation etc. The river was once potentially rich in fisheries. The headwaters of the Ganga system in the high altitudes are snow trouts which are commonly available, *Schizothorax* spp. *Schizothoraichthys* spp. *semiplotus* sp. and certain catfishes have been commonly encountered. Mahaseer (*Tor tor* and *T. putitora*) upto a size of 1067 m and Assamese, katli, *Acrossocheilus hexagonolepis* and *Bagarius bagarius* form the main food fishes. In the plains of 2230 km. stretch between Haridwar (U. P.) and Nabadwip (W.B.), the dominant species among major carps were *cirrhinus mrigala* (Ham), *Catla catla* (Ham), *Labeo rohita* (Ham) and *Labeo Calbasa* (Ham), followed by *Labeo Pangusia* (Ham), *Labeo bata* (Ham). *Labeo dero* (Ham), *Cirrhinus reba* (Ham), *Hilsa ilisha* (Ham), *Setipinna phasa* (Ham), *Gadusia chapra* (Ham), among clupids and *Macrobrachium malcolmsonii* (H. milne Edwards), *M. camarrei* (H. milne Edwards), as freshwater prawn. Among catfishes, the dominant species were *Mystus aor* (Ham), *M. Seenghala* (Sykes), *S. Silondia* (Ham),Wallago attu (schnsider), *pangasius pangasius* (Ham), *Bagarius bagarius* (Ham), *Rita rita* (Ham) followed by *Clupisoma garua* (Ham), *Eutropiichthys vacha* (Ham), *Aila coila* (Ham), *Ompak bimaculatus* (Bloch), *Ompok pabda* (Ham), among other group of fishes and *Notopterus notopterus* (pallas) and *Notopterus chitala* (Ham) among featherbacks.

Towards the end of 18th century around Calcutta it was freshwater zone and except Hilsa all other freshwater species were available in plenty. It was around 1943 considerable changes in the fish composition were noticed and in 1953 mainly brackishwater fishery extended upto Barrackpore. Subsequently the deltaic tidal plain which covered Nabadwip to Diamond Harbour, besides its usual flow was influenced by tidal waters. In sixties the catch was fairly high. The composition in the upper stretch was *H. ilisha*, freshwater prawn spp., *S. Phesa, Harpodon nehereus* etc. while in the lower region besides the above, some more forms viz. *Polynemus indica, tachysurus jella, Trechinrus* spp., *pama pama* etc. were dominating in commercial scale. Miscellaneous group formed on important rank in the catch and sometime their landing exceeded the Hilsa fishery.

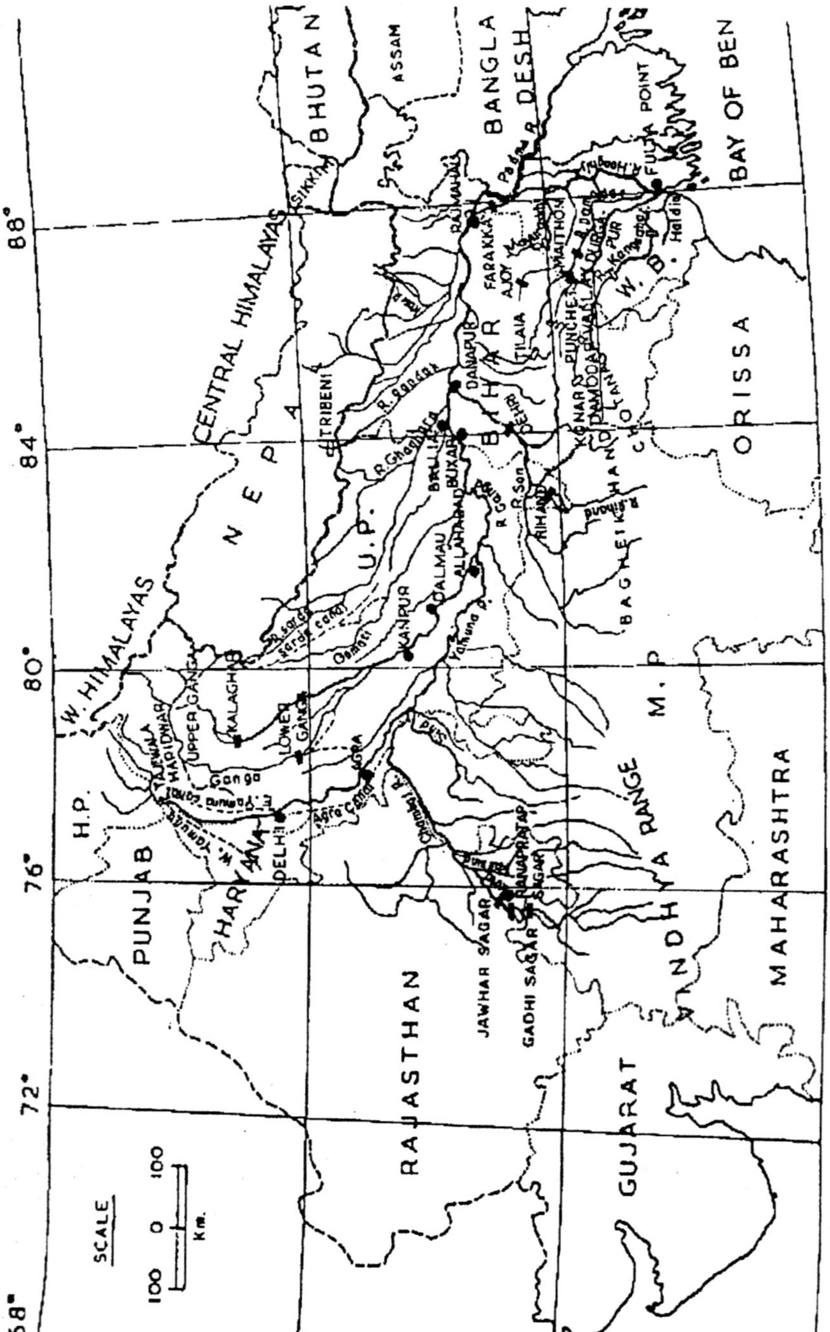

Fig 1 : The Ganga River System

With the coming up of numerous irrigation projects on the Ganga and its tributaries after independence have degraded the ecosystem by removal of water in an alarming fashion for irrigational use, industrial and domestic consumption etc. The later intern contributes various polluting materials in the river. The cumulative effect of such wastes create ecological disaster in the ecosystem. Further, due to rapid deforestation, the hydraulic characteristics and silt load in the most of the peninsular rivers have undergone rapid change over the years resulting in increased frequency of flash floods in these rivers and the main Ganga on one hand and much reduced flow during the dry season on the other. The fishery in the main river has thus considerably dwindled in recent years while in the Bhagirathi-Hooghly, with the additional release of water from Farakka further deterioration of the eco-system is checked to some extent with regard to flow only. Thus the ill effects of nature-induced-problems are further aggravated by the activities of human beings, their cumulative effects are thus heading towards ecological disaster in the entire Gangetic ecosystem. The present document provides information on some important aspects like

(a) Nature induced problems

(b) Human activities

(c) Ecological disaster

(d) Decline of fishery potential

(e) Stressed environment and biotic life

(f) Conservation and management of the Ganga river system.

2

Physiography and Hydrology

The Ganga and its twin, the Yamuna have their origin in the Garhwal portion of the Western Himalaya, which has an area of some 38,000 Km^2. The Great Himalayan range here comprises numerous snow peaks rising above 7000 meters. Here lie several peaks like Nandadevi (7,817 m), Kamet (7,756 m), Badrinath (7,138 m), Trisul (7,120 m), Satopanth (7,084 m) and Dunagiri (7,066 m), all exceeding 7000 m besides many other snow peaks rising to a height ranging from 6,000 m to 7000 m. Each of these high peaks of great Himalaya is surrounded by considerable snow-fields (known as himals) which send down large glaciers. It is from the snouts of these glaciers that the principal head stream of the Ganga and the Jamuna originated. Being supported by large snow fields and glaciers, all the Himalayan tributaries of the Ganga river are characterized by all regulated flow and assured supply of crystal clear melt-water throughout the year, though in the cold weather period the flow is considerably reduced. Further south of Middle Himalaya ranges is the Siwalik range which rises to a height of 900 m to 1000 m usually formed of younger sedimentary deposits comprising of sand stores and conglomerates. Behind the Siwalik range are the structural depressions (duns) which are places of human settlements, like that of Dehradun.

The Central Himalaya, which falls in Nepal is the home of several principal tributaries of the Ganga and a large share of water in the Gangetic basin comes from these snow fed streams like the Ghaghara and the Gandak (Fig. 1). Through Nepal also flows the major stream of the Kosi, which has one of its heads originating far across the Great Himalaya in the plateau of Tibet. In Nepal lie the Great Himalaya peaks which are mainly composed of metamorphosed limestones.

The Central Indian Highland within the Ganga basin area is mainly drained by numerous tributaries of the Yamuna, the principal ones being the Chambal, the Betwa and the Ken. The Son and the Damodar are the two other major streams. Several other medium and minor streams (some non-perennial) originate from the Central Indian Highlands (Fig. 1). The water flow in all these peninsular rivers is largely regulated by the protective cover provided by extensive forest belts. But due to rapid deforestation all through the Central Indian Highland, the hydraulic characteristics and the silt load in most of the peninsular rivers have undergone rapid change over the years, resulting in increased frequency of flash floods in these rivers on one hand and much reduced flow during the dry seasons on the other.

The Ganga, a snowfed perennial river, is formed by the confluence of two small rivers the Bhagirathi originating in the Gangotri glacier (3,129 m asl) and the Alaknanda from Saptal glacier (30^0 55' n, $79^0$7' E). After gorging 250 km in the Himalayas it dedouches in the plains at Haridwar and mendering over a distance of about 2,290 km in the plains to join the Bay of Bengal. Almost the whole of Northern India, North and East of Deccan plateau falls within the Ganga drainage basin. The Total

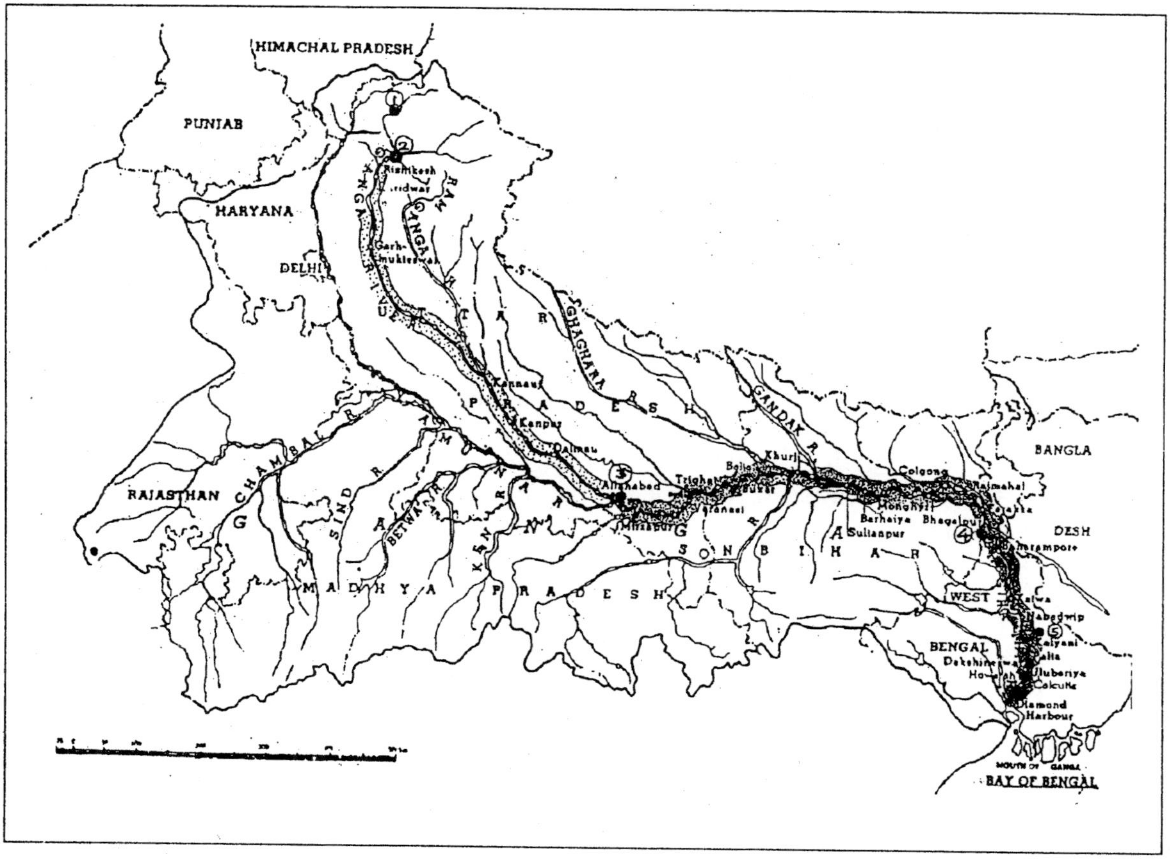

Centres	*Section*	*Length (Km)*	*Nature*	*Slope*	*Mean annual rate of flow (m³/sec)*	*Index*
1-2	Source-Rishikesh	250	Mountainous	1 in 67	850	
2-3	Rishikesh-Allababad	800	Upper plain	1 in 4100	850-1700	
3-4	Allababad-Farakka	1005	Middle plain	1 in 13800	4000-10200	
4-5	Farakka-Nabadwip	230	Deltaic non-tidal plain	1 in 23000	1000-1300	
5-6	Nadadwip-Outfall	240	Deltaic tidal plain	1 in 24000	Variable due to tides	

Fig. 2 : Physiographical and Limnological Conditions of the Ganga

drainage basin of the Ganga extends over a million sq. km. It is estimated that the basin's total surface flow of water is around 4,90,000 m^3 and that not more than a fraction of this can be used impairing ecological growth.

Due to much higher intensity of rainfall in the Himalayan region more than 60 per cent of the water flowing into the Gangetic basin comes from the Himalayan streams joining the Ganga from the North. The peninsular streams together contribute hardly 40 per cent of water through the catchment area of the peninsular streams extending well over 60 per cent of the entire Ganga basin. (861,404 sq. km, Das Gupta, 1984)

Hydrologically, the Himalayan rivers are of great importance compared to the peninsular streams of the Ganga basin. The weather in the Ganga basin is characterised by a distinct wet season receiving more than 70 per cent of the total annual rainfall (800-2500 mm; 50 per cent of the rain is in the form of snow in the higher reaches - 3000 m and above) during the period of South West monsoon (June-September), a dry spell during summer (March-May) and a short winter (December-February). In the greater part of the basin the rainy season spreads over 3 months—July, August and September and 70 to 85 per cent of the total annual rainfall occurs during this period. In the eastern part of the basin (West Bengal and Bihar) the wet season is longer, usually starting in June and continuing till the end of September or early October. (Das Gupta, 1984).

In the Gangetic plains, westwards of the Ganga the daily maximum temperature in May rises to 40^{0}C and above. During this month, the mean daily maximum air temperature shoots up as high as 42.1^{0}C at Allahabad followed by Agra (41.8^{0}C), Hissar (41.6^{0}C), Gaya (41.3^{0}C), Lucknow (41.20^{0}C) and Bareilly (40.5^{0}C). Usually, May is the hottest month in greater part of the basin except in the lower Bengal. Under such high temperature conditions prevailing over the basin for greater part of the years, a large amount of water is lost through evaporation. In fact in lower Bengal and the Himalaya, in the remaining part of the basin the total annual evaporation is much higher than the total annual rainfall received (Table 1), Das Gupta (1984).

Table 1 : Rainfall and evaporation at Selected Stations in the Ganga basin

State	*Centres*	*Mean Annual rainfall (mm) 1931-60*	*Mean Annual Evaporation (mm) 1959-75*	*Difference between Annual rainfall & Annual evaporation (mm)*	*Surplus Water in wet months (July-Sept) (mm)*	*Rainfall (Jul-Sept) expressed as percentage of annual rainfall*
	Dehra Dun	2313.7	1289.2	+ 1024.5	1548.8	78.3
	Bareilly	1068.3	1908.7	– 840.4	384.2	76.2
Uttar Pradesh	Gonda	1294.0	1803.5	– 509.5	620.2	75.4
	Agra	781.6	2232.2	– 1450.6	164.8	80.8
	Allahabad	1027.3	1843.6	– 861.3	407.1	78.2
Haryana	Hissar	389.9	2579.1	– 2189.2	Nil	81.5
Delhi	Safdarjang	714.2	2223.8	– 1409.1	41.3	74.7
Rajasthan	Kota	841.6	2591.4	– 1749.9	223.1	84.2
Bihar	Gaya	1097.3	2220.1	– 1122.8	408.9	77.0
	Patna	1109.6	1886.1	– 776.5	385.5	73.5
West Bengal	Asansol	1391.8	1456.1	– 62.2	676.1	65.7
	Dum Dum	1575.1	1401.9	+ 173.2	685.9	56.9

Source : Indian Meteorological Department (In Das Gupta, 1984).

Since greater amount of rainfall (70-85 per cent of the annual total) occurs in three months (July-September) only in most of the areas within the basin, it is during this period that the water available from the rainfall is usually more than what is lost through evaporation (30 per cent) and seepage (20 per cent). This is how the total amount of surface flow of water in the Ganga basin eventually comes to around 493.8 billion cubic metres of water. It has been estimated that some 293.8 billion cubic metres of water are lost through evaporation while another 212.8 billion cubic metres go down into the soil, as ground water recharge every year in the Ganga basin.

After a detailed investigation some interesting picture on the high fluctuation of water in the Ganga is available. At Rishikesh, the Ganga delivers 27 mm^3 of clean water per annum on an average. A large quantity of water is taken out by the upper Gangetic network off the headworks at Hardwar for irrigating a major portion of Ganga-Jamuna doab. This has reduced the main flow to 15 mm^3/ annum at Balawali. As it flows further downstream a few minor tributaries join adding to the volume of water which is now loaded with enough sediments. At Garhmukteshwar, the flow attains a greater volume (22 mm^3) in a year. About 240 km downstream of Haridwar till Narora is reached the annual flow rises further. Here another large irrigation network (Lower Ganga canal) takes off at a barrage which spans the 1164 m wide bed of the river leaving again a much reduced flow downstream. Ramganga, joining at Kanauj, annually adding substantially some 15.26 mm^3 of water to the main flow. At Allahabad, the Jamuna (76 mm^3) joins from the right. From Allahabad downstream the river receives several major tributaries at more frequent intervals, the Tons and the Son from the right and the Gomati, the Ghagra, the Gandak, the Burhi-Gandak and the Kosi from the left till the Ganga reaches the head of its delta at Farakka below Rajmahal. (Fig. 1)

The Bhagirathi originating from the Ganga at Sumati of Murshidabad district flowing down joins the sea. Below Nabadwip it is known as Hooghly and prior to Farakka barrage was influenced by tidal waters. Tributaries like Jalangi, Churni are joining the Bhagirathi-Hooghly from the left, while Mayurakshi, Ajoy, Damodar and Kangsabati, the seasonal rivers, contribute large volume of water to the Bhagirathi-Hooghly from the right bank during wet months.

The simultaneous action of addition of water through seasonal and perennial rivers to the master Ganga at different points and their destruction for various purposes and evaporation leaves the flow of water in an alarming condition in certain stretch of the Ganga in some months of the year. As for example, the total irrigated area in the Ganga basin is about 20 m hectares of which 50 per cent (10 m ha) lies in Uttar Pradesh alone and 40 per cent (4 m ha) of which is irrigated by canal water.

Monthwise 90, 50 and 10 percentile stream flow values of a few station on the Ganga are shown in Table 2.

Table 2 : Month-wise stream flow (m^3/Sec) in the Ganga at 90, 50 and 10 percentile levels

A. Rishikesh (1971-81)

Month	*90 percentile flow*	*50 percentile flow*	*10 percentile flow*	*Mean flow*
January	141	165	171	160
February	138	159	180	155
March	135	174	216	187
April	170	239	306	246
May	241	404	494	381
June	498	827	1,497	997
July	1,950	2,794	2,921	2,610
August	2,146	3,180	3,619	3,159
September	130	1,418	2,132	1,387
October	335	433	535	460
November	190	256	300	263
December	149	190	211	188

Contd....

B. Allahabad (1970-81)

Month	90 percentile flow	50 percentile flow	10 percentile flow	Mean flow
January	358	458	688	515
February	266	385	591	458
March	224	294	581	419
April	184	288	550	359
May	157	332	470	349
June	245	474	1,097	805
July	1,995	7,400	10,439	7,470
August	6,183	19,693	21,674	18,224
September	2,696	15,768	24,131	15,949
October	849	3,052	4,699	2,985
November	436	1,125	1,510	1,044
December	463	598	737	620
C. Patna (1965-81)				
January	860	1,321	1,683	1,335
February	540	1,186	1,437	1,152
March	579	1,019	1,414	1,014
April	677	932	1,399	997
May	609	1,065	1,582	1,108
June	976	1,595	3,167	2,043
July	4,749	11,011	19,356	12,458
August	13,536	27,992	37,424	24,751
September	7,835	21,817	34,673	33,823
October	3,493	6,611	12,163	7,911
November	1,100	2,705	4,666	2,816
December	1,083	1,691	2,365	1,747
D. Azamabad (1960-81)				
January	1,551	2,154	2,612	2,126
February	1,180	1,902	2,356	1,841
March	897	1,513	2,114	1,542
April	1,060	1,352	2,036	1,469
May	1,139	1,613	2,173	1,694
June	1,619	2,915	4,239	3,273
July	4,957	13,757	21,883	15,048
August	19,558	33,493	49,968	37,044
September	14,396	35,830	44,962	33,700
October	4,390	10,434	24,424	15,453
November	2,683	4,899	8,100	5,056
December	2,071	2,791	3,765	2,904

Contd....

E. Nabadwip (1975-81)

Month	*90 percentile flow*	*50 percentile flow*	*10 percentile flow*	*Mean flow*
January	922	1,026	1,188	1,048
February	438	1,021	1,163	957
March	393	875	1,063	838
April	350	884	1,081	829
May	596	1,105	1,159	1,013
June	1,123	1,266	1,402	1,263
July	1,392	1,824	2,237	1,817
August	1,663	1,980	2,235	2,000
September	1,380	2,120	2,600	1,995
October	1,428	1,519	2,407	1,734
November	927	1,005	1,132	1,044
December	1,106	1,221	1,293	1,203

Source : Central Water Commission (Das Gupta 1984).

It will be seen that the values of flow during the flood stage in July, August and September are several times higher than the rates in the lean flow period during the dry months.

Flow Characteristics

Variation of slope in the entire Ganga bed studied during the period 1960-81 (Fig. 2) indicate that in the hilly region upto Rishikesh the slope during 1971-81 is appreciable (1 in 67) with mean annual rate of flow as 850 m^3/s. Further, the leanest month is February with mean rate of flow as 155 m^3/s, while August has the highest mean rate of flow (3159 m^3/s), which is 20 times higher. But the corresponding 90 percentile volumes are much less, 138 m^3/s and 2146 m^3/s respectively.

From Rishikesh to Allahabad, a stretch of about 800 km, there is appreciable decrease in the slope (1 in 4100) with mean annual rate of flow as 850-1700 m^3/s. The slope being 1 in 4100 clearly indicates that considerable siltation has been effected in the river bed. (around Ghazipur silt deposition was 35000 t/yr, Sarkar - 1988). As a result, the bed was becoming shallower and wider, which means increase in shape factor (Tables 3 and 4), a clear sign of deterioration (width of Narora barrage is 1164 m).

Monthly mean flow rate at Allahabad during 1970-81 indicates that the leanest period was in May (349 m^3/s), while maximum flow was in August (18224 m^3/s). The corresponding 90 percentile flow were only 157 and 6183 m^3/s during the above months. Here also width of the river is 2000 m, a clear indication of higher shape factor (Tables 3-5).

Further down from Allahabad to Farakka a stretch of 1005 km. The slope is appreciable reduced, 1 in 13,800, indicating further deterioration due to silting (Tables 3–5). Although, the mean rate of flow in this zone has increased 10,200m m^3/s) at Azamabad, due to more number of tributaries like the Jamuna, the Tons, the Son from south and the Gomati, the Ghagra, the Gandak, the Burhi Gandak and the Kosi from the north all joining the master river. But since the bed is elevated, the cross section obviously has increased (width of the river at Patna is around 7000 m) resulting high value of shape factor.

At Patna the leanest month is April with a mean rate of flow as 997 m^3/sec and flood attaining a higher mark in September with a mean rate of flow as 33,823 m^3/sec; but the 90 percentile value shows different characteristics. Lowest flow (540 m^3/sec) is recorded in February against the highest value (13,536 m^3/sec) recorded in August. The increased flow could not even help eroding the bed. This shows that the river bed has extended on either side (width of Farakka is 1609 m).

Table 3 : Hydrological features of R. Ganga at some important centres (Modified from Jhingran & Ghosh 1978)

	Distance from source (km)	*Width (m)*		*Depth (m)*	
		Summer	*Monsoon*	*Summer*	*Monsoon*
R. Ganga					
Kanpur	541	100–150 (125)	1500	5–6 (5.5)	15–20 (17.5)
Allahabad	788	150–200 (175)	2000	5–6 (5.5)	15–20 (17.5)
Varanasi	992	200–300 (250)	1100	4–5 (4.5)	13–15 (14)
Ballia	1155	70–100 (85)	1500	10–15 (12.5)	25–30 (27.5)
Patna	1274	70–250 (160)	80–1000 (900)	5–7 (6)	10–15 (12.5)
Bhagalpur	1504	150–200 (175)	600–800 (700)	5–6 (5.5)	15–20 (17.5)
Rajmahal	1631	150–200 (175)	500–600 (550)	8–10 (9)	15–17 (16)
R. Jamuna					
Allahabad		80	160	5	16

Table 4 : Shape factor for the above centres during Summer and Monsoon

	Summer	*Monsoon*	*Remarks*
R. Ganga			
Kanpur	27	85	Higher values indicate deterioration due to silting
Allahabad	36	114	
Varanasi	66	78	
Ballia	8	54	
Patna	41	72	
Bhagalpur	36	34	
Rajmahal	19	34	
R. Jamuna			
Allahabad	16	10	

Table 5 : Slope characteristics in different stretches of the Ganga with corresponding mean rate of flow

Stretch	*Section*	*Length in Km*	*Av. Slope*	*Mean annual rate of flow (M^3/Sec)*
1. Source to Rishikesh	Mountainous	250	1 in 67	850
2. Rishikesh to Allahabad	Upper plain	800	1 in 4100	850–1700
3. Allahabad to Farakka	Middle plain	1005	1 in 13800	4000–10,200
4. Farakka to Nabadwip	Deltaic non–tidal plain	230	1 in 23000	1000–1300
5. Nabadwip to outfall	Deltaic tidal Plain	240	1 in 24000	Variable due to tides

Source : Das Gupta S.P. (1984).

At Rajmahal, the river bed recorded as high as 20 ft and at the same time the river bed in Uttar Pradesh being high, the central portion of the river bed cannot remain at a lower level. As a result the river bed in Bihar has every possibility of raising its bed by 10–15 ft. Thus for the same flow of water the river will proportionately raise its bed when at a later stage the flood water will make its way out of the embankment causing inundation of larger areas in Bihar (Table 6).

Table 6 : Areas (in ha) affected by flood in Uttar Pradesh and Bihar during 1950-1981

Period	*Uttar Pradesh (1953-81)*		*Av.*	*Bihar (1950-81)*	*Average*
First	5 years	9.1 – 41.3	21.2	7.9 – 25.0	14.7
Second	5 years	0.7 – 32.5	17.7	2.1 – 13.2	9.2
Third	5 years	0.8 – 34.3	12.7	2.7 – 15.3	9.2
Fourth	5 years	3.1 – 52.6	23.4	2.2 – 42.6 (*)	14.2
Fifth	5 years	13.0 – 36.3	22.2	7.3 – 31.4	20.6
Sixth	4 years	7.0 – 73.4 *	42.2	8.1 – 23.7	15.9

* Maximum 73.4 m ha in 1978 ; (*) Maximum 42.6 m ha in 1971

Modified from Table (Chaphekar S.B. and G.N. Mhatre, 1986)

The heavy floods of 1971, as reported by Bhattacharjya (1986), in Bihar and Maldah in West Bengal are the evidences. Available report states that at Rajendra Bridge near Patna, the maximum discharge recorded was about 32.5 lakh cusecs could allow the free flow of only 23 lakh cusecs. Even the mighty Kosi river can increase the flow to 40 lakh cusecs near Farakka after it joins the main river. Hence, the excess flow (40-23 cusecs) or 17 lakh cusecs caused flood havoc in Bihar and Maldah in North Bengal even though local rainfall was 5 per cent less than normal. Due to silting in the Ganga bed, its tributary like the Pun Pun, the Son, the Kamla, the Kosi, the Gandak and the Ghagra will find difficult for their normal discharge into the Ganga. As a result Bihar valley will experience prolonged flood (Table 6) which will have no escape. Further, in a river like Ganga due to excessive width the depth cannot remain uniform. At places huge islands are formed while at other places there are deep pools (Tables 7 A-D). Islands formed in the previous year may be converted into a deep pool. In such a wide river like the Ganga, the difference between high island and deep pool may exceed 20 ft. Earlier records (Sarkar 1988) state that around 1869, the depth was maximum between Allahabad and Rajmahal (843 km). At Rajmahal the depth ranged between 3.0 and 4.57 m and the velocity of water during summer was around 1.34 m/sec against 3.12 -3.57/m/sec. during other periods. Maximum discharge recorded in September was 15,00,000 cusecs.

Deep pools in river bed have considerable importance for harbouring large size carps especially during non-monsoon months. A survey work of deep pools in a 120 km long stretch of river Jamuna from Rajapur to Allahabad and at different stretches of the Ganga between Allahabad and Fatehpur (upstream of Allahabad) and Allahabad and Bhagalpur (down stream of Allahabad) was made by CICFRI, Barrackpore (Annual Report 1988-89, 1989-90) These are charted in Table 7.

Table 7 : List of Deep Pools

A. On the Jamuna (Rajapur-Allahabad)

	Name of Village	*Approximate length (km)*	*Approximate depth (m)*
1.	Khaujasa	0.5	30-4C
2.	Pratappur	1.0	40-50
3.	Mau	2.0	60–80
4.	Barwar	1.0	40–50
5.	Katri	0.5	30–40
6.	Rajapur	1.0	50–60

B. On the Ganga (Allahabad —Fatehpur, Upstream)

	Name of Village	*Approximate length (km)*	*Approximate depth (m)*
1.	Kurai	0.50	15
2.	Karaghat	0.50	20
3.	Ezuva	0.15	10
4.	Navasta	0.50	40
5.	Champutper	1.0	18
6.	Ujjanighat	1.0	25
7.	Badampurghat	1.0	10
8.	Palhanai	1.50	40
9.	Kanithi	1.0	40
10.	Kalakankar ghat	0.50	30
11.	Rajghat	0.50	15

C. Allahabad—Varanasi (Downstream)

	Name of Village	*Approximate length (km)*	*Approximate depth (m)*
1.	Manda Mukundapur	0.5	20–30
2.	Jeta	1.0	20—40
3.	Chakhata	1.0	20—30
4.	Nawar	1.0	50—60
5.	Umapur to Baruka	2.0	20—30
6.	Bhairupur Ghat	0.50	20—30
7.	Gogaun	0.50	20—30
8.	Khavia	0.15	20—30
9.	Pepra Dah	1.0	20—40
10.	Chandika	1.0	20—50
11.	Tekaur	1.0	30—50
12.	Chunar Ghat	0.5	20—40
13.	Raipuriya	0.5	25—35
14.	Narayanpur	0.5	25—40
15.	Chhoti Mirzapur	0.5	20—30
16.	Garhwa ghat	0.5	25—40
17.	Assighat to Rajghat	5.0	20—50

D. Patna and Bhagalpur

River Ganga was also surveyed in and around Patna and only one deep pool could be located near Sherpur. Near Bhagalpur, the following five deep pools were located but their sizes could not be determined. These are :

1. Narainpur
2. Ghogra
3. Titahi
4. Kahelgaon
5. Ekehari

Source : Annual Report 1988-89 and 1989-90, CICFRI, Barrackpore.

In the estuarine zone ((Bhagirathi–Hooghly) the condition prior to the construction of Farakka barrage (1975) were different. The diversion of the main flow of the Ganga into the Padma, the bed of Bhagirathi–Hooghly being 7.3 metre above the Ganga bed and shifting of the course of the Damodar further down and rejoining the Hooghly near Haldia at right angle, considerable deterioration of the Bhagirathi-Hooghly resulted (Fig. 3). The Bhagirathi which flows through Behrampur, recorded about 140,762 cusecs and still further down the stream near Jangipur it was 140,160 cusecs with velocity of water 1.8—2.31 m/sec is now considerable reduced during dry months with the result that Bhagirathi converted into a small spill channel (Sarkar, 1988).

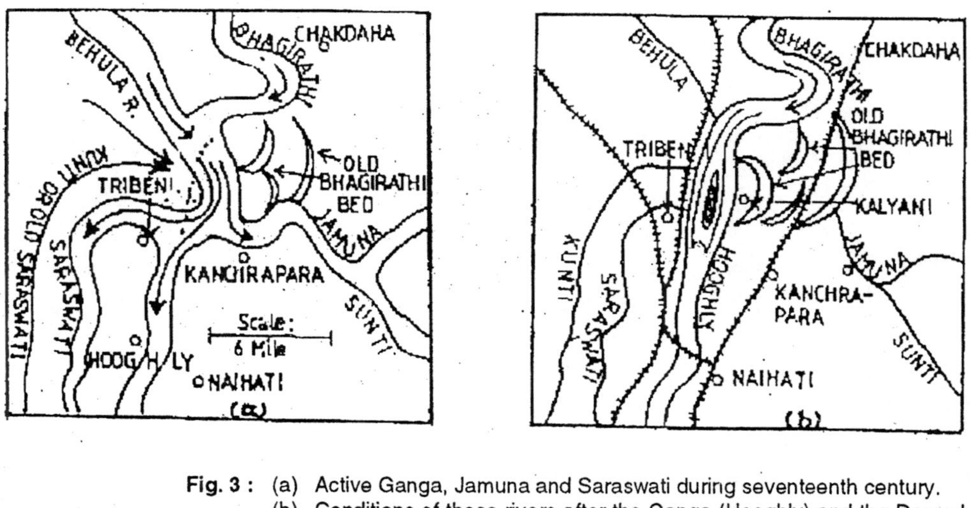

Fig. 3 : (a) Active Ganga, Jamuna and Saraswati during seventeenth century.
(b) Conditions of these rivers after the Ganga (Hooghly) and the Damodar (Behula) shifted their Courses.

Source : Ray, P. (1993)

This fact was also evidenced when people could even wade across the river during 1972. The inland water-ways Authority of India recorded least available water depth between Tribeni and Farakka to be two metres (Mukherjee, 1993).

Near Calcutta, the Hooghly is wider, (about 609.6 m) as a result velocity of water decreases. During monsoon huge quantity of alluvial soil is brought alongwith the water. Bera (1984) is of the opinion that if the flow of water is above 1.52 m/sec the silt moves along with the water, but if it is below, silt settles, further; if the velocity exceeds about 2.1 m/sec, effect of erosion takes place.

The discharge @ 40,000 f^3/sec comes to 345 crore f^3/day. According to Bera (1984) during

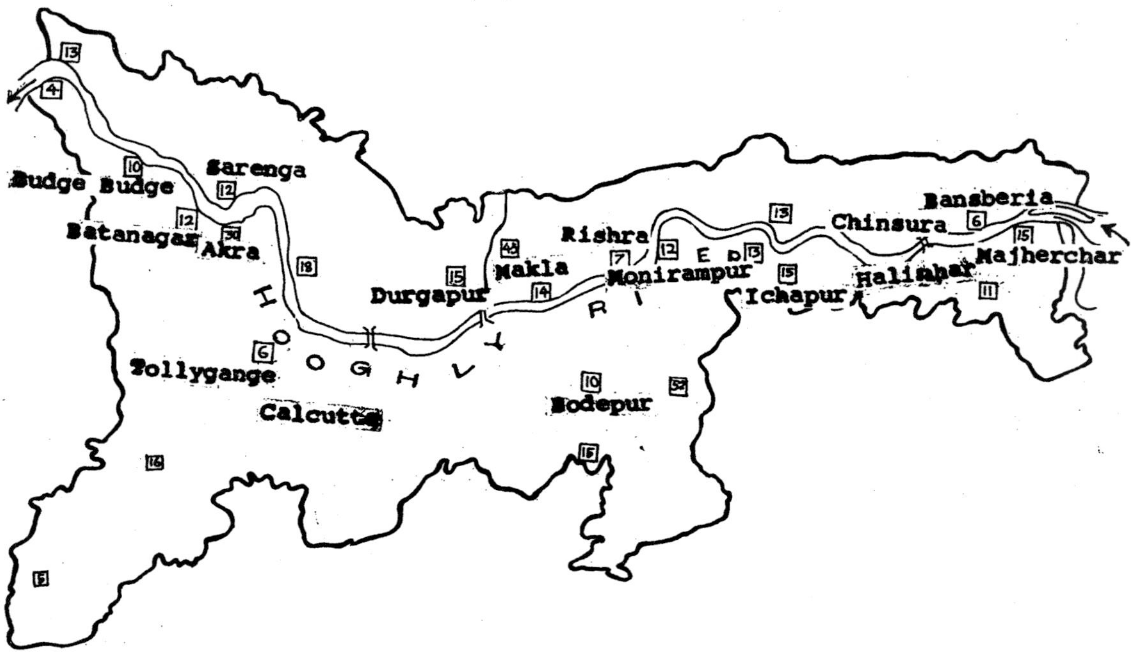

Fig. 4 : Location and number of brick field clusters flanked on both sides of the Hooghly estuary under Calcutta Metropolitan District.

Source : Bhattacharya, S.K and S.M. Chatterjee (1972)

monsoon the quantity of alluvial soil brought daily along with the water is about 17.25 lakh f^3. In four months of monsoon the quantity comes to 20 crore f^3. Earlier study (Chatterjee, 1972) revealed that 90—100 million f^3 of silt were carried alongwith the water every year into the Bhagirathi-Hooghly, three fourth of which settles in the river bed. This comes to 60-66 m f^3/per year. In the present observation 3/4 of 20 crore f^3 that settles in four months comes to 13.3 crore f^3 which is more than double the earlier value in course of 12 years. Also in 1947-48, the dredging quantity was limited to about 7 million tonnes and in 1967-68 it was in the order of 21 million tonnes. Again according to ministry of irrigation and Bower an estimate, about 2.8 million m^3 of silt (about seven million tonnes) enters the Bhagirathi-Hooghly channel annually. About 2/3 of silt is deposited below the Bansberia point. In peak rainy season it is carried further down and deposited 20 km. south of Calcutta. In addition the river receives about 8 million tonnes of sand from the sea which chokes up the stream flow (Mukherjee 1993). To be precise the quantum of silt deposited annually, is undoubtedly alarming. The effect of such huge silt load is shown in Tables 3-5. Also, 279 brick fields flanked on both sides of the river between Kalyani and Achipur (100 km) are the outcome of such silt load (Bhattacharjee and Chatterjee 1972) (Table 8 and Fig. 4.) The reduction of flow velocity near Calcutta causes such huge mass of silt to settle.

Table 8 : Brick field clusters of Calcutta Metropolitan District flanked on both sides of the Hooghly Estuary

	Area	*No. of Brick fields*		*Area*	*No. of Brick fields*
1.	Achipur	4	11.	Kalyani, Majherchar	15
2.	Budge Budge, Pujali	10	12.	Uluberia upto Bauria	13
3.	Batanagar, Nangi	12	13.	Sarenga, Manickpur	12
4.	Akra	30	14.	Podra, Radhadasi, Benkra	18
5.	Bhangakhal, Badartala	18	15.	Durgapur (Howrah)	15
6.	Tollygunge	6	16.	Makla, Raghunathpur	45
7.	Barrackpore, Manirampur	12	17.	Kotrung	14
8.	Garulia	3	18.	Rishra, Serampur	7
9.	Ichapur	15	19.	Champdani, Chandernagore Chinsura	13
10.	Halisahar, Kanchrapara	11	20.	Bansberia	6

Source : Bhattacharya S.K. and S.M. Chatterjee (1972).

Out of 25 Centres under Calcutta Metropolitan District having 377 brickfields only 20 Centres having 279 brickfields are shown here as they are very close to the Hooghly river and fed by its water.

In the last 10 years as stated by Bera (1984), the depth of water is reduced from 3-5 ft (0.9-1.51m) in the main channel. Decrease of depth of river bed results increase of tidal range. In 1947, the river experienced 70 bore days which in 15 years time rose to 161 days. (Banerjee, 1972). Fig. 5 indicates the gradual increase in annual frequency of bore tides in the river Hooghly.

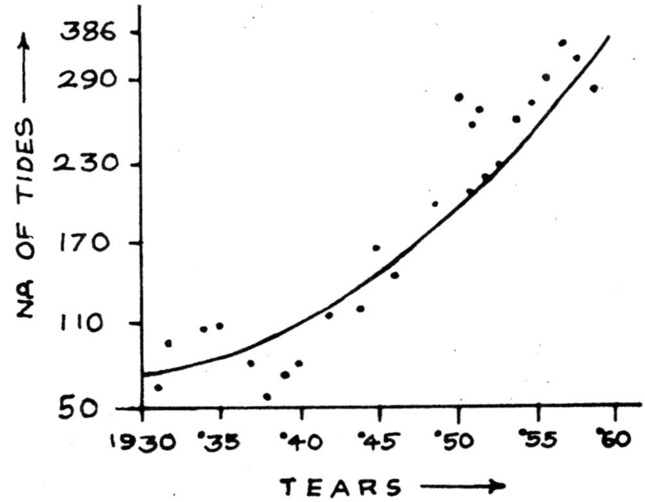

Fig. 5. Annual Frequency of Bore Tides in Hooghly River (1930-1960)
Source : Banerjee, B.N. (1972)

Chakraborty and Banerjee (1966) who studied the depth conditions of the Hooghly estuary between Bally bridge (down stream) and tribeni (up stream) occasionally observed uneven muddy and sandy areas at the bottom of the estuary. This confirms large variation of depth ranging between 4.5 and 14.5 m during low tides. The settled muddy bed shifts up and down within the river bed depending upon the velocity of water during flood and ebb tide conditions. Even tidal waters together with fresh water release from Farakka cannot create suitable velocity to erode the deposited silt. During the dry season, discharge of freshwater into the estuary decreases and the abstraction rate for various purposes increases which enhances the impact of flood tides over the ebb tides. This allows the saline water to penetrate deep into the freshwater zone. The average velocities of boretide and neap tide during different seasons *i.e.* winter, summer and monsoon are found to be 27.7, 28.3 adn 27.9 km/hr for the high tide and 10.16, 12.1 and 10.96 km/hr during low tide of bore tidal phase respectively. During neap tide, the corresponding values are 18.18, 18.5 and 20.3 km/hr at the time of high tide and 10.15, 12.6 and 12.5 km/hr during low tide. On an average, the bore tide phenomenon takes place for about 12 days/month, while the neap tidal phase lasts for about 14 days/month. The height of the bore tides both during high and low tides is always higher than that of neap tidal phase. The ebb tidal flow lasts for a longer period than the flood tide. Duration of ebb tide flow is 6.5-8.5 hr during dry months and 7.0-11.5 hrs in the wet months. The flood tide flows last for 4.0-6.0 hr and 2.5-4 hr in dry and wet months respectively. Both bore and spring tides have appreciable scouring effect on the littoral zones of the estuary. The mouth of the Hooghly-estuary is wide and deep (kulpi, 4,59 km, Diamond Harbour, 4.65 km. and Falta, 1.2 km). The corresponding maximum depth at high tide during monsoon are 16..9, 27.5 and 20 m. The maximum depth at low tide during winter being 10, 20 and 13.9 m respectively. The banks taper into a broad funnel admitting the tides to extend high up into the river. They often form 'bores' in the narrow part of the estuary and maintain average height of 4.0-4.6 m with strong scouring action. (Ray, 1981). Additional discharge from Farakka simply extends the freshwater zone further down.

3

Tributaries and Their Role

By and large all the tributaries have a tremendous role in keeping the main river alive. Snow–fed rivers are practically perennial while others are seasonal and contribute only during monsoon. Their drainage is dependent upon the catchment area. Since the total annual evaporation in the flowing stretch other than Himalaya and West Bengal is much higher than the total annual rainfall received, it has become necessary to hold the water received during monsoon by constructing dams, barrages, anicats, weirs etc. at suitable places both in the tributaries and the main river for use in agriculture, industries, human consumption, navigation, pisciculture *etc*. By doing so considerable volume of water is consumed which otherwise would have been wasted.

From Rishikesh, as the Ganga flows further down stream few minor tributaries join adding volume of water which is now loaded with enough of sediments giving a hue of yellow ochre to the Ganga water.

Ramganga joins the main stream at Kanauj adding annually some 15.28 billion m^3 of water. At Allahabad, the Ganga is united from the right to its counterpart twin the Jamuna which is fed by its tributaries the Chambal, the Sind and the Betwa, which contributes more water (79 billion m^3/year). It is from Allahabad and its downwards the river receives several major tributaries at more frequent intervals-the Tons and the Son from the right and The Gomati, The Ghagra, The Gandak, The Burhi-Gandak and The Kosi from the left till the Ganga reaches the head of its delta at Farakka below Rajmahal bend. The water quantity and the sediment load being modified at each turn according to the character of the contributing stream. The Ganga carries a lot of sediments. The sediment load often rises to a level of 2g/lit of water (Rao, 1975). By far the largest amount of water per annum in the entire Ganga basin is contributed by The Ghagra (94.4 billion m^3) followed by The Kosi (61.56 billion m^3), The Gandak (52.5 billion m^3), the Son (22.42 billion m^3), The Gomati (7.39 billion m^3), The Burhi-Gandak (7.1 billion m^3), and the Tons (4.93 billion m^3).

By the time it reaches Patna, the flow of the Ganga attains a sizable volume of 240.5 billion cubic metres per year reaching further to 320.4 billion m^3 at Azamabad. Table 9 & 10 show the contribution made by different tributaries, their catchment areas annual yield, mean rate of flow and seasonal variations etc.

In Jamuna, upstream of Delhi the winters are characterised by low flow, while on the down-stream of Delhi the minimum flow occurs in summer, and the river bed gets almost exposed. In the peninsular tributaries like the Son and the Tons, the lowest flow occurs in the Summar season. It is also be seen from the table that the flow of water in all the streams in the Ganga basin during long lean season from December to May remains very critical. The high land profile of Central India which

Table 9 : Catchment area*, annual yield and mean rate of flow of streams in the Ganga basin

River	Catchment area (Km^2)	Annual yield (mm^3)	Rate of flow m^3/sec
Yamuna	3,66,223	76,000	2,410
Son	71,259	22,420	711
Pun Pun	8,530	3,577	114
Ghagra	57,578*	94,400	2,993
Ramganga	32,493	15,258	484
Gomati	30,437	7,390	234
Tons	16,860	4,927	156
Kiul	16,580	5,900	187
Burhi-Gandak	10,150	7,100	225
Gandak	7,620*	52,200	1,655
Kosi	11,000*	61,560	1,952
Dwarka	8,850	4,687	149
Rupnarayan	8,530	4,400	140
Karmanasha and Adjoining streams	11,709	5,750	182
Ajay	6,050	3,207	102
Other streams	1,70,035	31,400	996
Haldi	10,210	5,300	168
Damodar	25,820	12,210	387

* Area figure relate to the part of basin falling within Indian territory only. *Source* : Rao K.L. (1975).

It is revealed from the above table that barring the 5 major tributaries of the Ganga all have poor flow rate ranging from as low as 100 to 500 m^3/sec. The seasonal variation of streams at a few selected centres on The Ganga, The Jamuna, The Son and The Tons are given in the Table 10.

Table 10 : Seasonal variation of stream flow in the Ganga basin

		Mean Seasonal flow (mm^3)			
Centres		Monsoon (June-Sep)	Postmonsoon (Oct-Nov)	Winter (Dec-Feb)	Summer (March-May)
R. Ganga					
(Uttar Pradesh)	Rishikesh (1971-81)	26,631	1,912	1,305	2,150
	Balawali (1976-81)	13,306	947	167	760
	Garhmukteswar (1967-81)	18,392	1,756	908	982
	Fatehgarh (1972-81)	17,300	1,266	264	281
	Kanpur (1960-81)	30,763	4,139	1,518	910
	Allahabad (1970-81)	112,278	10,703	4,148	2,987
	Mirzapur (1976-81)	97,142	9,362	4,529	3,362
	Varanasi (1960-81)	112,206	11,244	4,213	2,793
	Buxar (1960-81)	113,247	17,741	5,463	3,438
(Bihar)	Patna (1965-81)	192,625	28,488	11,044	8,341
	Azamabad (1960-81)	235,357	54,494	18,055	12,474
(West Bengal)	Baharampur (1975-81)	12,817	6,055	8,140	6,277
	Nabadwip (1975-81)	18,666	7,763	7,910	7,107
R. Yamuna					
Delhi & U.P.	Tajewala	6,800	1,400	900	1,400
	Delhi Rly. Bridge	6,500	800	200	400
	Etawah	7,300	1,000	300	200
	Pratappur	61,600	9,300	3,400	1,700
R. Tons					
(U.P.)	Meja Road (1960-81)	4,159	491	494	83
R. Şon					
(Bihar)	Chapan (1960-81)	14,867	2,105	1,915	1,534
	Koelwar (1960-81)	17,699	2,256	1,497	966

Source : Central Water Commission (in Das Gupta, S. P. 1984).

is restricted to only the Ganga basin are mainly drained by the numerous tributaries of the Jamuna, the principal ones are being the Chambal, the Betwa and the Ken. The Son and the Damodar are the other two major streams. Several other medium sized and minor streams (some non-perennial) originate from the Central Indian high lands. The water flow in the peninsular rivers are largely regulated by the protective cover provided by extensive forest belts. But due to rapid deforestation all through the central Indian high land, the hydraulic characteristics and the silt load in most of the peninsular rivers have undergone rapid change over the years resulting in increased frequency of flash floods in these rivers on one hand and the flow is much reduced during the dry seasons on the other hand (cf. Table 6 and 9).

The Ganga is the major stream and is the recepient of all water-lines in the region. The regional drainage can be studied through the sub-systems, (1) The Ghagra system, (2) The Gandak system, (3) The Kosi system-all situated in the north of Ganga; (4) The Son system and (5) the other tributary drainage to the Ganga from the South. (Fig. 1)

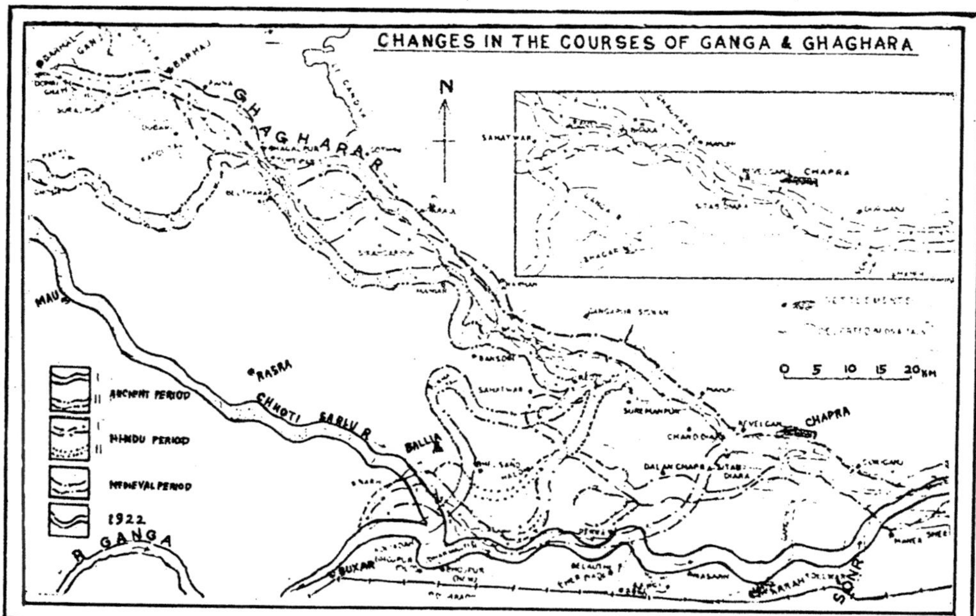

Fig. 6 : Changes in the courses of the Ganga and Ghagra
Source : Singh, R.L. (ed.) (1971)

The Ganga receives numerous tributaries from the southern uplands among which The Son is the largest. The river has been notorious for its changing courses in the past as is traceable from several old beds on its east (Fig. 7) but has been tamed squarely with the anicut at Dehri and now more so with the Indrapuri barrage, a few km. upstream.

In the northern Ganga plain, floods are recurring phenomenon. Large areas in this region get flooded when the tributaries join the Ganga, almost all the rivers in the region like The Rapti, The Ghagra, The Gandak, The Kosi, The Son and the Ganga itself spill over during the monsoon period and shift their courses frequently leaving remnants of various sizes behind them (Fig. 6-8) as is

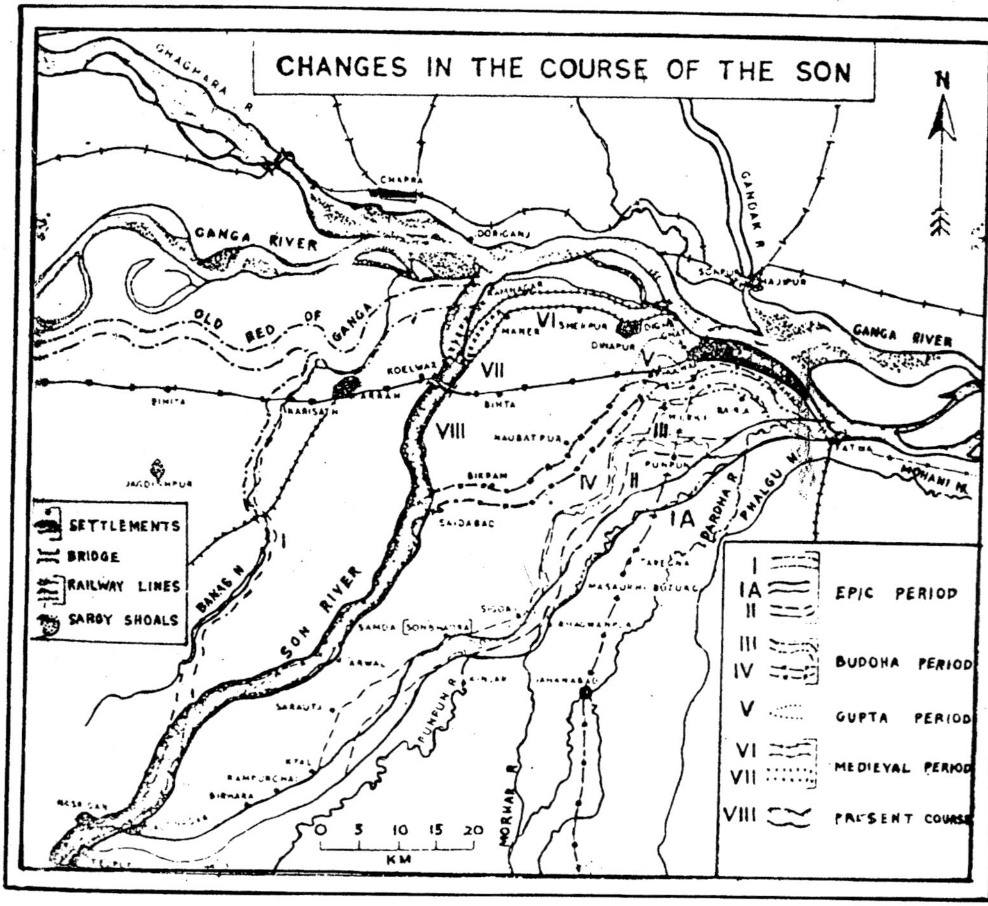

Fig. 7 : Change in the course of the Son.
Source : Singh, R.L. (ed.) (1971)

reflected by the remnants of their former beds in the form of Ox-bow lakes, meander loops, dead arms, chors, tals and ruins of settlements etc. For instance (1) The Ganga has shifted over 35 km between Bhojpur tal and Surha tal causing changes in the site of Ballia; (2) Town and river ports like Gola, Barhaj etc. have been affected by similar shifts of the Ghagra and (3) The Kosi the most dynamic among the rivers has marched about 120 km during the last 200 years from near the town of Purnea up to the Tiljuga within Durbhanga district (Fig. 6-8). Further, an important phenomenon is normally observed with the tributaries joining the main river. Mighty streams gathering all the runoffs where terrain is hilly, remain in copious and the river flow is torrential. During the rainy season the channel is either in spate or runs in trickle. The extremely flashy tributary creates a havoc as it dibouches almost parpendicularly into the main stream that flows below the danger level with lesser discharges. The flow of the main river is thus arrested (*e.g.* R. Damodar, R. Ajay and R. Mayurakshi in West Bengal; R. Kosi and The Son in Bihar etc. Fig. 6-8) for a certain period and around the confluence the

---

Ox-bow lakes, meander loops, dead arms, chors tals and ruins of settlements are named differently due to their different sizes and formed as remnants of a river.

rivers remain in spate and thwarts the flowing of the main river water causing flood. Suspensoids brought in along with the water by both the rivers settle around the confluence. This causes rapid shoaling of bars year after year and floods are thus a permanent phenomenon which gadually increases. Inumerable seasonal streams which can create havoc during monsoon and some perennial rivers are causing such problems and change the physiographical conditions of the river.

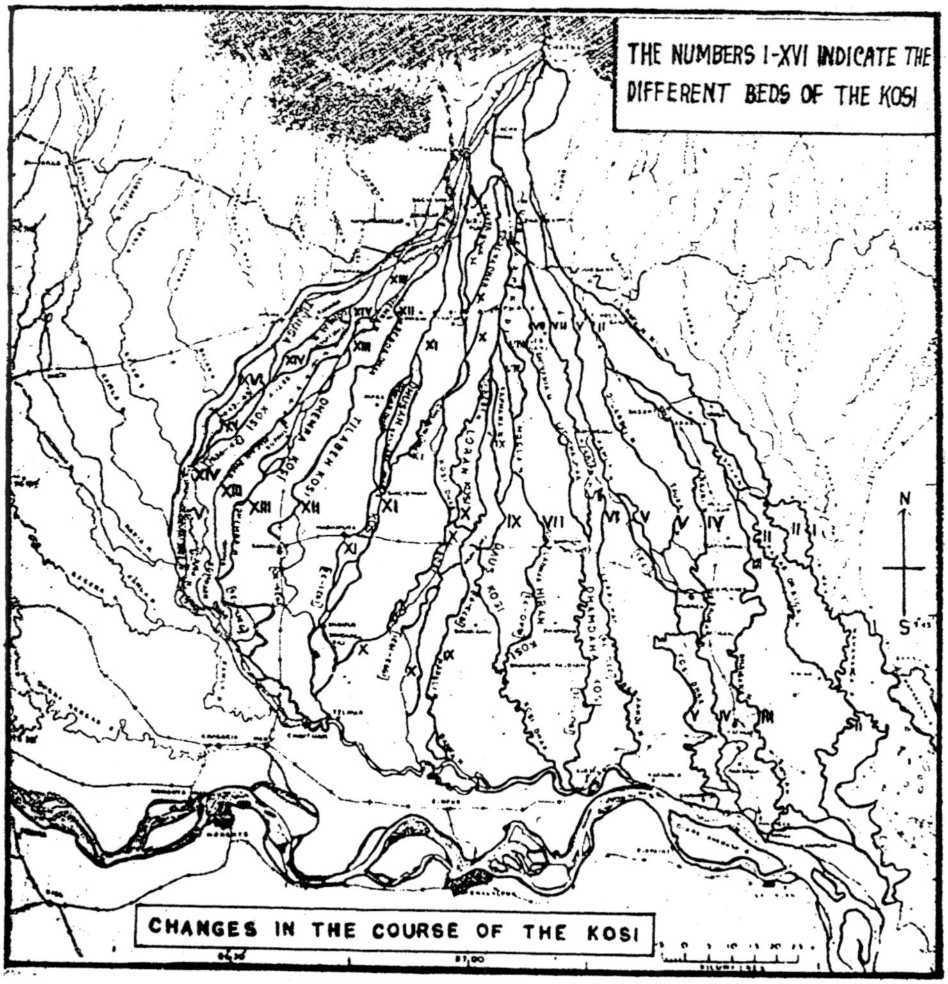

Fig. 8 : Changes in the course of the Kosi

Source : Singh, R.L. (ed.) (1971)

4

Major Canal Projects in the Ganga Basin

In the Ganga basin, canals provide the largest amount of water for irrigation. Some of these are even capable of irrigating few hundred square km. The main upper Gangetic canal drawn from Haridwar, is an example of a long 230 km stream of water, pouring forth at the rate of nearly 300 m^3/sec. The main canal together with principal branches have a total run off of about 900 km. Further down, the lower Gangetic canal has recorded discharge of more than 150 m^3/sec.

The main branches run for over 1000 km while the distributaries are over 5000 km. Further, to the east, Sarda canal drawing water from the mighty Ghagra river at the rate of 325 m^3/Sec. The total length of canal network presently work out to 9500 km., to be eventually augmented to a vast network of 13,680 km., of canal. Table 11 gives a list of 18 major Canal structures along with storage capacity of each.

Table 11 : Major Canal projects in the Ganga basin

State		*Canal Project*	*Live storage (mm^3)*	*Canal discharge m^3/sec.*
Haryana :	1.	Western Jamuna	—	106
Uttar Pradesh :	2.	Upper Ganga	2613	298
	3.	Lower Ganga	—	157
	4.	Eastern Jamuna	—	85
	5.	Agra Canal	—	92
	6.	Sarada	634	325
	7.	Matatila	780	Not available
	8.	Betwa	84	67
	9.	Ken	60	57
	10.	Ramganga	1937	141
Bihar :	11.	Gandak	—	147
	12.	Kosi	—	Not available
	13.	Son	8971	—do—
	14.	Trioeni	—	—do—
	15.	Damodar	4226	—do—
	16.	Mayurakshi	555	—do—
	17.	Kangsabati	972	—do—
Rajasthan/ Madhya Pradesh	18.	Chambal	8500	—do—

Source : Rao K.L. (1975) Ministry of Information and Broadcasting. (In Das Gupta, S.P., 1984).

From Table 12 an approximate consumption of water by man in cities, towns and rural areas of the Ganga basin can be calculated.

Table 12 : Distribution of towns and cities together with rural and urban population in the Ganga basin (1977)

State	*Total No. of towns and cities*	*Total No. of cities**	*Urban-population*	*Rural-population*	*Total population*
Haryana	28	4	999,291	3,303,429	4,301,720
Uttar Pradesh	256	16	14,042,237	84,178,854	98,221,091
Bihar	155	10	5,656,714	64,211,651	69,868,365
West Bengal	152	21	7,282,214	23,619,327	30,901,541
Rajasthan	25	6	2,097,528	10,389,100	12,486,628
Madhya Pradesh	69	4	3,734,054	16,194,902	19,928,956
Himachal Pradesh	6	0	98,163	397,780	495,943
Delhi	1	1	5,378,000	561,000	5,940,000
Ganga Basin	269	62	39,289,201	202,856,043	242,145,244

* Each with population 100,000 and over estimated from census of India, 1971, 1981.

On account of high concentration of population and location of factories in the urban areas, the towns and cities serve as the major source of demestic and industrial products which eventually find their way into the aquatic system. The quantity of water used for various purposes in the Ganga basin has not yet been fully assessed. However, from the estimate so far available, it appears that a total quantity of 142.6 billion m^3 of fresh water is used annually to meet the requirement of various sectors, Table 13 shows the estimated quantity of water used in irrigation, domestic purpose and other industrial activities for each state in the Ganga basin on the basis of available information so far. The table also gives a balance sheet of annual consumption of fresh water and generation of waste water in the Ganga basin in mm^3.

Table 13 : Balance sheet of annual consumption of freshwater and the corresponding waste-water generated (mm^3).

State	*Consumption of freshwater*				*Generation of waste-water per cent*			
	Irri-gation	*Dome-stic*	*Indus-trial & other use*	*Total annual consum-ption*	*Irrig-ated field*	*Other source*	*Total annual gener-ation*	*Con-sumption*
Haryana	8706	56	63	8825	1741	21	1762	80
Uttar Pradesh	74907	1692	3042	79641	14981	528	15509	80
Bihar	21625	793	289	22707	4325	128	4453	80
West Bengal	11267	680	459	12406	2253	317	2570	79
Rajasthan	10465	187	51	10703	2093	74	2167	79
Madhya Pradesh	6898	296	69	7263	1380	108	1488	79
Himachal Pradesh	135	6	1	142	27	2	29	78
Delhi	481	443	26	950	96	350	446	53
Ganga Basin	134484	4153	4000	142637	26896	1528	28424	80
Per cent Consumption	94.28	2.91	2.80	100.00	18.86	1.07	19.93	

Source : Centre for Study of Man and Environment (1984) (In Das Gupta S.P. 1984).

The Table 13 reveals that Uttar Pradesh is the biggest consumer of water using 55.8 per cent of the basins total consumption. Also the largest share of water (94.3 per cent) within the basin is claimed for use in irrigation alone.

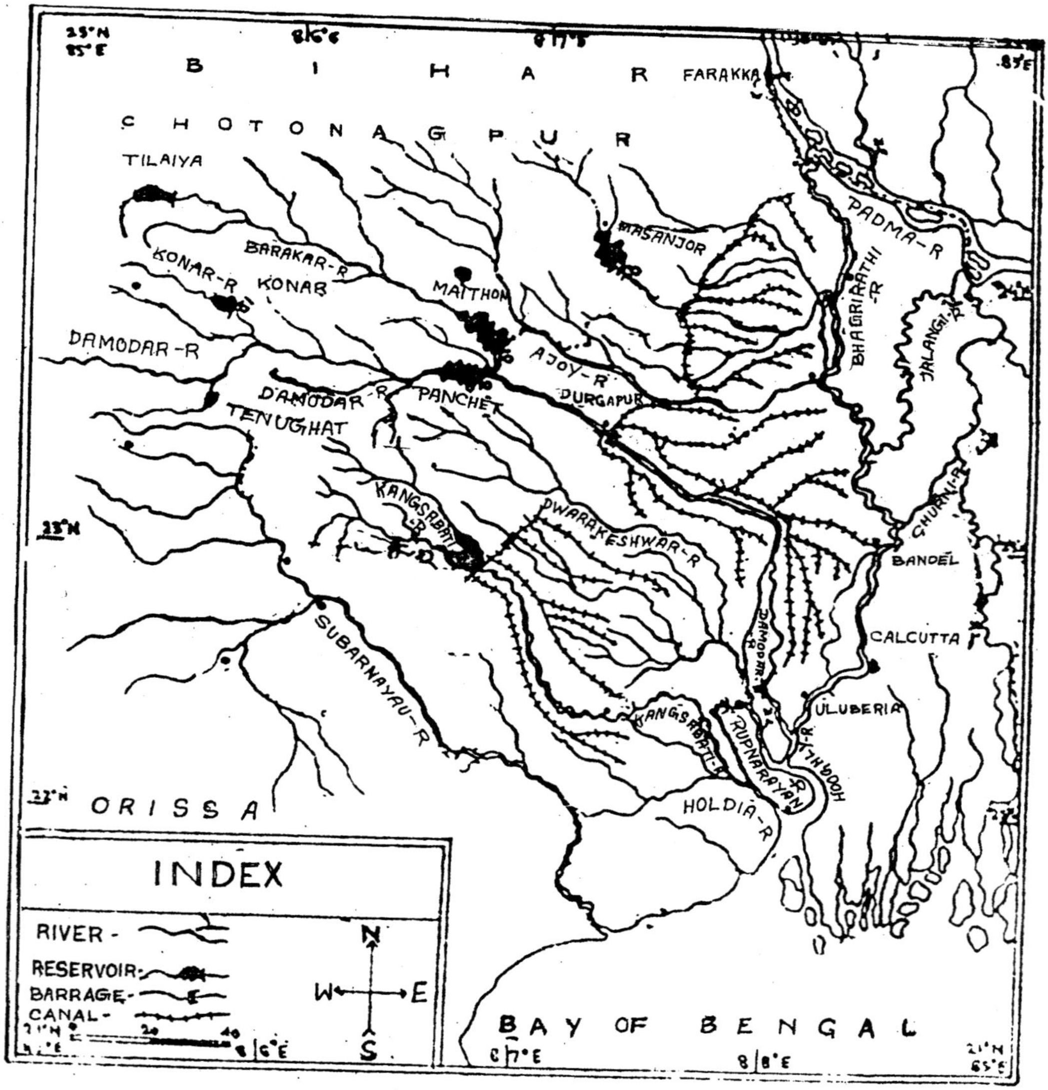

Fig. 9 : Drainage system of different rivers along with Bhagirathi-Hooghly in West Bengal.
Source : Bera, S (1984)

In the process of use of freshwater while greater part of the total quantity is actually consumed, a sizable residual amount is let off as waste water containing various pollutants. Nearly 19.9 per cent of the total quantity of freshwater used is finally discharged as waste water effluent amounting to some 28.4 billion m^3 in a year within the Ganga basin. Abstraction of such huge volume of water reduces the flow velocity so much so that eroding the river bed becomes almost impossible even during the wet months when shape factor shows a high value (cf. table 4) stream velocity recorded at various centres like Haridwar (0.35m/sec or 1.55 ft/s), Kanpur (0.59 m/s or 1.78ft/s), Allahabad (0.90 m/s or 2.97 ft/s), Varanasi (0.40m/s or 1.32 ft/s), Farakka (0.44m/s or 1.29 ft/s, Nawabganj (0.42 m/s or 1.6 ft/s), Dakshineswar (0.27m/s or 0.89 ft/s) and Diamond Harbour (0.22m/s or 0.49 ft/s) during July clearly indicates that even though there is an increase in the discharge figure (August and September) shape factor being high, the water will spread on either side of the river some time spils resulting loss of water velocity of the river. The conditions thus becomes favourable for settling the suspensoids. Besides, tributaries originating from mountains having greater slope and joining the main river parpendicularly further arrests the velocity of the flow of the main river and allows precipitation of suspensoids. A clear example is Bhagirathi-Hooghly where @ 17.25 lakh f^3 of silt brought per day during monsoon causes 20 crore f^3 of silt to settle in 4 months time. Here Ajoy and Damodar Rivers join the main stream (R. Hooghly) in a perpendicular direction. Even tidal water associated with fresh release from Farakka cannot create suitable velocity to erode the cumulative deposited silt from the river bed. It can only push the saline water further down (Ray, 1980). (Fig. 10).

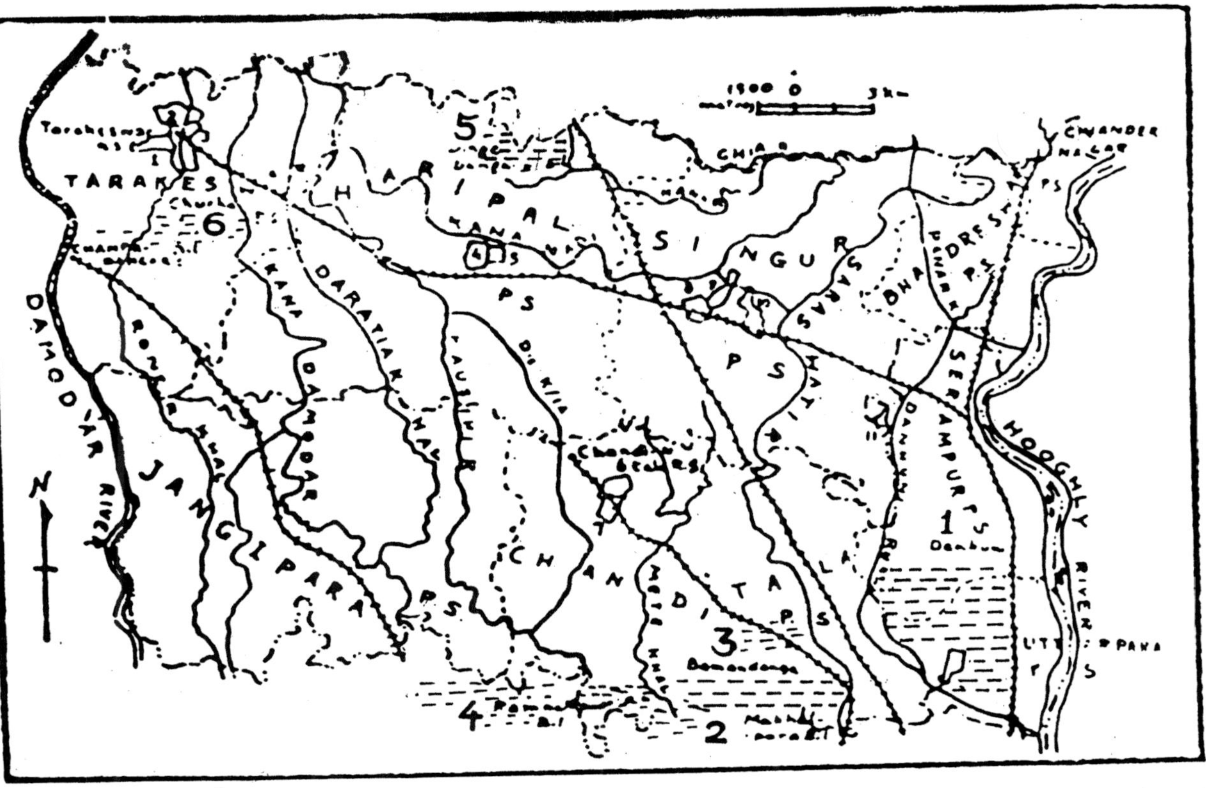

Fig. 10 : Nos 1-6 indicates formation of wetlands in between Damodar and Hooghly rivers. Bhattacharya, R. and Sankar De. (1972)

5

Irrigation Projects and Abstraction of Water

It is now an established fact that development is associated with the destruction of ecology. Construction of dams, running of irrigation projects and abstraction of water for various purposes reduce the flow appreciably resulting in an inadequate flow. A number of dams have come up in the basin to provide water for irrigation, hydropower development and flood control. In the Ganga basin about 33.5 billion m^3 of water is presently stored in reservoirs, the largest one is the Rihand reservoir (Govind Ballab Pant Sagar), holding nearly 9 billion m^3 of water.

In all there are 25 dams having 30 m height and above, built on the Ganga river, in different States for the purpose of irrigation, hydel power, flood control, navigation, water supply and recreation. These are as follows:

State	*Dams Completed upto 1979*	*Dams Under Construction*	*Total Dams*
Uttar Pradesh	7	7	14
Bihar	9	1	10
West Bengal	1	Nil	1

Source : Register of Large Dams in India (1979).

The major projects now in operation in the sub region are- The upper Ganga canal (taken out from the right bank of the Ganga at Haridwar where a dam is built across the river) and the lower Ganga Canal (it takes off on the right bank of the Ganga at Naraura (Bullandshahr) by construction of dam across the river) utilizing over 9 mm^3 of water. The Rohilkhand group of canals from Ramganga, the Sarda canal, the Tribeni canal taking off from the Gandak and the Kosi canal, all add up to the massive withdrawal from the Ganga system. Recently, a large irrigation scheme was started to divert Ghagra water to feed the existing Sarda system, where demand has gone up leading to an acute scarcity of water. Besides this there are also irrigation projects of the Chambal and The son and their subtributaries. Table 14 indicates abstraction points on streams in various States.

The utilization rate of river water varies from industry to industry. It is estimated that for per ton production, the requirement of water for pulp and paper is about 20,000 – 1 million litres and for Textile it is 50,000 – 2,00,000 litres. Demand for drinking water and other domestic uses is to the tune of 45.5 litres/day/head in the rural areas and 137 litres in the urban areas. Adding these figures for all industrial cities located in and around the Ganga basin, the total amount of water required would be staggering. (Ray, and Jhingran, 1987).

Table 14 : Abstraction points on streams in different states

Sl. No.	State/District	Town	Abstraction from stream
Uttar Pradesh			
1.	Pauri Garhwal	Srinagar	Alakananda
		Kotdwara	Ramganga
2.	Almora	Ranikhet	Ramganga
		Almora	Ramganga
3.	Nainital	Kathgodam	Ramganga
4.	Pithoragarh	Pithoragarh	Sarda
5.	Lucknow	Lucknow	Gomati
6.	Jaunpur	Jaunpur	Gomati
7.	Rai Bareli	Rai Bareli	Sai
8.	Mirzapur	Robertsganj	Belan
9.	Kanpur	Kanpur	Ganga
10.	Varanasi	Varanasi	Ganga
Madhya Pradesh			
1.	Satna	Nagod	Tons
		Satna	Tons
2.	Rewa	Rewa	Tons
3.	Jabalpur	Murwara	Son
4.	Shahdol	Umaria	Son
		Shahdol	Son
5.	Sidhi	Sidhi	Son
6.	Surguja	Ambikapur	Rihand
Bihar			
1.	Rohtas	Adhaura	Karmanasha
2.	Gaya	Gaya	Falgu
3.	Hazaribag	Ramgarh	Damodar
4.	Munger	Munger	Ganga
		Jamalpur	Ganga
5.	Bhagalpur	Bhagalpur	Ganga
		Kahalgaon	Ganga
6.	Santhal Parganas	Singharsi	Gumani
		Barhati	Gumani
		Barharwa	Gumani
		Dumka	Mayurakshi
		Deoghar	Ajay
		Jamtara	Ajay
7.	Dhaubad	Jamadoba	Damodar
		Sindri	Damodar
West Bengal			
1.	Murshidabad	Murshidabad	Bhagirathi (Ganga)
		Behrampur	Bhagirathi (Ganga)
2.	Birbhum	Siuri	Mayurakshi
		Sainthia	Mayurakshi

(Contd...)

Sl. No.	State/District	Town	Abstraction from stream
3.	Barddhaman	Asansol	Damodar
		Raniganj	Damodar
		Durgapur	Damodar
		Barddhaman	Damodar
4.	Nadia	Sautipur	Bhagirathi
5.	Purulia	Purulia	Kasai
6.	Hugli	Chunchura-Hugli	Hugli
		Srirampur	Hugli
		Uttar Para	Hugli
7.	24 Parganas	Palta	Hugli
		Kamarhati	Hugli
		Baranagar	Hugli
		Garden Reach	Hugli
8.	Haora	Haora	Hugli
9.	Medinipur	Medinipur	Kasai
		Kharagpur	Kasai

Source : Das Gupta, S.P. (1984).

6

Deforestation Vis-a-Vis Soil Erosion

Associated with deforestation Heavy rainfall encourage high rate of silt erosion and are brought down to the main Ganga basin alongwith the water. Table 15 indicates that forest area in the Ganga basin is highest in Himachal Pradesh (59.37 per cent). The low values recorded in Haryana (2.4 per cent), Bihar (7.54 per cent) and West Bengal (7.55 per cent) of the geographical area are quite alarming.

Table 15 : Forest area (Sq. Km.) in the Ganga basin

State	*Total reporting area*	*Forest area*	*% of forest area*
Haryana	21265	516	2.4
Uttar Pradesh	294413	43238	14.68
Bihar	138296	10438	7.54
West Bengal	50717	3833	7.55
Himachal Pradesh	5799	3443	59.37
Madhya Pradesh	201705	46536	23.07

Source : Centre for Study of Man and Environment (1984) (In Das Gupta S.P, 1984)

Forest area in the Ganga basin is under severe stress and strain under the present system of forest management. Even in the better forested state of Himachal Pradesh (59.4 per cent) Madhya Pradesh (23.1 per cent) and Uttar Pradesh (14.7 per cent), the proportion of land actually under dense tree cover within the Government forest tracts is very low due to extensive felling of trees carried out in recent decades. According to Deoras (1973) the rains wash away from bare lands nearly 40 tons of loose soil out of about 2 acres, but when covered by forests and grass this wash is only 1 to 0.3 tons for the same area.

7

Sediment Load and Nature of Soil

Susceptibility to severe erosion of mountain soil, submontane soil is dominant in Uttar Pradesh followed by Himachal Pradesh, alluvial soil, in Uttar Pradesh, Bihar, and West Bengal, Rajasthan, Haryana, Madhya Pradesh, Delhi and Himachal Pradesh (in order of abundance), shallow black soil, (dominant in Madhya Pradesh followed by Rajasthan and Uttar Pradesh); Laterite and Lateritic soil (dominant in West Bengal followed by Bihar, Madhya Pradesh and Uttar Pradesh) are very high. These are followed by red soil (dominant in Madhya Pradesh followed by Bihar, Uttar Pradesh, Rajasthan, West Bengal, Haryana and Delhi); red and yellow Soil (Dominant in Bihar, Rajasthan, West Bengal, Madhya Pradesh and Uttar Pradesh) and mixed red and black soil (Madhya Pradesh, Rajasthan and Uttar Pradesh) indicate low to moderate erosion. Das Gupta, 1984 (Table 16).

Table 16 : Soils in the Ganga basin and their susceptibility to erosion (area in square Kilometer)

State	*Mountain Soils*	*Sub montane soils*	*Alluvial Soils*	*Red Soils*	*Red & Yellow Soils*	*Mixed Red & Black Soils*	*Deep Black Soils*	*Medium Black Soils*	*Shallow Black Soils*	*Laterite & lateritic Soils*	*Total area*
Haryana	—	—	33,516	684	—	—	—	—	—	—	34,200
Uttar Pradesh	12,428	32,586	209,491	16,457	740	4,438	10,502	3,621	3,962	188	294,413
Bihar	—	—	91,881	32,338	19,333	—	—	—	—	858	144,410
West Bengal	—	—	57,323	704	7,271	—	—	—	—	6,712	72,010
Rajastan	—	—	40,768	12,166	16,827	9,810	9,332	9,394	9,085	—	107,382
Madhya Pradesh	—	—	17,272	39,075	5,990	11,168	14,251	79,881	33,470	598	201,705
Himachal Pradesh	1,296	4,280	223	—	—	—	—	—	—	—	5,799
Delhi	—	—	1,225	260	—	—	—	—	—	—	1,485
Ganga Basin	13,724	36,866	451,699	101,684	50,161	25,416	34,085	92,896	46,517	8,356	861,404
Per cent of total area	1.59	4.28	52.44	11.80	5.83	2.95	3.96	10.78	5.40	0.97	100.00
Susceptibility to erosion	Very high	Very high	Very high	High	Moderate	Moderate	Low	Low	Very high	Very high	

Source : Das Gupta S.P. (1984)

Alluvial fill covers more than 52 per cent of the basin. The alluvial deposits of the basin has not only covered the gangetic trough but also extend over a sizable portion of the peninsular foreland in the form of a layer, which is less than 3 m thick.

8

Remnants of Riverbeds Due to Shifting of Its Course

Floods area recurring feature of the Ganga plain. Not infrequently, but when the Ganga itself remains inspate and thwarts the incoming water from its numerous tributaries in this region, large areas get flooded. Almost all the rivers in the region develop a capacity to spill over in the monsoon period and are notoriously dynamic in their character, particularly the Rapti, the Ghagra, the Gandak, the Kosi, the Son and the Ganga itself (Singh, 1965) and also to a large extent the Damodar of West Bengal. These rivers have shifted their courses frequently covering wide areas throughout the historic times as is reflected by the remnants of their former beds in the form of ox-bow lakes, meander loop, dead arms, chars, tals and ruins of settlements.etc. (Fig. 6-8)

These water areas were known to be potentially rich source of fish supply. Further, the recent taming through construction of dams and barrages, embankments and roads has accentuated the problem of drainage, particularly in West Bengal during excessive rains both in the catchment area and in the lower reaches. This is mainly due to non-completion of drainage scheme in the lower valley. The reservoirs release water at a time when the channels are filled with drainage water from neighbouring areas and thus create water logged areas. Some typical instances of the remnants of river beds resulted from shifting of river courses in all the states are discussed below.

Uttar Pradesh

River Ganga has shifted over 35 Km between Bhojpur Tal and Surha Tal causing changes in the site of Ballia. Towns and river ports like Gola, Barhej etc. have been affected by similar shifts of the Ganga (Fig. 6). Information on the number of ox-bow lakes, chars and mans of varying sizes resulted from silting due to shifting of R. Ganga are not available.

Bihar

The Kosi, the most dynamic among rivers has marched about 120 km in course of last 200 years from near the township of Purnia to the Tiljugo in Darbhanga district (Fig. 7). Historically, it has oscillated from Brahmaputra to about the Ganga-Gandak confluence. The Gandak and the Kosi river system form are extensive network in N-Bihar in the district of Saran, Champaran, Muzaffarpur, Saharsa, Darbhanga, Monghyr and Purnea. Both the Gandak and the Kosi rivers have changed their courses even so often especially the latter leaving behind varying stretches of waters, the ox-bow lakes. Some of these continue to retain their continuity with the original rivers through some narrow channels atleast during the rainy season, while some others are completely cut off and remain

Table 17 A : Industries Causing Pollution in the Ganga

Upper Ganga Plain :

Antibiotic and heavy electrical factory, Textile mills, Tanneries, Woollen mills, Jute mills, Indian Explosive Ltd., Thermal power station and a number of Chemicals, Metallurgical and other factories, Indian Farmers Fertilizer Corporation Limited (IFFCO) manufacturing urea etc.

Middle Ganga Plain

Varanasi has besides large diesel locomotive (Maruadih), Chemicals (Sahapuri) and glass (Ramnagar), Asia Cycle Works, a number of suger factories and a dozen of various Industrial units.

Dalmianagar on Son at the tip of Rohtas plateau in the Sahabad district is a heterogeneous industrial complex with cement, asbestos, chemicals, paper and pulp, card board, plywood, sugar, vegetable ghee factories. Barauni has a large oil refinery and associated petrochemical complex, Bata shoe factory and Mc Dowell distillery at Mokamah (Barauni) in North Bihar Plain.

Table 17 B : Industries causing Pollution in the Bhagirathi-Hooghly.

Lower Ganga Plain

Pulp and Paper
- Sulphite Process
- Kraft Process
- Soda rocess

Textile
- Natural Cotton
- Synthetic rayon

Tannery
- Rubber
- Distillery
- Power (thermal)

Oil
- Petroleum
- Refinery

Miscellaneous
- Paints and Varnishes
- Polythene
- Hydrogenated vegetable oil and soap
- Safety match
- Chlorine
- Rubber Chemicals
- Insecticides
- Phosphate fertilizer
- Shellac
- Pesticides
- Antibiotics
- Chloro-alkyl plant

isolated bodies of water-shed. These are locally termed as 'man'. In addition to these, there are certain low lying extensive areas locally termed as 'chars' which got inundated from the flood waters of the adjoining rivers and retain the water for several months. In the Kosi basin one encounters long stretches of old river beds and over flow channels (Dhars) which remain dry during the greater part of the year. All these water bodies were reportedly yielding a good quantity of dish till recently. It is only in this State that a thorough districtwise survey has been made on a number of 'mans' and chars and their characteristics. (CIFRI, Survey Report 7, 1983). Captain Abraham Kuruvilla, Chairman, Dredging Corporation of India stated that the Kosi river is said to have risen by as much as 6 m above the normal level which explains why its banks overflow too early every monsoon season (Statesmen Report 13th June, 1988).

A large number of 'mans' of varying sizes and a few 'Chars' exist in the Champaran district having water spread areas ranging from 4-405 ha. According to DFO, Champaran, the production during 1965-66, 1966-67 and 1967-68 were 2883.30 t, 2672.30 t and 2537.40 t respectively. In Muzaffarpur district, the problem of decline of 'man' fisheries was reportedly not so acute excepting a few cases. The production figure of these is not available. Of the remaining districts information reveals that there were only a few 'mans' in the northern part of Monghyr district and 38 mans covering an area of only 60 ha in Saharsa district. Fishery was sharply on the decline.

With the commissioning of Gandak and Kosi Valley Project involving the raising or extension of existing bundhs and the construction of new ones, the continuity of some of the 'mans' and 'Chars' with the river was thus lost resulting in appreciable fall in their fish production.

West Bengal

The Hooghly and the Damodar, notorious for shifting the course are the only important rivers of West Bengal together with other streams which are either old courses of Damodar e.g. The Kana-Damodar, the Kana Nadi etc. or distributaries of the Bhagirathi-Hooghly like the Saraswaty, have created a number of marshes. These are principally the Dankuni bil, the Makhalpara bil, the Bamandanga bil, the Ramnathpur bil, the Dingebhanga bil, the Churka bil, and these are the depressions located mainly between the two principal rivers mentioned (Fig. 9 &10). The criss-cross channels are some evidences of a decaying distributary system which are no longer capable of carrying much silt or water of the marshes are without any outlet. The Damodar and other channels through which the flood waters drain, overspill, quite frequently and create water logged areas which is difficult to drain out (Bhattacharyya and De, 1972).

According to Master Plan for Fisheries Development report, (Part—I, 1974) there are about 2.0 lakhs acre of beels (this includes brackishwater also) in West Bengal which were once a potential source of fish supply. The areas being the beds of defunct rivers, in most cases have seasonal or perennial connection with nearby flowing rivers or canals. But there is no report about the number and extent of water areas under different size group of all these different categories of water sheets. Besides, the number of these water areas and to what an extent they are in derelict condition and how many of them are seasonal or perennial have also not been mentioned. Thus picture of these potential fishery resources of the state is not clear. Like Bihar, no systematic survey of such water areas or marshes have been carried out in this State.

9

Industrial Wastes—Their Pollution Load Disposed into the Ganga

The upper stretch of the Ganga is subjected to pollution by the effluents drained from antibiotic and heavy electrical factories at Rishikesh (9.02 MLD) and tanneries, woollen mills, textiles, jute mills etc. (167 MLD) at Kanpur.

The middle stretch suffers from pollution through drainages of effluent from fertilizer plant (5.50 MLD) at Allahabad. Important industrial units like Diesel, Locomotives, Chemicals, glass, cycle unit and dozen of various industrial units between Varanasi and Mughal Sarai also contribute their wastes appreciably. Besides, oil refinery associated with petrochemical complex, fertilizer, Bata shoe factory (11,000 gallon/day 49500L/d) distillery (2, 00,000 lit/d) also individually contribute this wastes in the Ganga in South Bihar. The quantum of effluents from these sources are quite high.

Besides discharging of wastes into the river, dumping of dead bodies in the Ganga also poses great danger to the river. It is estimated (Chaphekar and Mhatre, 1986) that around 30,000 dead bodies are burnt along the ghats at Varanasi every year and over 40,000 dead bodies are cremated annually at different ghats (Bhagalpur, Sultanganj, Monghyr and Sinaria) of south Bihar. Between half and two tons of wood is needed to burn a dead body. Such system is in vogue in other big cities. The flyash emitted can cover the river bed and adversely affect the growth and survival of fish food organisms.

The lower Ganga in the deltaic tidal stretch encounters heavy discharges from industrial effluents (430 MLD) from more than 96 factories such as, pulp and paper, distillery and yeast, viscose rayon, cotton textile, tannery, paints and varnishes, rubber and cycle rim, metal and steel, safety match, Chemicals, Thermal power plants, pesticides (BHC) unit, hydrogenerated vegetable oil and soap, phosphate fertilizer, oil storage, Chloro-alkyl plant, antibiotic etc. flanked on both the banks between Nayasarai and Birlapur (about 92 km) and oil refinery and pesticide (dimethoate) manufacturing unit at Haldia (150 km down stream of Calcutta) beside pesticide residues through urban and agricultural run off. (Magnitude to pollution with stretch is indicated in Table 18. Organic components of industrial wastes (240 MLD) in the tidal stretch make a total BD_5 demand of 55.1 t/d of which lignocellulosic matter makes a BOD demand of 28.6 t/d (Ghosh *et. al.* 1973). Further Total pollution load separately, their broad classification and chemical characteristics of some industrial domestic and municipal wastes are also indicated in Table 19, 20 and 21. The stretch also receives oil (0.31t), heavy metal like zinc, copper and chromium are drained from viscose rayon (8mg/l), Cotton Textile (19 mg/l) and Tanneries (28 mg/l) respectively. Wide spectrum of heavy metal contaminants from various industries released into the basin are given in Table 22.

Table 18 : Magnitude of Pollution with stretch in the Ganga river system investigated or surveyed

State	River/Dam	Place	Stretch	Magnitude of Pollution				Source	References
				Severe	Moderate	Slight	Total		
Delhi	Najafgarh nala (R.Jamuna)	New Delhi	Entire Stretch	√				DDT Factory	George, M.C., N.K. Kaushik and S.K. Srivastava (1965).
		New Delhi	20 km.		√			Municipal Waste	
		ITO (Delhi), Agra, Mathura and Allahabad	City side of Delhi	√				Heavy metal pollution	Jhingran, A.G. and H.C. Joshi (1987)
Haryana	R. Jamuna	Jamunanagar (Ambala)		√					
Uttar Pradesh	R. Suvaon	Balarampur	3.2 Km	√				Sugar mill	Banerjee, S and M.P. Motwani (1960)
	R. Gomati	Lucknow	20.8 Km.			√		Paper Mill, City sewage	Bhaskaran, T.R., R.N. Chakra-borty and R.L. Trivedi (1965)
	R. Dajora	Bareilly	8 – 9.6 Km.		√			Synthetic Rubber factory	Survery C.I.F.R.I., Barrackpore (1966)
	R. Kali	Bulandshahr	160 Km.	√				Sugar, Paints, dis-tillery, Soap, Glycrine, Rayon, Silk and yarn etc	George, M.G., S.Z. Qasim and A.Q. Siddiqui (1966)
	R. Ganga	Kanpur	25 Km.		√			Municipal waste, Tanneries, Textile Mill,	1. Ray, P. and A, David (1966) 2. Saxena, K.L., R.N. Chakraborty, A.Q. Khan, S.N. Chattopadhya and H.Chandra. (1966)
	R. Deoha	Pilibhit				√			Survey, C.I.F.R.I. Barrakpore.
Bihar	R.Son	Dehri-on Son	1.6 Km.	√				Caustic soda, chlorine etc.	Motwani, M.P., S.M. Banerjee and S.J. Karamchandani (1956)
		(Dalmianagar)	20.4 Km.		√			Paper Mill, Chemicals, Cement Factory.	

Contd....

State	River/Dam	Place	Stretch	Magnitude of Pollution				Source	References
				Severe	Moderate	Slight	Total		
	R. Daha	Siwan	1.6 Km.	√				Sugar and Distillery	1. David, A. and P. Ray (1966)
				34.4 Km.		√			2 Bhaskaran. T.R., R.N. Chakraborty and R.C. Trivedi (1965)
	R. Damodar	Between Bokaro and Panchet	90 Km.	√				Fly ash from Coal washeries, thermal power plant, cromate Cyanides, Phenol, ammonia, alkalies and Napthalene from Fertilizer Corporation	Gopalakrishanan, V., R. N. Pal and P.K. Chakraborty (1966)
	R. Ganga	Barauni		√				Oil refineries	Sunderesan, B. B., P. V.R. Subrahmanyan and A.D. Bhinde (1983)
West Bengal	R. Damodar	Entire Damodar (538 km)		√				Industries under Asansol-Durgapur agglomeration	Banerjee, R.K. (MS, 1988)
		Konar-Panchet (45 Km.) Maithan-Durgapur	(35 Km.)	√				Industries under Durgapur agglomeration.	Dhaneswar, R.S., C.S. G. Rao, M.K. A. Kutty D. Ray and R. C. Trivedi (1972).
	R. Churni	Mathabhanga-Churni 65 Km.		√				Darsana Sugar Mill (Bangladesh)	Ghosh B.B., M.K. Mukhopadhyay and M.M. Bagchi (1990)
	Bhagirathi-Hooghly (estuary)	Bandel – Baj Baj	72 Km.					Various complex industrial and Municipal wastes	Ray, P. and G. Laha (1981)

Table 19 : Pollution Load Contributed by Various Industries, Estates, and Municipalities with their Corresponding Percentages

Nature of waste water	*Vol. of waste Water (m³/day)*	*BOD₅ tonnes/day*	*Total solids tonnes/day*	*Suspended Solids tonnes/day*	*Dissolved Solids tonnes/day*
Industrial Wastes					
Pulp & Paper Industry	178612.34 (15.5%)	28.59 (26.90 %)	547.40 (23.61 %)	299.20 (28.29 %)	248.20 (19.68 %)
Textile Industry	20366.08 (1.77 %)	7.26 (6.83 %)	71.80 (3.10 %)	41.82 (3.95 %)	29.98 (2.38 %)
Distillery & yeast	675.08 (0.06 %)	10.29 (9.69 %)	5.242 (0.22%)	0.402 (0.04%)	4.84 (0.38 %)
Tannery	11365.00 (0.99 %)	4.54 (4.27 %)	86.70 (3.74 %)	13.50 (1.28 %)	73.20 (5.81 %)
Rubber	10910.40 (0.95%)	0.43 (0.4%)	8.70 (0.38%)	1.70 (0.16 %)	7.0 (0.56 %)
Oil Refinery	6600.00 (0.49 %)	0.158 (0.15 %)	9.976 (0.43 %)	0.215 (0.02 %)	9.761 (0.07 %)
Miscellaneous	129728.29 (11.24 %)	14.92 (13.95 %)	164.235 (7.08 %)	55.305 (5.23 %)	108.93 (8.64 %)
Domestic and Municipal Wastes	79555.00 (68.95 %)	40.08 (37.72 %)	1424.36 (61.44 %)	645.40 (61.03 %)	778.96 (61.78 %)
Total	1153807.19	106.268	2318.413	1057.542	1260.871

Source : Ray P. (1981).

Table 20 : Broad Classification of the waste waters and their Pollution Load/contributed into the Hooghly Estuary with Percentages

	Vol. of waste water (m³/day)	*BOD₅ tonnes/day*	*Total solids tonnes/day*	*Suspended Solids tonnes/day*	*Dissolved Solids tonnes/day*
A. Wastes Containing Chiefly Organic Materials :					
Biological wastes[1] (wastes from processing of biological materials and/or from biological process)	196471.30 (73.49 %)	45.91 (85.10 %)	680.642 (87.77 %)	346.522 (93.75 %)	334.11 (82.33 %)
Miscellaneous organic chemical wastes[2]	32770.41 (12.26 %)	2.598 (4.82 %)	41.281 (5.32 %)	9.520 (2.58 %)	31.761 (7.83 %)
Hydrocarbon wastes[3]	10910.40 (4.08 %)	0.43 (0.80 %)	8.70 (1.12 %)	1.70 (0.46 %)	7.0 (1.72 %)
B. Wastes Chiefly Mineral in Nature or Partly Mineral and Partly Organic[4]	27185.08 (10.17 %)	5.01 (9.29 %)	44.83 (5.78 %)	11.90 (3.22 %)	32.93 (8.11 %)

1. Pulp and paper, textile (cotton), tanneries, distillery, and yeast.
2. Paints and varnishes, shellac, oil refinery and hydrogeneted vegetable oil and soap.
3. Rubber.
4. Metal and Steel, textile (rayon), and match.

Source : Ray, P (1981).

Table 21 : Chemcial Characteristics of some Industrial, Domestic and Municipal Wastes Expressed as average Values During the Period 1968-1971

Nature of wastewater	pH	Alkalinity	Sp. Cond. m mhos	DO	BOD5	KMnO$_4$ OC mg/l			Nitrogen			
						3 mt	4 hr	30 mts 100°C	Free & saline	Alb Amn	Nitrite-N	Lignin
(1)	(2)	(3)	(4)	(5)	(6)	(7)	(8)	(9)	(10)	(11)	(12)	(13)
Pulp and Paper Industry Sulphate Process	6.5–9.4	50–940[b]	479–8779	Nil–6.8	12–700	Nil–8.0	90–225	4–228	Tr.–0.9	Tr.–1.6	Nil	265[d]
	(7.4)	(286)	(2191)	(2.3)	(144)	(4.0)	(145)	(120)	(0.4)	(0.33)	Nil	
Sulphite process	5.3–6.9	30–450	654–79993	Nil	400–1920	6–985	145–2144	240–3300	0.5–1.4	0.11–1.36	Nil	1000[d]
	(6.2)	(169)	(9095)		(986)	(370)	(1201)	(1899)	(0.8)	(0.89)		
Soda process[a]	7.1–9.5	126–2070	287–3789	Nil-7.7	13–450	3–180	14–360	5–520	–	–	–	200[d]
	(8.5)	(6.08)	(1214)	(3.45)	(60)	(34)	(84)	(140)				
Textile Industry Natural	5.6–10.5	160–1896[c]	514–7199	Nil–8	17–1335	1.2–75	4–250	6–1060	1.4–2.35	0.45–0.5	Nil	
		(564)	(2946)	(2.1)	(380)	(30)	(98)	(162)	(1.87)	(0.47)		
Synthetic	2.0–4.9	Nil–14	912–10284	Nil–2.3	26–1175	18–32	4-23	40–438	–	–	Nil	
	(2.8)	(1.0)	(3697)	(0.3)	(331)	(24)	(16)	(141)				
Distillery	5.0–6.1	502–040	7199–10284	Nil	5760–21600	1600–2000	1040–8500	1025–31300	Nil–0.1	Tr.–0.25	Nil	
	(5.5)	(1135)	(5160)		(15687)	(1806)	(4918)	(8502)	Tr.	Tr.		
Yeast	5.4–7.7	432–650	1565–2399	Nil	1460–5760	25–85	79–1200	130–1793	0.6–19.2	4.5–2.9	Nil	
	(6.6)	(564)	(1866)		(3387)	(45)	(790)	(1216)	(9.63)	(16.75)		
Tannery	5.2–8.1	42–500	2057–21838	Nil–3.2	60–1280	25–65	65–750	88–1020	5.6–25.5	0.9–8.0	Tr.–0.12	
	(7.2)	(292)	(7173)	(0.6)	(404)	(44)	(120)	(339)	(10.03)	(2.9)	(0.04)	
Rubber	6.6-8.0	240–320	604–835	1.5–6.7	10–106	2–72	10–40	9–89	1.48–3.3	0.80–0.8	Nil	
	(7.3)	(271)	(697)	(4.8)	(40)	(11)	(19)	(51)	(2.2)	(0.34)		
Oil refinery[e]	5.5–9.0	110–142	2181–3999	2.6–6.0	7-30	–	–	–	–	–	–	
Miscellaneous Paints and varnish	6.1–10.5	104-920	757–4499	Nil–9.6	8–457	Nil	6–25	8–376	0.3–6.0	03–1.5	NIl–56	
	(8.1)	(370)	(1513)	(3.00	(137)		(11)	(88)	(2.0)	(0.43)	(.12)	
Safety match	6.8–8.1	187–396	3428–4388	Nil–6.1	16–76	–	–	20–118	2.4–6.22	.85–1.5	NIl	
	(7.5)	(308)	(3345)	(3.0)	(39)	–	–	(79)	(4.30)	(1.17)		
Shellac	5.0–7.3	832–1500	1285–19452	Nil	310–1600	50–250	225–800	320–1420	–	–	Nil	
	(6.3)	(1137)	(9250)		(963)	(75)	(375)	(900)				
Metal and Steel	6.1–80	11–250	299–4234	Nil–6.2	10–74	–	–	8–76	Nil–0.1	Nil–0.4	Nil	
	(7.2)	(194)	(1161)	(2.4)	(46)	–	–	(40)	(Trace)	(Trace)		

Contd...

Nature of wastewater	pH	Alkalinity	Sp. Cond. m mhos	DO	BOD_5	$KMnO_4$ OC mg/l			Nitrogen			
						3 mt	4 hr	30 mts 100°C	Free & saline	Alb Amn	Nitrite-N	Lignin
(1)	(2)	(3)	(4)	(5)	(6)	(7)	(8)	(9)	(10)	(11)	(12)	(13)
Combined Wastes (Bagh Canal)	6.9–8.2 (7.8)	360–480 (407)	4056–9506 (6969)	Nil–2.4 (1.6)	126–156 (136)	10–35 (25)	32–65 (55)	54–118 (80)	0.12–4.24 (2.78)	Tr.–8.4 (2.9)	Tr.–0.2 (0.12)	
Domestic and Municipal Wastes	6.7–8.6 (7.7)	204–660 (412)	1531–8999 (4048)	Nil–7.4 (0.4)	24–560 (187)	15–95 (45)	20–156 (120)	31–540 (198)	0.12–12.5 (6.6)	0.14–4.0 (0.83)	0.02–0.2 (0.15)	

Results are expressed in mg/1 except otherwise stated
a–Data pertain to 1970–1973
b–Contains carbonate alkalinity
c–Contains hydroxyl alkalinity
d–Basu (1966)

e–Data provided by B.B Ghosh, Central Indland Fisheries Research Institute, Barrackpore
Further data :
TSS (mg/1) : 18–55
TD (mg/1) : 1200–1420
Sulphide (mg/1) : 0.2–2.0
Phenolic Compound (mg/1) : Trace–1.0
Oil (mg/1) : 2–12
Ghosh, Ray and Gopalakrishnan (1973)

Source : Ray, P. (1981)

Table 22 : Wide spectrum of heavy metal contaminents released into the environment

Industry	Zn	Cu	Fe	Mn	Pb	Cr	Ni	Cd	Sn	CO	Sn	CN	Hg	As	Se	(1) Ag	(2) B	Phenol
Electroplating	*	*				*	*	*					*					
Textile Waste	*	*				*	*	*										
Iron and Steel			*	*														
Iron and Ferro-manganese	*	*	*	*	*													
Synthetic Rubber PVC	*				*			*	*									
Soap detergents																	*	
Pulp and Paper	*	*			*	*	*						*					
Organic Chemical and petrochem	*		*		*	*		*			*		*			*		
Fertilizer	*	*	*	*	*	*	*	*					*	*				
Petroleum Refinery	*	*	*		*	*	*	*										
Leather tanning and finishing						*												
Alkalies, Chlorine Inorganic Chemicals	*		*		*	*		*			*		*					
Paints and Varnishes	*				*	*		*										
Perfumery, drug pharmaceuticals & Chemicals			*		*	*		*					*					
Galvanising	*							*										
Cosmetics	*																	
Battery/Battery Plant.	*				*		*	*					*			*		
Fly ash	*				*	*	*	*					*	*	*			
Cement factory	*																	
Domestic Waste	*	*			*	*	*	*						*		*		
Spoil Heaps								*		*		*		*				*
Photography development & X-ray plates.								*								*		
Fish Processing	*	*				*	*	*										
Meat Processing	*	*				*	*	*										
Bakery	*	*				*	*	*										
Brewery, soft drinks and flavouring	*	*				*	*	*										
Ice Cream	*	*				*	*	*										
Laundry						*	*											
Soft drink																		

1. Highly toxic to bacteria, soil flora and plants. 2. Powerful herbicide.

Apart from industrial and municipal wastes, there is another source which contributes substantial metallic pollutants from the washings of images at the time of immersion in the Ganga every year. Ghosh (1993) has revealed some valuable information on this aspect in the article "The alarming news of the Ganga pollution." His data is based on for the last three years during October to January when all the immersion takes place. According to him normally the water contains 0.1/mg chromium, 0.05/mg lead, 0.1/mg zinc and 0.1/mg copper per litre. During October to January, the values of chromium and lead increases 10 to 20 times and zinc and copper to 1.5 mg per litre. He has concluded that such progressive increase of the metallic pollutants is capable of upsetting the ecological balance of the Ganga and requires to be controlled immediately. Ghosh (1993) however did not indicate the stretch he studied.

Pollution from Oilspill

The Ganga experienced an ugliest incident of human indifference when it was set ablaze near Monghyr, Bihar in 1968. (Sunderasan *el. al.* 1983). The discharge of oil from refinery effluent led to the fire and resulted in colossal loss of fishery in the river. A similar case of heavy discharge of furnace oil spill in the river Damodar from an industrial unit in Bihar (1987) which was also caught fire in a 1.5 km stretch.

10

Ecological Imbalance of the Ganga

Quite a substantial portion of water is withdrawn from the Ganga by the draught hit states upstream—Uttar Pradesh, Haryana and Bihar for the purpose of irrigation is changing the regime of the stream flows considerably. For example, at Haridwar head works of the upper Gangetic canal net work nearly 60 per cent of the total available water is drawn off for the purpose of irrigation, reducing the Ganga to a mere trickle with an average flow of hardly 48 m^3/sec let off for all other uses. A sort of balance should have to be established for various competitive uses of water in the Ganga basin with a view to adequately controlling the pollution levels along different reaches of the Ganga and its tributaries.

During seventies average of mean seasonal flow of the Ganga indicated below for Uttàr Pradesh, Bihar and West Bengal clearly states that in Uttar Pradesh river condition remained worse for long eight months from October to May, followed by West Bengal and Bihar.

Seasonwise ratio of waste to river water per month clearly states that the Ganga in Uttar Pradesh throughout the year viz. the dilution ratio is maintained at 1 : 1 ratio both during monsoon and post monsoon period, while in winter and summer the waste exceeds the flow. West Bengal suffers during postmonsoon (1 : 8), winter (1: 9) and summer (1 :7) while Bihar suffers only during winter (1 : 7) and summer (1 : 5) Table 23 A, B, C.

Table 23 (A) : Average of mean seasonal flow mm^3/month of the Ganga in different states

	Monsoon *June-Septem.*	*Postmonsoon* *Oct.-Nov.*	*Winter* *Dec.-Feb.*	*Summer* *March-May*
Uttar Pradesh (Av. of 8 centres)	13219	1291	532	444
Bihar (Av. of 3 centres)	45102	8393	2880	2021
West Bengal (Av. of 2 centres)	3935	1727	2006	1673

(B) Seasonwise ratio of waste to river water in different states

Uttar Pradesh	1:1	1:1	x	x
Bihar	1 : 121	1: 22	1:7	1:5
West Bengal	1:18	1:8	1:9	1:7

Source : Modified, based on the data of Das Gupta (1984).

x indicates water exceeds the flow of the Ganga.

Contd....

(C) Monthwise generation of waste water mm31 month

	Monsoon June-Septem.	Postmonsoon Oct.-Nov.	Winter Dec.-Feb.	Summer March-May
Uttar Pradesh		15509 ÷ 12 = 1291		
Bihar		4453 ÷ 12 = 371		
West Bengal		2570 ÷ 12 = 214		

* Based on annual generation of waste water (Das Gupta, 1984)

Table 24 : Water Quality of Ganga during Summer 1977 in the upper stretch around some important Cities

Centers		Water temp °C	D.O. mg/l	B.O.D. mg/l	Coliform MPN No/100 ml
Rishikesh	u/s	18	8.0	4.4	100
	d/s	18	8.0	6.7	2400
Haridwar	u/s	18	8.0	8.0	2400
	d/a	20	8.0	7.2	1600
Narora	u/s	30	7.5	4.2	130
	d/s	29	7.4	7.2	63
Kannauj	u/s	28	6.0	4.6	240
	d/s	29	5.9	7.0	130
Kanpur	u/s	27	7.5	10	35
	d/s	29	6.0	27	2400
Allahabad	u/s	31	6.5	6.0	350
	d/s	30	6.4	6.8	2400
Varanasi	u/s	27	6.4	4.0	49
	d/s	31	5.8	10.0	2400

Source : Bhargava D.S. (1981A); u/s —upstream, d/s—downstream of the town.

Table 25 : Water quality of the Ganga during extreme Summer condition (May) of 1981 and 1982 (average value)

Monitoring station	Catchment area (Km^2)	Mean flow in Ganga m^3/S.	DO	BOD	Total Coliform MPN/100ml.	Stream velocity m/Sec May	Stream velocity m/Sec July
1. Haridwar	22.685	381	5.7	2.0	2,400*	0.6	0.35
2. Garhmuktesar	9.798	170	6.8	5.1	1,430	—	—
3. Kannauj	29.364	52	6.2	7.3	NA	—	—
4. Kanpur	4.274	86	6.5	15.3	NA	0.5	0.59
5. Allahabad	12.065	133	7.4	9.3	31,250*	—	0.90
6. Varanasi	11.799	285	6.4	9.7	12,075	—	0.40
7. Patna	33.265	1108	7.4	1.3	28,315	—	—
8. Monghyr	16.187	—	6.4	2.2	24,200	—	—
9. Farakka	22.895	—	8.1	0.03	125,500	0.35	0.44
10. Nabadwip	6.048	1013	7.5	0.6	58,500	0.56	0.42
11. Dakshineswar	3.947	—	6.0	2.0	240,000	0.27	0.27
12. Diamond Harbour	1.207	—	7.1	8.0	8,250	0.09	0.22

* In case of non-availability of May figures, data for the nearest month have been taken,

Source : State Water Pollution Control Boards of Uttar Pradesh, Bihar and West Bengal, reported by Das Gupta, S.P. (1984).

Table 26 : Seasonal variations of the mean concentrations of indicators of organic pollution in the Ganga during 1980-1982

	BOD (mg/1)			*Do (mg/1)*			*Coliform bacteria (MPN) no/100 ml.*		
	Post monsoon	*Winter*	*Summer*	*Post monsoon*	*Winter*	*Summer*	*Post monsoon*	*Winter*	*Summer*
Upper Ganga Plain	9.76	8.48	10.05	6.67	7.25	5.89	9920	14663	10669
Hardwar- Nagbasuki temple (Allahabad)	2.50-17.70	1.70-18.60	2.50-16.30	5.25-8.05	5.26-8.92	4.35-10.10	43-24333	1677-41000	1416-25488
Middle Ganga Plain	3.55	3.34	3.41	7.12	7.85	6.88	25133	193286	26429
	0.47–7.30	0.54–6.50	0.72-8.37	5.20–8.27	4.67–11.75	6.57–8.84	3400–80913	5505-844667	5910–80534
Below Allahabad-Farakka.									
Lower Ganga Plain	1.0	2.82	3.07	6.26	8.27	6.46	28859	271579	260509
	0.53–1.23	0.35–15.23	0.45–15.58	4.53–7.80	7.47–9.10	5.40–6.90	2110–53533	8833–916667	22600–902333

Source : State Water Pollution Control Board U.P., Bihar and West Bengal reported by Das Gupta S.P. (1984). Table modified.

Table 27 : Organic load (BOD) disposed in the Ganga basin (1977)

States	*Urban (Kg/day)*		*Rural (Kg/day)*	*Total*
	Domestic	*Industrial*	*Generated from human and cattle*	
Haryana	47,250	29,857	100,138	177,245
Uttar Pradesh	455,859	873,815	2,017,991	3,347,665
Bihar	99,798	83,064	1,340,167	1,515,029
West Bengal	310,751	131,885	478,101	920,737
Rajasthan	59,400	15,654	410,393	485,447
Madhya Pradesh	118,790	24,513	527,488	670,791
Himachal Pradesh	4,138	—	18,966	23,104
Delhi	250,500	7,452	11,505	269,457
Ganga basin	1,338,486	1,166,240	4,904,749	7,409,475

Source : Das Gupta, S.P. (1984).

Table 28 : Seasonal fluctuations (u/1) of planktons at different centres of R. Ganga (1978-79)

Sampling centres	*Summer*		*Monsoon*		*Winter*	
	P	*Z*	*P*	*Z*	*P*	*Z*
Mokamah	—	34	—	7	—	23
Barauni	4137	17	530	7	1298	20
Monghyr	3513	20	405	8	1326	26
Sultanganj	1484	14	337	8	1107	21
Bhagalpur	2614	27	377	6	1041	19
Sahibganj	1605	25	318	5	870	21
Rajmahal	4720	27	343	8	3091	18
Farakka	2404	31	298	9	2867	14

P : Phytoplankton; Z : Zooplankton

Source : Bilgrami K.S. and J. S. Dutta Munshi (1979).

Location of industrial establishments, sewage generation points in the main Ganga as well as in the estuary surveyed by different workers are shown in Fig. 11-14. Some pollution parameters indicating water quality criteria above and below of some big cities, bank and mid-river during extreme summer and also during different seasons and stretches of the Ganga organic load disposed from different states are indicated in Tables 24–27.

Seasonal fluctuation of plankton studied between Barauni and Farakka shows comparatively higher values during summer followed by winter (Table 28). Unpolluted bank and mid-river concentrations were always showing less values than polluted banks, mostly by 80-89 per cent. Coliform bacteria an indicator of organic pollution are presented in Tables 26,29 & 30. Its conc. in the bank and also in the mid-river between Nabadwip and Kalna superseded the permissible limit for drinking water (5000 MPN/100ml.). Seasonal variation of coliform bacteria in a more wider area viz. upper Ganga plain (Haridwar-Allahabad), Middle Ganga plain (Allahabad-Farakka) and lower Ganga plain, during monsoon, winter and summer were also studied during 1980-82. Middle Ganga plain and Lower Ganga plain and also winter and summer were indicating much higher values (Table 26). In this case, at all the centres, the concentration exceeded the permissible limit for drinking water although DO was indicating optimum level and BOD reasonably low values (indicating not

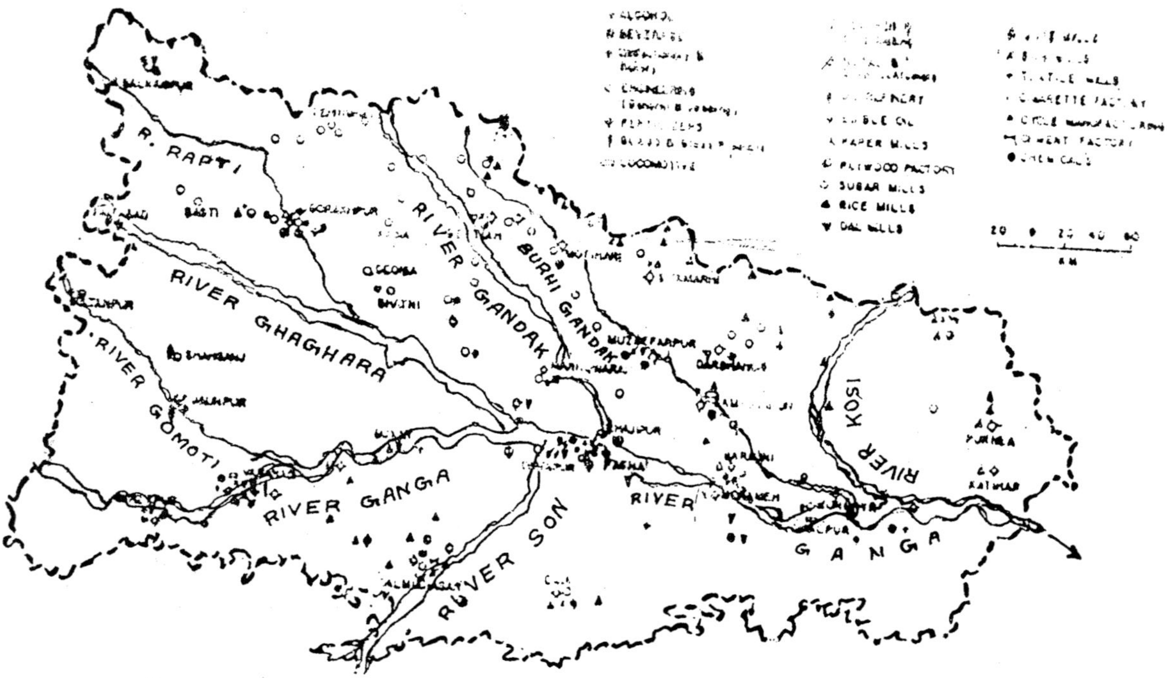

Fig. 11 : Indicating Location or various industries in the middle Ganga Plain

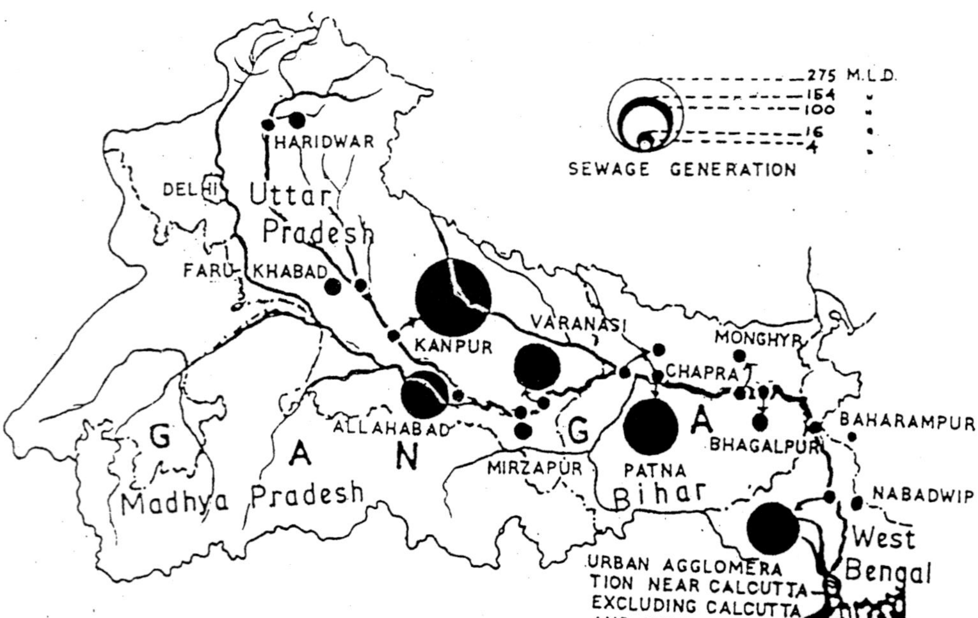

Fig. 12 : Estimated Sewage Generation from Class-1 Cities on the Bank of River Ganga

Source : Das Gupta, S.P. (1984)

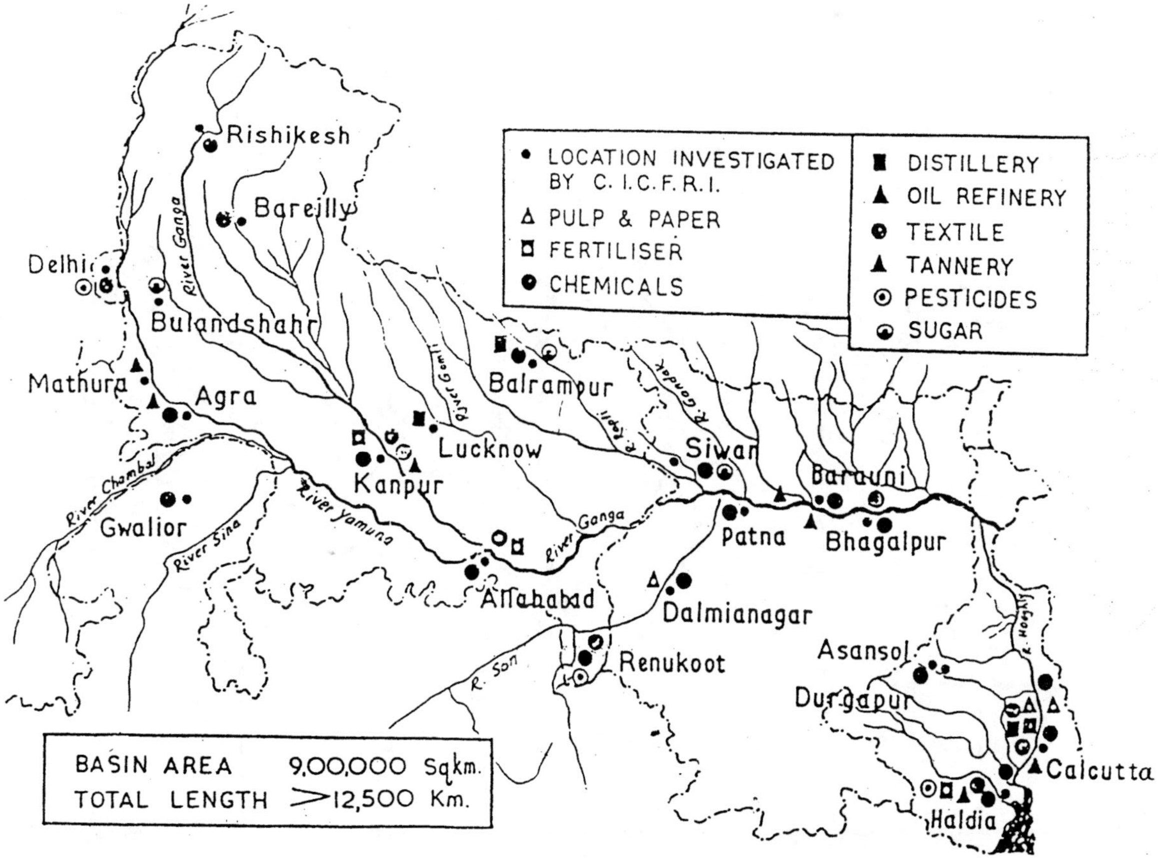

Fig. 13 : Location of Industrial Establishments Affecting the Fishery Resources of the Ganga River System

Source : Das Gupta, S.P. (1984)

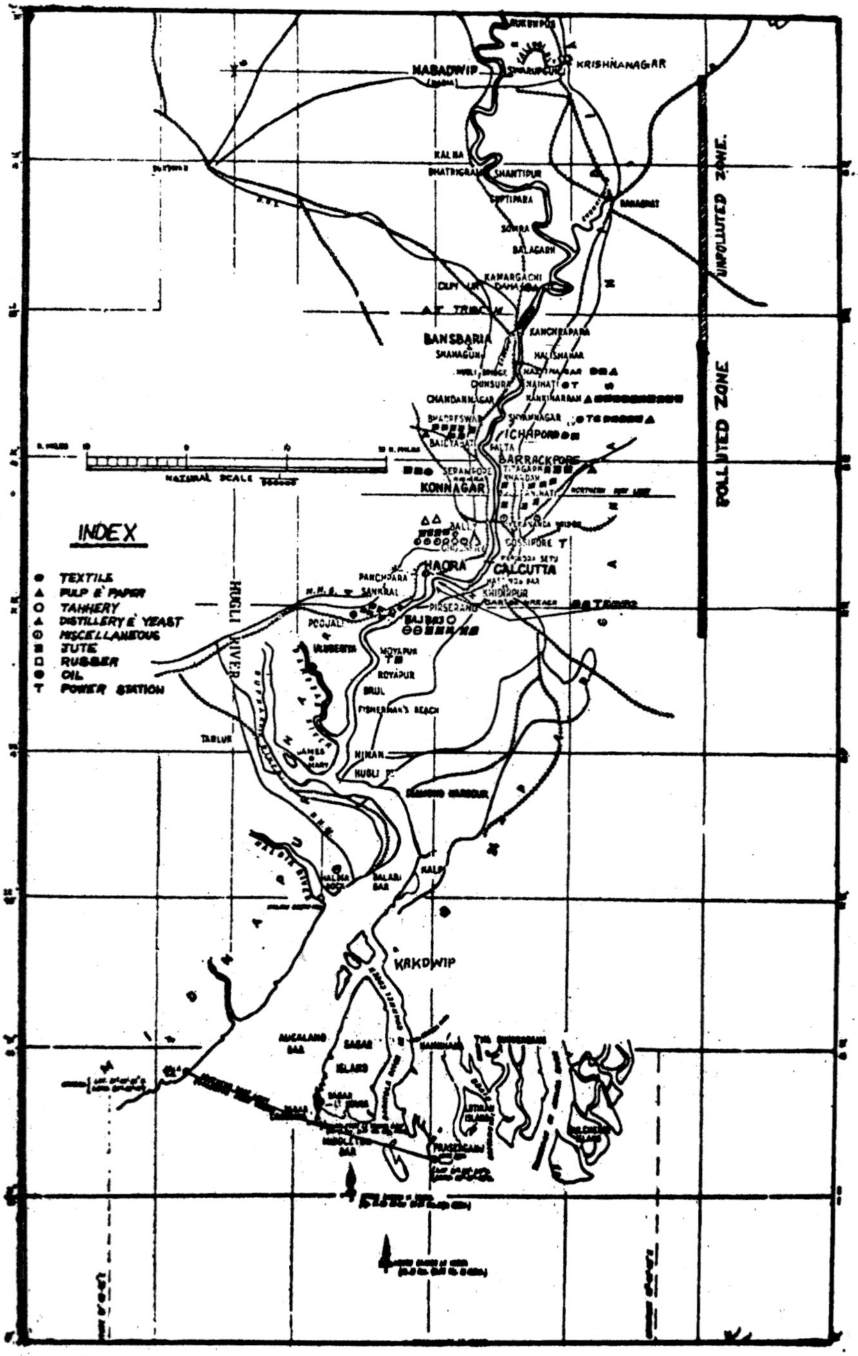

Fig. 14 : Industries Causing Pollution in the Hooghly Estuary
Source : Ray, P. and G.C., Laha, (1981)

Table 29 : Comparison of some selected Pollution parameters between bank and mid river of the Ganga in a stretch Monghyr— Katwa during 1981

Centre	*Period of sturdy*		*BOD (mg/l)*	*COD (mg/l)*	*Total Coliform count (MPN) (no/100 ml.)*	
Monghyr	**January to June**	Bank :	0.5 – 2.3 (1.7)	24 – 28 (25.3)	460 – 10670 (4510)	—
		Mid River :	0.9 – 1.5 (1.2)	16 – 20 (18.7)	23 – 1600 (866)	(80% less.)
Rajmahal	—do—	Bank :	0.5 – 2.2 (1.3)	8 – 24 (17)	1600 – 18000 (10235)	—
		Mid River :	0.4 – 0.9 (0.7)	8 – 24 (15.7)	93 – 2400 (1076)	(89% less)
Kahalgaon	—do—	Bank :	1.3 – 1.8 (1.6)	8 – 20 (16)	10680 – 21000 (16272)	—
		Mid River :	0.9 – 1.4 (1.1)	11.6 – 16 (15)	43 – 2400 (1045)	(93% less)
Bhagalpur	—do—	Bank :	1.0 – 2.3 (1.4)	15 – 24 (20.3)	1600 – 13680 (8637)	—
		Mid River :	0.6 – 0.9 (0.8)	8 – 17 (12.1)	93 – 2400 (118)	(98% less)
Nabadwip	**April & May**	Bank :	—	11000 – 21000 (16000)	21000 – 110000 (65500)	—
		Mid River :	—	4600 – 110000 (57300)	4600 – 46000 (25300)	(62% less)
Katwa	—do—	Bank :	—	240000 – 240000 (240000)	110000 – 240000 (175000)	—
		Mid River :	—	110000 — 240000 (175000)	21000 – 46000 (33500)	(80% less)

Source : **Water Pollution** Control Board, Bihar and West Bengal reported by Das Gupta, S.P. (1984).

Values in the parenthesis are mean values.

Table 30 : Seasonal fluctuation of Coliform bacteria (MPN) per 100 ml of water in the lower stretch of Middle Ganga Plain 1978-79

	Summer	*Monsoon*	*Postmonsoon*	*Winter*
	March—May	*June—September*	*October—November*	*Dec.—Feb.*
Barauni	134	187	299	146
	94 — 161	141 — 221	253 — 345	120 — 177
Mokama	44	83	146	80
	28 — 63	32 — 177	120 — 172	76 — 84
Monghyr	103	222	280	121
	24 — 177	109 — 345	212 — 348	94 — 148
Sultanganj	212	427	385	179
	161 — 253	278 — 542	345 — 426	141 — 221
Bhagalpur	148	183	227	117
	109 — 172	94 — 278	177 — 278	94 — 148
Sahebganj	63	149	126	98
	35 — 84	48 — 345	109 — 148	79 — 120
Rajmahal	67	76	86	71
	40 — 92	39 — 141	79 — 94	53 — 84
Farakka	42	89	71	59
	33 — 53	39 — 148	48 — 94	49 — 76

Source : Modified from Bilgrami, K. S. and J. S. Dutta Munshi, 1979.

much of organic pollution). Finally in another study made during extreme summer of May for consecutive two years, 1981 and 1982 (Das Gupta 1984) the entire stretch of the Ganga showed quite high values between Allahabad and Dakshineswar 24,000–2,40,000 MPN). The DO ranged between 5.7 and 8.1 mg/l and BOD mostly below 10mg/l (Table 25). Around Calcutta, bankwise bacterial density studied by NEERI during 1975-76, in a stretch Pathak to Chintamony (about 20 km) during low and high tide of summer and monsoon indicated, right bank to be more polluted (Left bank : LT. 26.7×10^5; HT : 29×10^6 during dry season against the right bank LT : 37.26×10^5 HT : 11.253×10^6 during monsoon). Some of the values mentioned by Das Gupta, 1984 (Table 26) quoting report of Uttar Pradesh, Bihar and West Bengal Govts. appear to be quite high (higher values : 24,333-9, 16,667) unlike Bhargava 1981a (Table 24) and Bilgrami and Dutta Munshi, 1979 (Tables 29,30) who have indicated higher values to range 350-2400 and 148-542 MPN (No/100ml). In none of the cases mentioned above, the reason for such high values is not indicated.

11

Impact on Biota Around the Outfall and the Main River—Their Pollution Status

Physiochemical and biological studies around the effluent discharge points of industries are important in determining implication of pollution. The effluents discharged have different types of chemical composition. The composition include mineral acids, bases and many other toxic substances of non-biodegradable and biodegradable in nature (Ray, 1981). In the main river pollution, studies

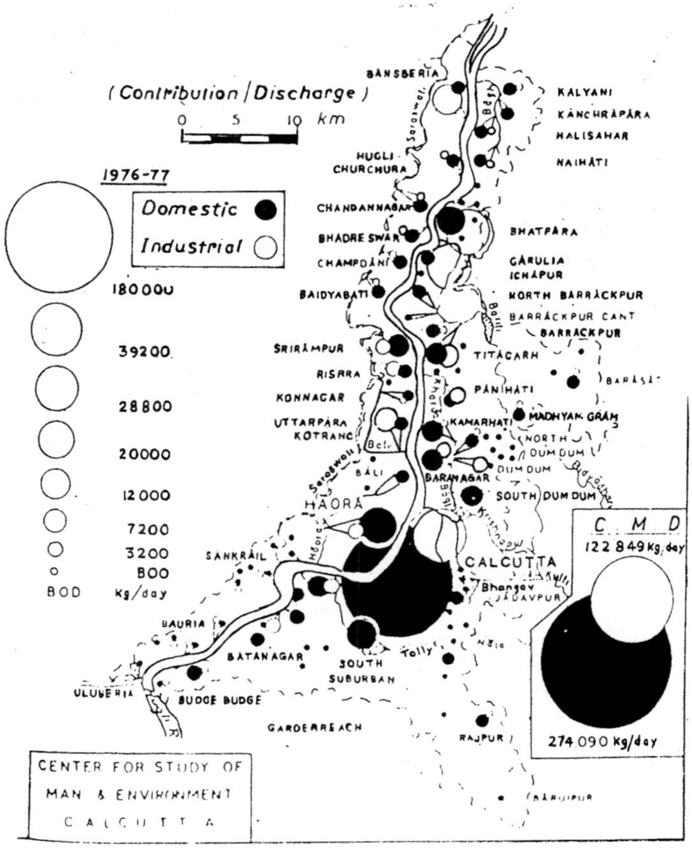

Fig. 15 : Generation and Discharge of Wastewater

(Tannery, industrial and sewage effluents) made are by Ray and David (1966) and Saxena *el. al,* (1966) around Kanpur, and waste from fertilizer factory producing urea by Chandra *et. al.*, (1981) at Phulpur (Allahabad); From domestic and industrial wastes by Rai (1978) around Varanasi; From industrial and oil refinery by Bilgrami and Datta Munshi (1979) in a stretch of 256 km from Barauni to Farakka and finally in the estuarine stretch of Bhagirathi-Hooghly by the effluents from synthetic rayon industry by Ghosh *et.al.*, (1976); Pulp and paper mills (soda process) by Basu *et. al.*, (1973) Ray *et. al.*, (1979) and Ghosh *et. al.*, (1980b); Rubber industrial complex by Ghosh *et. al.*, (1979 a); Pulp and paper industry (sulphite process) by Basu (1966); Ghosh *et. al.*, (1977) and Ray and Mitra (1980); Distillery by Ray *et. al.*, (1977); Tannery by Ghosh *et. al.*, (1980 a) etc. (Most of these studies were made for 1-$1^1/_2$ hour prior to high tide while in few cases, observations were made during the high tide also). (Besides these, additional studies were made for heavy metal pollution by Ghosh and Bagchi (1981) and Ghosh *et. al.*, (1983), Joshi (1985-1986) further investigated accumulation of pesticide residue specially DDT in fish flesh of the estuary.

Any alteration in turbidity, PH, DO, CO_2, SS, immediate oxygen demand (IOD), BOD and heavy metal concentration in the water affecting the survival, growth and reproduction of aquatic organisers or that which impaires the legitimate use of water is generally referred to as aquatic pollution. Sand, clay, silt, chalk, gypsum and highly resistant organics like cellulose, liguin, fly ash are conservative pollutants as they do not decompose easily. Domestic sewage and effluents from tannery, distillery, breweries, yeast, shellac, pulp and paper, rubber, oil, fat etc. are biodegradable. These substances contribute towards the pollution. The impact of pollution depends on lateral, vertical and longitudinal mixing of river water which is determined by turbulence. The turbulence is dependent on discharge, slope depth canal roughness, configuration and wind action.

In the main Ganga, pollution studied so far is mainly from organics, contributed by various industrial and domestic effluents. Modern industrial wastes are highly complex in character containing various pests and metallic pollutants. Organic wastes although complicates the ecosystem by upsetting the oxygen balance through disintegration of easily biodegradable material, have tremendous capacity in fertilizing the water which intern helps in enriching aquatic biomass having different characters.

Pollution study at Kanpur in a stretch of 35 km by the effluent of textile, tannery and municipal sewage reveal that fast homogenous dissipation of raw sewage effluents occurs within 7 to 8 km below the outfall. It also serves to fertilize the river to a great extent (Fig. 16) by stimulating growth of aquatic organisms where major Indian river carps like *Catla, L. rohita, C. mrigala, L. Calbasu, L. bata* and catfishes—like *Mystus aor, M. seenghala, W. attu, Silonia Silondia, Bagarius Bagarius* and other species are captured round the year. Near the sewage outfall, for 4-5 km below, only some of the above and medium sized catfishes together with minor *Aspidoperia morar, Oxygaster* spp and *Crossocheilus latius* are found in large quantities. The river, Ganga in general did not show any septic or highly polluted zones that might affect the fish life. A complete deoxygenation caused by disintegration of putrescible settled sludge is not possible due to fast flowing currents on the right bank (being fairly deeper) which replenish the oxygen deficiencies (150 mg/ℓ of BOD in the sewage sample could not bring forth undesirable effects by lowering the D.O. values). But Saxena *et. al.* (1966) claims that in most of the area the D.O. content was less than 4 ppm and the coliform count ranged from 10^5–10^9 MPN 100 ml. The fish do not appear to be subjected over prolonged period to environmental stress even during the low flow in summer months because of sufficiently high dilution (1:40-64) attained and fast currents. Thus complete mixing of wastes is not possible and mild pollution effect is limited to the right bank only. Fish feed upon the abundant Tubifex and part of sludge along with worms generally found in areas of gross organic pollution and escape into clear

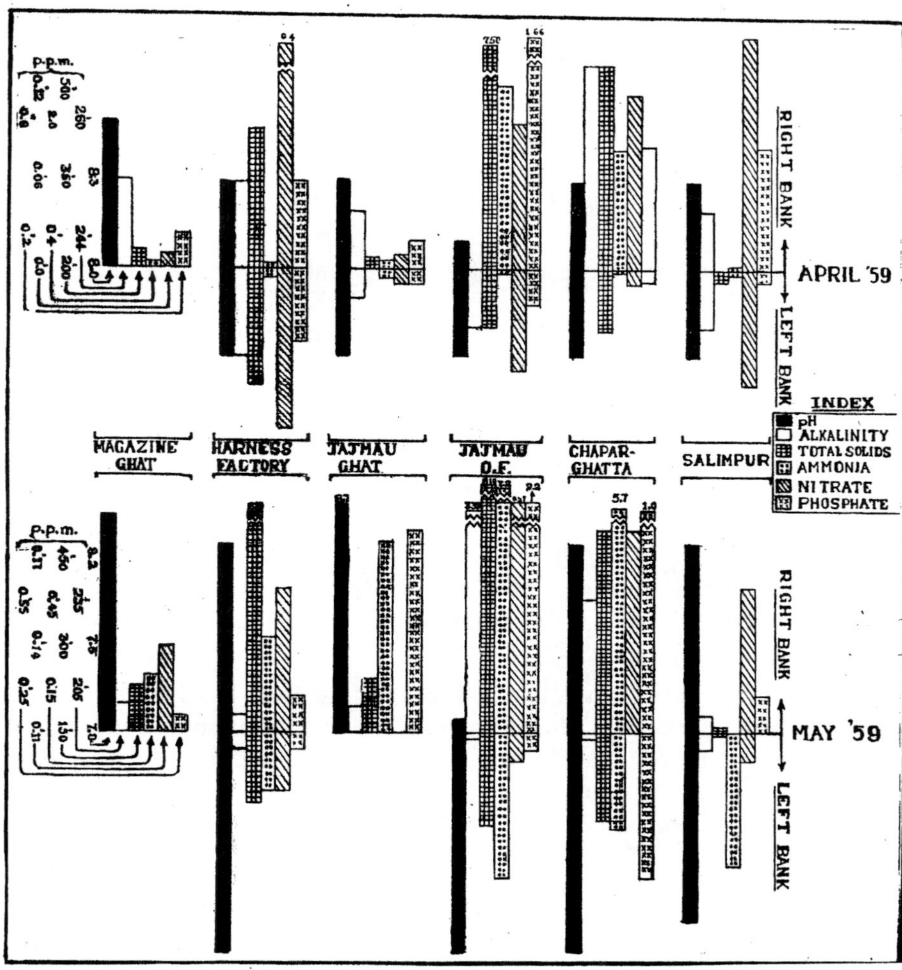

Fig. 16 : Fertility effect of wastes during extreme summer conditions.
Source : Ray, P. and A. David, (1966).

waters adjoining the polluted zone. Diversion of major wastes into the trunk sewer lower down, reduced almost completely the toxic effects of textile and tannery wastes because of high dilution attained in sewage. Diversion of the wastes into agricultural fields especially during summer further reduces the organic load led normally into the river. Precipitation of suspended matters and their decomposition do not influence fish breeding, as, such areas are limited and the monsoon floods annually wash off signs of pollution. (Ray *et. al.* 1966).

Coliforms were found to be higher in summer (about 8 million MPN/100 ml) Saxena *el. al* (1966).

Effluents released (@ 5500m^3/d) from IFFCO complex manufacturing urea and ammonium sulphate, into R. Ganga at Phulpur, Allahabad, revealed that the composite effluent characterised by high free ammonia (100.60 – 112.20 ppm), alkalinity (300 – 580 ppm), Nitrate – N (54 – 72 ppm), sp. conductance (1218 – 1862 micro mhos/cm), fairly high ph (8.6 – 8.9) and oil (21.5 mg/l) when first released caused extensive biological damage (Chandra *et. al.* 1981). About 50 quintals of fish consisting *C. catla* of 1.5 – 4.5 kg, *C. mrigala* of 0.5 – 1.5 kg, *L. rohita*, 2.0–3.5 kg, *M. aor*, 450 – 650 gms,

M. seenghlaa, 1.5 – 2.54 kg. *Clupisoma qarua* (110–150 gm) and damage two farm animals were reported. The toxic effect of waste was observed up to 300 m in downstream. 500 m below the outfall region plankton population which was around 4550 μ/1 prior to discharge of effluents, disappeared completely except a few blue green algae. Likewise benthic community was erased up to 300 m. According to the author free ammonia at fairly high pH was responsible for fish kill. Appreciable recovery however was noticed at 500 m below the outfall (phytoplankton 46% and Zooplankton 94%).

The Ganga at Varanasi an oldest city of the country receives increased domestic and industrial wastes every day. In his studies of algal communities of the Ganga, Rai (1978) reported presence of Stigeoclonium and Schizomeris in some areas of the river (Rajghat) which indicates organic pollution. His study further revealed that species like *Cylindrotheca, Gyrosigma* etc. reported to be the inhabitants of less polluted waters were not present in that region. Besides it is estimated (Chaphekar and Mhatra, 1986) that around 30,000 dead bodies are burnt along two ghats at Varanasi every year. Between $^1/_2$ to 2 tons of wood is needed to burn each dead body. This produces considerable amount of ashes (about 37,500 t, taking average as 1.25 t/dead body) that goes into the river and flows down to a considerable distance and settle to make the river bed sterile by covering it. There are innumerable ghats (bathing place) like this located on the banks of the Ganga producing large quantities of flyash which possibly cover a large area of river bed. During pre-monsoon months when the demand of Ganga water increases considerably the flow of water is reduced, the ashes can settle and render the bed unproductive.

The stretch of 256 km of Ganga between Barauni to Farakka is being charged regularly by effluents from industries like Bata shoe factory, distillery (on the southern Bank) and oil refinery, thermal power station and fertilizer factory (on the northern side) of the river. Three years study (1976-79) of the stretch by Bilgrami and Dutta Munshi (1979) reveals that total density of phytoplankton varied with seasons. (Table 28). Phytoplankton, 1484 – 4720 units in summer, 298 – 520 units during monsoon and 870 – 3091 units/l during winter zooplankton values were much lower in concentration, 14 – 34 units, 5 – 9 units and 14 – 26 units/1 during the above seasons.

Bacteria and fungi contributed major group of decomposers. The most probable number of bacteria (MPN) per 100 ml indicated by Bhargava (1981a) and Bilgrami and Dutta Munshi (1979) in Tables 24 and 30 appears to be much lower than the values indicated by Das Gupta (1984) in Tables 25, 26 & 27. Further mid-river values are lower by 80 – 93 per cent than the margin (Table 25). The Ganga experienced an ugliest incident of human indifference when it was set aflame near Monghyr, Bihar in 1968 (Sunderasan *et. al.* 1983). The discharge of oil from a refinery effluent led to the fire and resulted in colossal loss of fishery in the river.

In the estuary however conditions are different because of different tidal influences. The effluents discharged into the estuary have extremely variable chemical composition. They include mineral acids, bases and many other toxic substances of non-biodegradable and biodegradable nature. Some studies on the impact of industrial effluents on the Hooghly have been done and are discussed below. Generation and discharge of waste water (domestic and Industrial) around the Hooghly estuary are indicated in Fig. 14 and 15.

About 240,152 m^3 of organic waste is discharged every day which amounts to 89.82 per cent of the total liquid effluents discharged into the estuary. The corresponding loads are 90.72 per cent for BOD_5; 94.21 per cent for TS; 96.79 per cent for SS and 91.88 per cent for DS. Pollution around the outfall of different factories releasing organic wastes such as effluents from pulp and paper mills (sulphite, sulphate and soda process) textile, tannery, distillery, yeast, shellac, rubber, etc. contains highly decomposable organic matter along with non-biodegradable materials like sand, silt, clay, chalk, gypsum, lignin, etc. Lignin content in pulp and paper waste is rather high, 200 ppm and 1000 ppm for soda and sulphite wastes respectively which corresponds to 2.2 t and 16.6 t per day (Basu, 1966). Continuous discharge of these substances raise the bed of the estuary, reduces plankton, eliminate food producing riffles, spawning grounds, fish eggs and larval populations. High turbidity in and around the outfall areas is another important contributory factor.

Finely suspended particles, possibly lignin clog the gills of fish and have synergestic effect due to asphyxia and poisoning of gastro-intestinal tract (Ray, 1961). Filtered effluents did not show any ill effect on the test fish. Experiments done using china clay (particle size 0.13 - 5 microns) and diatomaceous earth in 270 ppm concentration resulted in more than 50 per cent mortality of the test fish within 2-12 weeks (DSIR, 1961). Biopsy of the test fish revealed pathological changes in the gills, thickening of epithelial cells and after fusion of the adjacent lamellae. In some individuals even fishrot was observed. Reduction in the supply of food due to the adverse effect of effluents on plankton and other biota, destruction of spawning grounds and direct and indirect effects (by avoidance) of suspended solids affect fisheries adversely. Coloured effluents, particularly from pulp and paper, textile and paint industries and also from distilleries affect the growth of plankton population adversely. Decomposition of biodegradable materials in the vicinity of outfall promotes dense growth of sphaerotilus, protozoa and zooglea, covering the stream bed and smothering benthic biota. Increased oxygen demand by organic sediments and high ambient temperature result in a marked deficiency of dissolved oxygen beneath the slime, though it may not be absent in the surface waters. Outfall areas in and around Tribeni Tissue and Pulp Mill at, Hazinagar, represent the above condition (Ray, 1980; Ray and Mitra, 1980).

Cage culture experiments, were conducted during monsoon at the main and second outfall points of Tribeni Tissue Mill, Gastropods (*Lymnea sp.*) as test animal for 72 hr. showed 50 per cent and 60 per cent survival respectively. In winter, the mortality was total at the main outfall, and 5 per cent at the second outfall. The soil water interface remains in a reduced state as a result of high oxygen demand. Continuous emission of CO_2 gas was observed in the littoral zone during the dry months, silting reduce the harmful effects to a certain extent (Ray *et. al*, 1979).

The effect of suspensoids of sulphite pulp and paper mill waste on the biota in a stretch of 4 km around the outfall area was studied by Ray and Mitra (1980). High temperature was found to enhance decomposition of organic matter in the sediments which reduces the amount of dissolved oxygen at the soil water interface. Carbon dioxide is evolved more vigorously during March-May but ceases completely with the monsoon floods due to the settling of the suspended materials. The estimated carbon dioxide value in water ranged between 3.5 and 15.5 mg/1 and 13.5 and 25. 5 mg/1 above and below the outfall point respectively.The sediment few mm below the surface were completely black generating H_2S and CO_2 within the top few mm of the sub-stratum. H_2S gets oxidised leaving behind an anaerobic zone of unsaturated inorganic and organic (carbonaceous) material at the soil water interface. This process continues even during the low tide when large areas below the outfall are exposed. These changes alter the equilibrium of the system. The survival of invertebrates is seriously affected. Benthic organisms are wiped out completely and the equilibrium of the system changes drastically. However, the onset of monsoon changes the situation by diluting the pollutants. Common organisms of the estuary were conspicuously absent from the outfall areas. The impact of pollution was more severe during summer. The numbers of species recorded were 2.4 and 6 at 1.5 km above the outfall in summer, monsoon and winter respectively 3,4 and 5 at 0.7 km above the outfall, 3,2 and 4 at 1.5 km below the outfall and 4, 8 and 7 at 2.5 km below the outfall physical and chemical characteristics of the substratum in relation to living organisms are given in Fig. 17. The following species formed the biotic community *Macrobrachium* spp, *Melanoides Lineatus* (Grey), *Melanoides granifera* (Lam), *Assiminea francessii* (Wood), *Stenothyra deltae* (Bensen), eel, *Acetes* sp, *M. tuberculata* (Muller), *Thiora scabra* (Muller), *Vivipara dissimiles* (Muller), *Vivipara bengalensis* (Lam), *Lymnea Luteola f: Ovalis* (Gray), megalopa larvae, chironomid and oligochaetes.

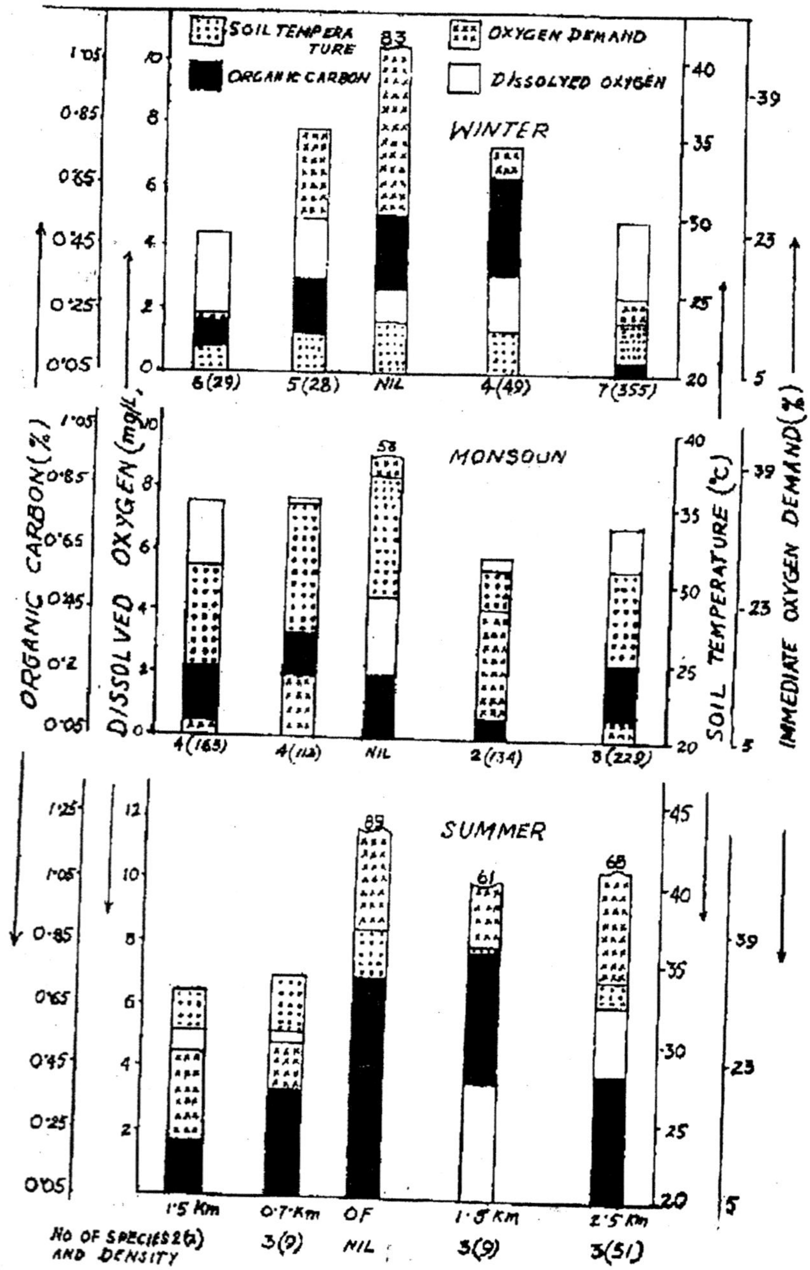

Fig. 17 : Physical and chemical parameters of soil and water in relation to biotic life.
Source : Ray, P. and E. Mitra (1980).

Emergence and propagation of plankton from the soil samples collected from 1.5 and 0.7 km above the outfall, and 1.5 and 2.5 km below the outfall using pollution free water of the Hooghly estuary were studied. The dominant species recorded at each location are as follows :

In the summer season

Ankistrodesmus, Oscillatoria, Euglena and *Closterium* (1.5 km above outfall). *Navicula, Oscillatoria, Cymbilla* (0.7 km above outfall)

Englena (Outfall)

Oscillatoria (1.5 km below outfall)

Oscillatoria (2.5 km below outfall)

In the monsoon season

Euglena, Pinnularia, closterium (1.5 km above outfall).

Cymbella, Closterium (0.7 km above outfall).

Closterium, Navicula (outfall)

Closterium, Euglena (1.5 km below outfall)

Navicula, Oscillatoria (2.5 km below outfall)

In the winter season

Euglena (1.5 km above outfall)

Oscillatoria, Euglena (0.7 km above outfall)

Euglena (Outfall)

At 1.5 and 2.5 km below the outfall plankton density was poor. Maximum plankton count was 20,360 and 227, 240 units/1 during summer and monsoon respectively after an incubation period of four months in the jars containing soil from the outfall point. Although in the laboratory relatively few species could be identified in the jar containing soil collected from the outfall but their densities were the highest indicating high nutrient content in the substratum (Ray and Mitra, 1980).

During winter the survival of 10-15 mm size young *Lymnea* after exposing them at soil-water-interface in cages for seven days was 100,10, 60 and 100 per cent at 1.5 km at the up stream, near the outfall and 1.5 km and 2.5 km down stream respectively.

Other industries discharge their effluents into the main current where water is fairly deep. Accordingly plankton are likely to suffer adversely but, the substratum generally remains free from any adverse impact.

Further, the scouring effect of tidal water flushes out the settled material which otherwise is supposed to be detrimental for the benthic biota. The tidal wave carries the young gastropods to comparatively near the calm upstream areas (Ray and Ghosh, 1976). The 96-h LC 50 values for *Microbrachium* spp. as 31-52 per cent and 1.75 – 7.5 per cent for *Daphnia similes* with synthetic rayon waste, 19.0 – 20.2 per cent for *Microbrachium* spp. and 6.0 per cent for Dephnia similis with paints and varnishes. The 24-h Lc 50 values recorded for *Puntius sophore* with distillary; *Puntius sophore* and *Mystus vittatus* with pulp and paper; *Puntius sophore* with tannery; *Puntius sophore* and *Mystus vittatus* with textile and *Daphnia lumholzi* with oil refinery were respectively 6.7, 20.6–34.7, 6.7–10.03,

Table 31 : Toxicity Tests of Some Selected Industrial Effluents

Industry	*Test animal*	*Period of experiment*	*Temperature range °C*	*LC_{50} (%) by Vol.*	*References*
1. Synthetic rayon	*Macrobrachium* sp	96 hr	27.5 – 34	31 – 49	
	—Do—	—Do—	—Do—	52	
	—Do—	—Do—	—Do—	33	Ray, Ghosh
	Daphnia similis	—Do—	—Do—	1.75	Bagchi
	—Do—	—Do—	—Do—	5.6 – 7.5	and Mazumder (MS)
2. Paints and Varnishes	*Macrobrachium* Sp.	—Do—	29.5 – 32.0	19.0 – 20.2	
	Daphnia similis	—Do—	—Do—	6.0	
3. Rubber	*Macrobrachium*	—Do—	29.0 – 31.0	32	
4. Distillery	*P. Sophore*	24 hr	19.0 – 25.0	6.7	
5. Pulp and paper	—Do—	—Do—	21.0 – 25.0	25.4	Ray (1961)
(a) Pulp	*M. vittatus*	—Do—	—Do—	34.7	
(b) Paper	*P. sophore*	—Do—	—Do—	24.5	
	M. vittatus	—Do—	—Do—	26.5	
(c) Combined	*P. sophore*	—Do—	—Do—	20.8	
	M. vittatus	—Do—	—Do—	20.6	
6. Tannery					
(a) Sulphide	*P. sophore*	—Do—	21.5 – 22.0	7.7	
	—Do—	—Do—	30.6 – 32.8	10.2	
(b) Vegetable	—Do—	—Do—	21.5 – 22.0	10.03	David and Ray (1960)
	—Do—	—Do—	30.6 – 32.8	6.9	
(c) Chrome	—Do—	—Do—	21.5 – 22.0	9.55	
	—Do—	—Do—	30.6 – 32.8	6.9	
7. Textile					
(a) Kier	—Do—	—Do—	21.5 – 22.0	3.9	
	—Do—	—Do—	30.6 – 32.8	5.0	
(b) Dye	—Do—	—Do—	21.5 – 22.0	10.45	
	—Do—	—Do—	30.6 – 32.8	11.75	
(c) Bleach	—Do—	—Do—	21.5 – 22.0	2.55	
	—Do—	—Do—	30.6 – 32.8	0.07	
	M. vittatus	—Do—	—Do—	0.11	
Combined	*Macrobrachium* sp.	96 hr	26 – 30	6.5	Ray, Ghosh, Bagchi and Mazumder (MS)
8. Oil Refinery	*Daphinia Lumholtzi*	24 hr	28.0	10	Ghosh and Bagchi (1979)

0.07 – 11.75 and 10.0 percent (Table 31). Effluents from these industries should be diluted or adequately treated before releasing into the receiving water. On account of high oxidizable organic matter the immediate oxygen demand becomes high. This results in nil or low Do, high BOD and OC (3 m and 30 m at 100°C by permaganate) in the receiving water. Such a condition becomes unfavourable for the growth and propagation of plankton below the outfall (Table 32). Major damage to plankton results during the transifional period of low and high tide and vice versa at each tidal phase, particularly during the neap tide in summer when dilution ratio of effluents to receiving water was minimum and effluents spread over a much wider area. The condition deteriorated further by the nature of toxin, dilution, mixing pattern, duration of tides, flow of water, tidal phase, seasonal influence and retention period of the effluents (Ray and Ghosh, 1976). Self-purification capacity was greatly affected. On several occasions, around the outfall points plankton were either absent or had very low density. Rapid fluctuation in the chemical characteristics of waste (rise or fall of PH) also affect plankton adversely.

Table 32 : Effect of Industrial Wastes Upon the Plankton Population at the Outfall Region in Different Seasons

Industry	Location	SEASON	
		November-February	*March-June*
1. India Pulp & Paper (Sulphite process)	Above the outfall	339 (P—259; Z —80)	59 (P—37 ; Z—22)
	At the outfall	77 (P—41 ; Z—36)	7 (P—3; Z—4)
	Percentage reduction	77 (P—84 ; Z—55)	88 (P—92 ; Z—82)
2. Tribeni Tissue (Soda process)	Above the outfall	657 (P—55 ; Z —102)	171 (P—133 ; Z—38)
	At the outfall	20 (P—7 ; Z—13)	87 (P—54 ; Z—33)
	Percentage reduction	97 (P—99; Z—87)	49 (P—59 ; Z—23)
3. Distillery & yeast	Above the outfall	89 (P—48 ; Z—41)	70 (P—45 ; Z—25)
	At the outfall	57 (P—32 ; Z—25)	55 (P—30 ; Z—25)
	Percentage reduction	36 (P—33 ; Z—39)	21 (P—33 ; Z—0)
4. Rubber Industry	Above the outfall	123 (P—70 ; Z—53)	78 (P—44 ; Z—34)
	At the outfall	80 (P—57 ; Z—23)	55 (P—40 ; Z—15)
	Percentage reduction	35 (P—18.2 ; Z—56)	26.6 (P—9 ; Z—55)
5. Kesoram Rayon	Above the outfall	236 (P—180 ; Z—56)	195 (P—140 ; Z—55)
	At the outfall	130 (P—96 ; Z—34)	49 (P—37; Z—12)
	Percentage reduction	42 (P—46 ; Z—39)	74 (P—73; Z—78)

P = Phytoplankton Z = Zooplankton

Source : Ray. P. (1981).

Value of primary production (gross and net) was estimated by light and dark bottle techniques are given in Table–33A (Ghosh *et al.* 1976) and subsequently by Babu Lal (1987, 1990), Table 33 B and C by C^{14} technique. During the summer and monsoon seasons productivity was low around the outfall of pulp and paper mill (Soda process). It however, improved appreciably during winter. Ghosh *et al.*, (1977) have further shown that primary production was significantly affected at the outfall of pulp and paper mills (Sulphite and Soda process (Table–34 A & B). In case of tannery outfall, gross and net production (Ghosh *et al.*, 1980 a) was 71 and 81 per cent, 87 and 74 per cent and 43 and 41 per cent respectively during the summer, monsoon and winter seasons. Considerable oil was discharged into the estuary along with other effluents mainly from ICI industries (709 m^3/d) at Rishra and from tannery (18,800 m^3/d) Batanagar. Petroleum refinery also disposed 6600 m^3/d or oil on an average ; about 4491 m^3/day of oil is disposed into the estuary, (Ghosh and Bagchi, 1979).

The impact of oil refinery effluent on the biota around the outfall have not been evaluated. However, the value obtained from toxicity tests (LC_{50}) for 24–h using *Daphnia lumholtzi* was 10 per cent by volume (Ghosh and Bagchi, 1979). It shows that the effluents are toxic to microorganisms.

Table 33 : A Primary production of the Hooghly estuary

Ghosh B.B., M.M. Bagchi and P. Ray, (1979) in Ray, P., (1981) (by light and dark bottle technique) mg/C/m^3/hr)

	Seasonal Variation					
	Summer		*Monsoon*		*Winter*	
Centres	*Gross*	*Net*	*Gross*	*Net*	*Gross*	*Net*
Kalna	69.0	52.1	37.7	50.2	60.6	54.5
Tribeni	7.3	2.6	31.5	8.3	150.5	92.1
Hazinagar	145.5	58.1	10.4	31.8	56.7	27.1
Kakinada	13.5	0.9	7.0	1.7	23.8	0
Barrackpore	46.9	24.7	29.3	8.4	24.7	20.1
Titagarh	13.8	12.6	5.2	5.2	12.1	12.3
Batanagar	56.2	27.4	11.2	9.8	12.9	14.7
Diamand Harbour	30.9	25.5	25.5	51.6	74.3	51.4
Kakdwip	57.5	15.5	43.0	44.5	79.2	42.2

B. Babu Lal (1987) (by C^{14} Technique)

Before Farakka discharge. Ranged 0.375 to 0.562 mg/c/m^{-3}/d^{-1}

Post Farakka discharge.

Nawabganj	—	38.00 to 95.64	(av. 64.85)
Barrackpore	—	22.62 to 95.16	(av. 55.35)
Diamond Harbour	—	37.66 to 75.05	(av. 53.89)
Fresherganj	—	44.51 to 57.40	(av. 45.55)
Kolaghat	—	29.16 to 53.01	(av. 43.15)

Minimum values of Primary production were recorded during monsoon and maximum during post monsoon period.

C. Annual variation in the range and average rates of carbon assimilation in the Hooghly estuary before and after the construction of Farakka Barrage (1965-86)

Year	*Rate of carbon assimilation* *Range*	*mg C m^{-3}/d^{-1}* *Average*	
1965	0.375 — 0.562	0.468	By Light and dark bottle method (1965–1985)
1969	Nil — 292.0	—	
1975	24.0 — 106.8	65.4	
1977	80.4 — 117.6	99.4	
1983	78.12 — 163.3	120.71	
1984	40.80 — 187.2	119.0	
1985	37.44 — 150.72	94.08	
1986	29.16 — 95.64	62.50	(By C^{14} technique)

Source : Babu Lal (1990).

Table 34 A: Effect of Pulp and Paper Mill Effluent (Soda Process) on the Primary Productivity

Location	Summer		Monsoon		Winter	
	Gross	Net	Gross	Net	Gross	Net
Above Outfall						
Neap tide	0 – 14.6 (7.3)	0 – 52 (2.6)	0 – 52.5 (31.5)	0 – 33.3 (8.3)	138.5 – 162.5 (150.5)	65.6 – 118.7 (92.1)
Spring/Bore tide	10.2 – 35.9 (22.6)	6.2 – 7.9 (7.0)	0 – 10.7 (9.0)	0 –7.3 (2.4)	12.5 – 91.7 (52.1)	0 – 12.5 (6.2)
Out fall						
Neap tide	0 – 18.4 (5.2)	Nil	0 – 3.1 (2.2)	0 – 5.2 (1.3)	43.7 – 65.8 (54.1)	21.9 – 65.6 (43.7)
Spring/Bore tide	7 – 12.5 (10.2)	0 – 21.6 (5.4)	0 – 14.6 (6.6)	Nil	22.9 – 25.0 (23.9)	Nil
1.5 Km Below Outfall						
Neap tide	0 – 8.3 (4.2)	Nil	0 – 8.5 (4.7)	Nil	55.2 – 109.3 (82.2)	26.0 – 72.0 (49.4)
Spring/Bore tide	14.1 – 28.1 (18.4)	0 – 198.1 (29.5)	5.2 – 39.6 (18.3)	0 – 5.2 (1.7)	12.5 – 104.2 (60.8)	12.5 – 47.1 (19.8)

(Ray. P. 1981)

Table 34 B : Effect of Sulphite Pulp and Paper mill waste on the Primary Production in the Vicinity of the Outfall area

Location	Summer		Monsoon		Winter	
	Gross	Net	Gross	Net	Gross	Net
Above Outfall 750 m	118.8 – 197.7 (145.5)	33.3 – 72.9 (58.1)	66.7 – 187.5 (104.0)	0 – 80.2 (31.8)	28.2 – 89.9 (56.7)	0 – 60.6 (27.1)
Out fall	0 – 84.4 (41.4)	0 – 9.4 (3.1)	Nil	Nil	0 – 36.3 (18.1)	Nil
Below Outfall 1.2 Km.	51.8 – 103.1 (71.6)	7.8 – 107.3 (54.6)	132.5 – 154.0 (140.5)	33.5 – 100 (62.1)	49.9 – 152.1 (89.3)	0 – 33.4 (15.0)

(Ray, P. 1981).

Mineral wastes contain some amount of organic matter (synthetic rayon waste) and are highly acidic (pH 2.0 – 4.9). Plankton species were adversely affected at the outfall where they were reduced by 42 per cent in winter (November – February) and 74 per cent during summer (March – June). A sudden fall in pH (from 8.2 to 6.1), alkalinity (from 191 to 96 mg/l and DO (from 4.9 to 3.5 mg/l) observed on several occasions had a detrimental effect on the ecosystem. Near the outfall the amount of zinc was 2 mg/l in the waste and 3.84 mg/l in the bottom sediment at the outfall region. In the vicinity of the outfall, shrimp (*Macrobrachium roserbergi*) was found to accumulate 0.2 mg/l of zinc Although some recovery took place up to 250 m below the outfall, it appears to be carried further down in the estuary along with suspensoids. About 3 km below the outfall the littoral zone on the right bank was devoid of any benthic life. The LC_{50} determined for 96h using *macrobrachium* sp. and *Daphina similis* ranged between 31 – 52 per cent and 1.75 – 7.5 per cent by volume respectively. These results indicate the need for appropriate treatment or dilution of the effluents before they were discharged into the estuary for protecting the biotic community.

Ghosh *et. al.* (1983) while studying the pollution status in the Hooghly estuary with reference to heavy metals disposed through industrial wastes have revealed valuable informations on the behaviour of metals in the ecosystem. For example, metal content were studied in the industrial effluents, their concentration in sediments and water, bioaccumulation in some fishes and in fish tissues, shrimps, molluscs and crabs, biomagnification of metals (based on lowest ambient concentration of some metals). DDT resides in fish and their biomagnification, impact on plankton and benthos etc. The results are given in Tables 35 – 38 & (Fig. 18 – 21).

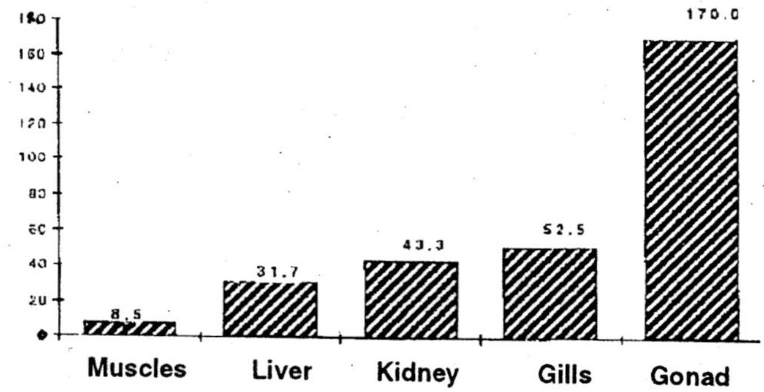

Fig. 18 : Bioaccumulation of zinc (in mg/g) in tissues of Rita rita in Hooghly estuary.
Source : Research Highlights, CICFRI, Barrackpore, 1989-90.

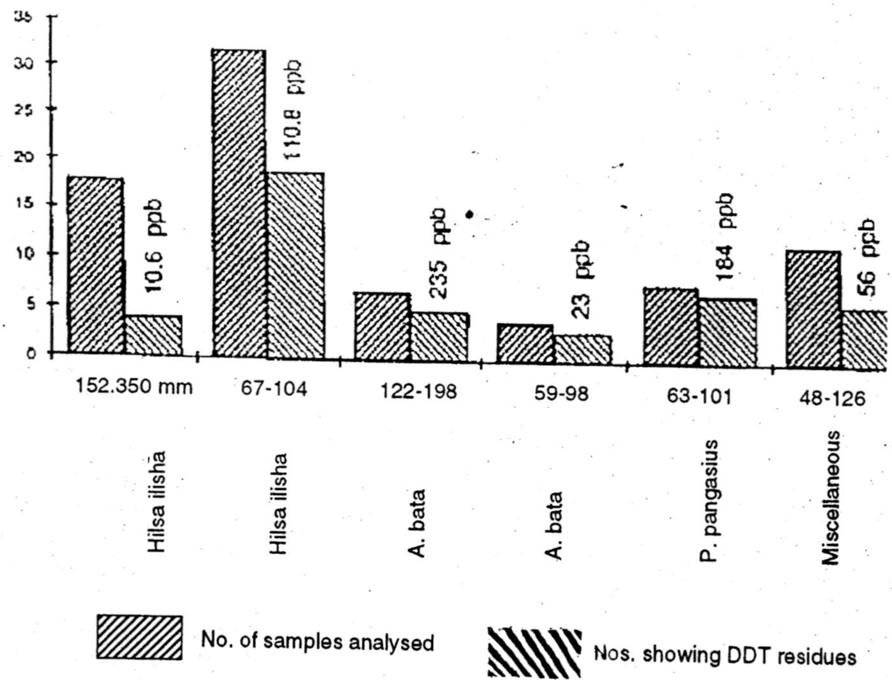

Fig. 19 : DDT Residues In Fishes In the Hooghly Estuary
Source : Joshi, H.C. (1986).

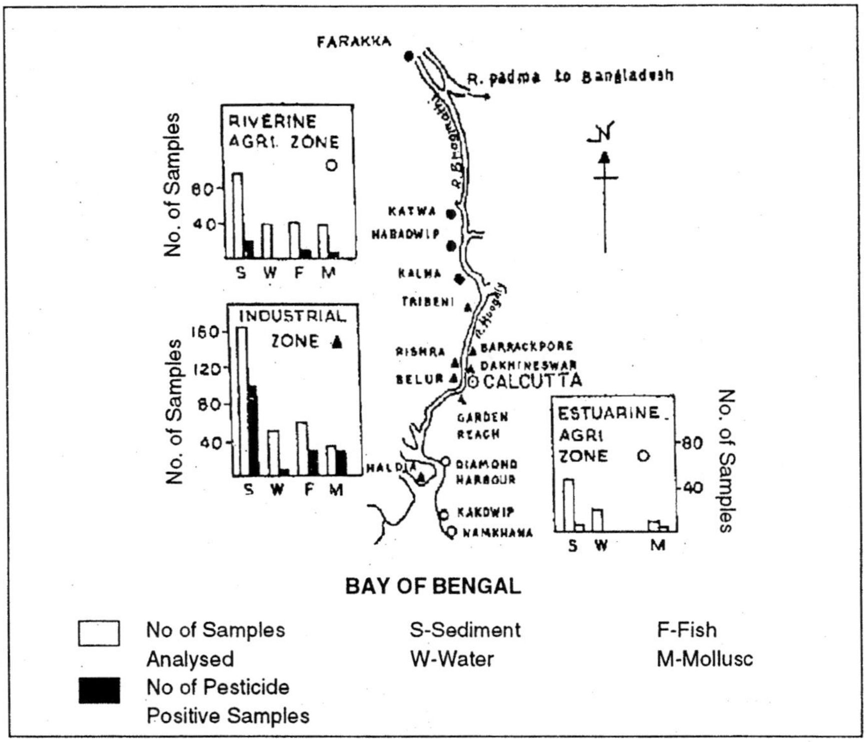

Fig. 20 : Pesticide Monitoring profile in the Hooghly Estuary.
Source : Joshi, H.C. (1985).

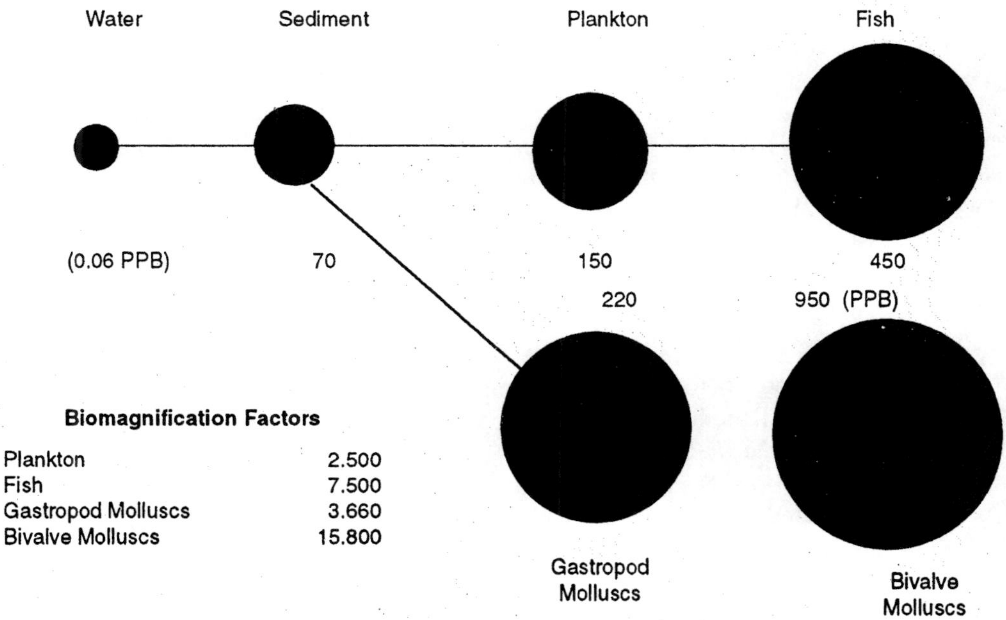

Fig. 21 : Biomagnification of DDT In Aquatic Food Chain In Hooghly Estuary
Source : Joshi, H.C. (1988).

Table 35 : Concentration of some heavy metals estimated (mg/l) in various industrial wastes. Ghosh *et. al.* (1983)

	Zinc	*Copper*	*Chromium*	*Lead*	*Remarks*
Rayon (Kuntighat)	0.06 – 8.50 (3.22)	ND	ND	ND	ND—Non detectable
Rubber & Cycle Rim (Sahaganj)	ND – 0.35 (0.12)	ND	ND – 1.92 (0.68)	ND	Fig. in the parenthesis denote average values
Paints & Varnish (Garifa)	ND – 0.42 (0.22)	ND	0.03 – 0.30 (0.14)	ND	
Paints & Varnish (Shalimar)	ND – 5.59 (1.01)	ND – 0.20 (0.03)	ND – 0.60 (0.14)	ND	
Cotton Textile (Garden Reach)	ND – 0.22 (0.12)	ND	ND – 1.23 (0.28)	ND	
Tannery (Batanagar)	ND – 0.50 (0.25)	ND – 0.50 (0.26)	0.2 – 18.50 (6.85)	ND	
Oil Refinery (Haldia)	ND – 0.10 (0.01)	ND	ND – 0.12 (0.02)	ND	

Table 35 indicates presence of Zinc, Copper and Chromium in the effluents of tannery, and paints and varnishes; (Shalimar) Zinc and Chromium in rubber and cycle rims, Cotton textile and oil refinery effluents, Zinc in Rayon effluents, while led was non-detectable. As high as 8.5 ppm of zinc in Rayon effluents (during summer), 18.5 and 0.5 ppm of Chromium and copper respectively in tannery effluents were estimated.

Outfall of Rayon industry registered the highest zinc accumulation (381.6 ppm) in the sediments followed by paints and varnishes (160ppm) and oil refinery (80ppm), chromium level to the extent of 9.6 ppm was estimated at the outfall of paints and varnishes while outfalls of the upstream registered a non-detectable amount of chromium and copper and ND – 10.7 ppm zinc. Zinc contaminated stretches extended up to 500 m below the outfalls of Rubber, paints and varnishes and oil refinery, and chromium below the outfall of paints and varnishes. Copper was non-detectable both in sediment and the water. Higher values of zinc were recorded in the surface water of Rayon outfall, and chromium at the outfall of Tannery.

Ghosh *el. al* (1983) have intensified the work further by estimating the accumulation of metals in different species of fish of small and larger sizes, shrimps, crabs and molluses etc. around the outfall of some industries disposing wastes on the Hooghly estuary (Table 36).

High accumulation of copper was observed in shrimps; mollusc registered more zinc and copper accumulation than crab, while all the fishes showed zinc accumulation. Accumulation of copper was registered in fishes like *Hilsa ilisha,, R. Corsula, P. pama, S. Panijas, R. rita, G. giuris E. macrodor* and *M. menoda.* Chromium accumulation was indicated in a few fishes like *R. rita* and *O. rubicundus.* In shrimps, copper was estimated in *M. rude, M. Mirabile* and *P. sculptilis* while highest zinc was estimated in *M. Villosimanus.* Lead was not detected in all the test arrivals. Marked bioaccumulation of zinc (maximum 120.4 ppm) and copper (maximum 20 ppm) in fish like *E. macrodon* and chromium (maximum 4.5 ppm) in *O. rubicundus* was estimated. Although copper was non-detectable in rubber effluents fish and shrimps caught around the outfall of rubber effluents

Table 36 : Accumulation of metals in some organisms of the Hooghly estuary at different outfall areas

Industry	*ORGANISM*		*Size*		*Concs. of metals (ppm)*			
			Length (mm)	*Weight (g)*	*Zinc*	*Copper*	*Chromium*	*Lead*
Rayon	*Fish* :	*Ailia Coila, Mastacembelus armatus * Apocryptes bato, Rita rita, Setipina phasa, Clupisoma garua Glossogobius giuris.*	60 – 185	1.9 – 16.5	2.5 – 105.8	—	—	ND
	Shrimp :	*Macrobrachium Villosimanus Macrobrachium rude*	51 – 80	2.5 – 4.0	29.2 – 112.2	—	—	—
Dunlop Rubber (Sahaganj)	*Fish* :	*Puntius ticto, Odontamblyopus rubicundus, R. rita, Mystusmenoda, G. giuris, Cynoglossouslingua, S. Phasa, Eleotris-macrodon.*	54 – 115	2.0 – 5.8	12.4 – 120.4	2 – 20	ND – 4.5	ND
	Shrimp :	*Macrobrachium - mirabile, Leander fluninicola, M. rude*	40 – 85	0.3 – 5.5	14.8 – 40.4	2 – 10	ND	ND
Paints & Varnishes (Garifa)	*Fish* :	*O. rubicundus*	115	5.0	7.2 – 62.4	—	ND– 0.26	ND
Paints & Varnishes (Shalimar)	*Fish* :	*A. bato, Silonia silondia, C. garua, Pama Pama, Pangasius pangasius, P. ticto, Colisa Fasciatus, O. rubicundus.*	50 – 149	1.5 – 30.4	8.4 – 46.4	ND	ND – 4.5	ND
	Shrimp :	*M. rude, M. mirabile*	42 – 85	1.8 – 5.0	20.4 – 40.0	14 – 38	ND	ND
	Crab :	*Scylla seratta*			24.4	14.0	ND	ND
Cotton Textile (Garden Research)	*Fish* :	*S. Phasa, Chela bacaila,* Hilsa ilisha*	98 – 116	5.0 – 5.2	26.4 – 64.6	ND	ND	ND
Tannery (Batanagar)	*Shrimp* :	*M. Villosimanus*	74 – 80	2.5 – 4.0	24.0 – 54.0	—	—	—
Oil Refinery (Haldia)	*Fish* :	*Boleophthalmus boddacrti, Boleophthalmus viridis, P. Pangasius*	82 – 100	5.0 – 5.0	10.0 – 24.0	ND	ND	ND
	Shrimp :	*Parapenaopsis sculptilis*	66.0	1.2	27.0	8.4	ND	ND
	Mollusc :	*Pila globosa*	—	—	83.0	20.0	ND	ND

*Adult fish; ND — Non-detectable :
Source : Ghosh B.B., M.M. Bagchi and D. K. De (1983).

Table 37 : Bioaccumulation of metals in fish, shrimp, crab and mollusc around the outfall of the Hooghly estuary

Industry	*Zinc*	*Chromium*	*Copper*	*Lead*
Rayon (Kuntighat)	(F. S) +	ND	ND	ND
Rubber and Cycle rim (Sahaganj)	(F.S)+	F+	(F.S)ND	ND
Paints and Varnish (Garifa)	F+	F+	ND	ND
Paints and Varnish (Shalimar)	(F.S.C.)+	F+	(S.C)$^{+}$	ND
Tannery (Batanagar)	S+	+	+	ND
Oil Refinery (Haldia)	(F.S.M.)+	+	(S.M)ND	ND
Cotton Textile (Garden Reach)	F+	+	ND	ND

\+ Indicates presence and ND not detected of metals in the Corresponding wastes of industries disposed in the river. Metals are found to be accumulated in Fish (F), Shrimp (S), Crab (C) and Molluscs (M) as the case may be.

Source : Based on data of Ghosh, B.B., M.M. Bagchi, and D.K. De (1983).

indicated its presence. This was possibly brought down from the effluent of Tribeni Tissue (Saha, 1980) located further up or partly due to migration of fish *E. macrodon* and shrimp, *M. rude.* The entire body of some juvenile fishes registered higher levels of zinc copper and chromium as compared to those in the adult fishes like *M. armotus* and *C. bacaila.*

Table 37 shows bioaccumulation of copper in fish and shrimp at the outfall of Rubber and cycle rim where copper is absent. Bioaccumulation of copper in shrimp and crab at the outfall of Shalimar paints where copper was not present. Zinc was present in fish at outfall of cotton textile were zinc found absent. Finally in outfall of oil refinery, copper was found to accumulate in shrimp and molluscs where copper was absent. Such anomalies were possibly due to different types of tidal influence during different seasons that might be the cause of movement of organisms like fish, shrimp, molluscs and crab.

Ghosh *et. al.* (1983) further intensified the study by estimating the metals in fish tissues like, gill, muscle, liver, kidney, gut (intestine) and gonad of some fishes like *Silago panijus, Rita rita* and *Hilsa ilisha* collected from in and around Barrackpore (Table 38 for the four metals studied). Accumulation of zinc was common in all the three species *Silago Punijius, Rita rita* and *Hilsa ilisha,* highest value was estimated in the gonad of *Rita rita.* Copper was found to accumulate in muscle, liver and gut (intestine), while in *Rita rita* it was present in muscle and kidney and in case of *Silago Panijus* only gut content showed its presence. Chromium was found to be accumulated only in the gill of *Rita rita.* Lead was non-detectable either in Rita or hilsa fish. From the available data it is apparent that the problem of metallic pollution was mainly from zinc.

Estimation of Biomagnification of metals, based on the lowest ambient concentration of zinc (0.02 ppm) and chromium (0.03 ppm) show that the former was of the order of 7340 in gonad, 6020 (whole fish), 5610 in shrimp, 4300 in molluscs and 1400 in crab. For chromium, the values were 640 in gill, 150 (whole fish) and 8 in case of shrimp. In earlier observation of Ghosh and Bagchi (1981) the highest zinc concentration factor for the whole fish and shrimp was reported as 3527 and 3740 respectively.

Apart from industrial and municipal wastes there is another source which contributes substantial metallic pollutants from the washings of images at the time of immersion in the Ganga very year. Ghosh (1993) has revealed some valuable information on this aspect in the article "The alarming tale of the Ganga pollution." His data is based on three years during October to January around Calcutta (Between Barrackpore and Diamond Harbour) when all the immersion of major seasonal images

Table 38 : Accumulation of heavy metals in tissues of some fishes of the Hooghly estuary around Barrackpore

Location	Species of fish	Size		Nature of tissue	Concentration of heavy metals (ppm)			
		Length (mm)	Weight (g)		Zinc	Copper	Chromium	Lead
Barrackpore	*Silago panijus*	272 – 274	115 – 120	Gill	40 – 60 (48.5)	ND	ND	—
(Dhobighat)				Muscle	12.5 – ND (15.7)	ND	ND	—
				Liver	—	—	—	—
				Kidney	30 – 40 (5.5)	ND	ND	—
				Gut (intestine)	25 – 65 (47.5)	ND – 30 (11.5)	ND	—
				Gonad	—	—	—	—
	Rita rita	420 – 430	900 – 920	Gill	11 – 15 (13.3)	ND	15 – 19.2 (17.4)	ND
				Muscle	10 – 13 (11)	2 – 4 (3)	ND	ND
				Liver	30 – 60 (28.7)	ND	ND	ND
				Kidney	17.5 – 25 (21.5)	5 – 15 (10)	ND	ND
				Gut (intestine)	17.5 – 25 (18.7)	ND	ND	ND
				Gonad	137.5 – 146.8 (114.5)	ND	ND	ND
Nawabganj	*Hisa ilisha*	379 – 493	650 – 1475	Gill	20 – 24* (21.4)	ND	ND	ND
					22.4 – 26 (24.1)	ND	ND	ND
				Muscle	7.5 – 9.5 (8.0)	ND	ND	ND
					6.4 – 8.0* (7.2)	8 –11 (8.4)	ND	ND
				Liver	13.8 – 16.8 (15.1)	2 – 8 (5)	ND	ND
					17 – 24*(21.4)	11 – 17 (14.4)	ND	ND
				Kidney	—	—	—	—
				Gut (intestine)	14 – 19 (18)	2 – 8 (5)	ND	ND
					4.5 – 8.5* (7.4)	ND	ND	ND
				Gonad	5 – 32 (10)	ND	ND	ND

*Smaller Size group (379 mm/650g); ND — Non-detectable ; Figures in the parenthesis denote average vlaues.
Source: Ghosh, B.B, M.M. Bagchi and D. K. De (1983).

takes place.* According to him normally the water contains 0.1 mg chromium, 0.05 mg zinc and 0.1 mg Copper per litre. During October to January the values of Chromium and lead increases 10 to 20 times and Zinc and Copper to 1.5 mg per litre. He has concluded that such progressive increase of the metallic pollutants is capable of upsetting the ecological balance of the Ganga and requires to be controlled immediately.

Under such ecological condition planktonic concentration were poorly represented. Similar observations were recorded by Ray (1981) during March to June in the Hooghly estuary. In the present case upstream of various outfall regions indicated a range 2-43 U/1 of plytoplankton and nil - 16 U/l Zooplankton while at the outfall their ranges were lowered viz. nil-14 and nil-8 U/l respectively. (Ghosh *et. al.*, 1983). The Chief organisms being *coscinodiscus* sp. *Melosira* sp and *spirogyra* sp among phyto-plankton and *Keratella* sp., *Brachionus* sp. and *Cyclops* sp among Zooplankters. Comparing their data with Ray (1981) it is distinctly clear that concentration of plankters in the later period is much less than the former years. This indicates that even after Farakka release the concentration declined sharply when stability was attained.

Of late Joshi (1985) in his pesticide residue monitoring in the Bhagirathi-Hooghly as made some interesting observations in water, sediment, fish, plankton, mollusc and suspended particulate matter. According to the author DDT, BHC, Dimethoate, Endosulfan residues were detected in 190, 57, 11 and 3 nos. of samples. DDT and BHC residues were not detectable in water samples. Molluscs showed maximum pesticide detection rate PDR (50 per cent) followed by suspended particulate matter (50 per cent), fish (43 per cent), sediment (37.7 per cent), plankton (28 percent) and minimum in water (3 per cent). Pesticide positive samples showed presence of DDT in mollases (65-963), fish (31-460), plankton (15-130) and sediment (17-89 ppb) while lower values were observed in case of BHC – y in molluscs (40–86), Fish (46 – 210) and sediment (21 - 70 ppb). Juvenile *H. ilisha* (67 - 104 mm) registered greater accumulation of DDT than those of the larger specimens (122-198 mm) while in case of estuarine fish like *Apocryptes bato* it was reverse (Joshi, 1986). Industrial zone registered highest pesticide residues (PDR 51 percent) PDR's separate values for riverine and estuarine agricultural zones were 13 per cent and 9 per cent respectively. Pre-monsoon (February- April) values for PDR were always higher than the monsoon (July–September) values. Biomagnification of DDT in plankton, fish, gastropods and bivalves has been 2,500, 7,500, 3, 600 and 15,800 times of its ambient level in water (0.06 ppb) Joshi (1985, 1986a and 1988).

GLC Estimation of Pesticide residues in sediment and biotic samples

Type of Samples	*No. of samples analysed*	*Pesticide residues in ppb*	
		*DDT**	*BHC – V*
Sediment	17	17 – 89	21 – 73
Fish	57	31 – 460	46 – 210
Molluscs	5	65 – 963	40 – 85
Plankton	3	15 – 130	Not detected

Source : Joshi, H.C. (1985) * DDT – DDT + DDE + DDD

Joshi (1985) remarks that as a food commodity these levels of DDT and BHC in fish are far below the tolerance limits prescribed by WHO (1976). However their adverse effects on the health of fish have remained to be seen.

*From an estimate it is stated that the number of important deities (Durga[1] and Kali[2]) of Calcutta immersed in the Hooghly will be around 20,000 each. From this it may be presumed that in the Gangetic West Bengal alone their number will be several thousands.

*1. T. V. Talk by D.G. of Police, Govt. of West Bengal (18.10.96)

2. T.V. News Belletin (Bengali).

12

Limnological Studies

Except some stray localised work, no systematic qualitative and quantitative studies of physio-chemical parameters and plankton were carried out in the entire stretch of the Ganga basin. Available information reveals that detailed studies were first carried out at Allahabad on the hydrology and plankton of both Ganga and Jamuna only in a short stretch during 1958-59 (Ray, *et al.*, 1966) and during 1962-64 (Devassy, unpublished). Similar investigations were also carried out at 7 Centres covering a 1090 km stretch of the Ganga from Kanpur to Rajmahal during 1960 (Pahwa and Mehrotra, 1966), Chandra *et al.*, (1981) studied the pollution around the outfall of IFFCO Complex at Phulpur while Bhargava (1981 b) not only indicated the pollution load and its control strategy in a stretch between Rishikesh and Varanasi but also indicated the purifying capacity of the Ganga. Chaphekar and Mhatre (1986) have tried to assess the human settlement, their activities and impact on the entire Ganga river in his document, "Human impact on Ganga river eco-system". Metal, pesticide and detergent residues of the Ganga river system was also studied by Joshi (1986 a, b and 1988). CICFRI, located at Barrackpore have made an attempt to study the various factors relating to environmental changes on biotic communities, pollution load and energy transformation, photosynthetic efficiency, a profile on fish community breeding and recruitment, estimation of fry and fingerlings etc. their declining nature in fishery of rivers Ganga and Jamuna system during 1986-91. (Annual Report, CICRI, 1988-89, 89-90).

Brackish Water Zone : Substantial work is also done in this stretch by CIFRI both before and after the commission of Farakka barrage. Aspects covered are qualitative and quantitative studies of physicochemical factors, phytoplankton and other biological features and their seasonal pattern in a small stretch of the river near Palta initiated by Roy (1949 and 1955). This was followed by a survey (Dutta *et. al*; 1954) conducted in a 108 Km. stretch from Palta to Diamond Harbour. Later, Shetty *et al.* (1961) and Saha *et al.* (1975) reported variations in the plankton of the Hooghly-Matla estuarine system throughout the 290 km tidal stretch of Hooghly. Rupnarayan estuary and upper zone of Matla estuary. Subsequently, extensive studies were made related to the effects of industrial and domestic effluents (Seth and Bhaskaran, 1950; Basu 1966; Ghosh and Basu 1968; Basu *et al* . 1970; Basu and Ghosh 1970; Gopalakrishnan *et al*, 1970 and 1973; Ghosh *et al.*, 1973, 1976, 1977, 1979 a, 1979 b; 1980 a, b; Ray *et al.* 1977, 1979 and Ray 1980) in and around the outfall of some industries and their overall impact in the river. The above findings are based on the studies made before Farakka barrage was commissioned in 1975. Subsequently after Farakka was commissioned ecological changes in the Hooghly was studied by Nandy *et al.* (1983), for a period of two years from 1975 to 77. The overall impact of industrial and municipal wastes on the trend of fisheries was highlighted by Ray and Laha

(1981). This was followed by an attempt to study the heavy metals in the estuarine water, bottom sediments, fish and shrimp together with their bioaccumulation in different organs by Ghosh *et al.* (1983), and pesticide pollution by Joshi (1985). Carbon assimilation in the estuary was studied by Babulal (1990), and finally the spawning activity of hilsa (De and Sehgal, 1989).

A. Physico-chemical Characteristics

Water temperature

Chakraborty *et al.* (1959); Pahwa and Mehrotra (1966) and Ray *et al.* (1966) recorded water temperature in a stretch of 1090 km. between Kanpur and Rajmahal ranging between 16.5-31.5°C. The lower level of temperature ranged between 16.5 and 18.5°C while the range of upper level was 30.5-31.5°C. Bhargava (1981 a) recorded the range of water temperature between 18 and 31°C during summer of 1977 at a few points above and below of some big cities from Rishikesh to Varanasi (Tables 39, 40, 41).

The water temperature recorded in the estuary in the different seasons during 1967–71 (Table 18) was 21.6-23.9°C (lower stretch) and 21.9-24.3°C (upper stretch) in winter; 27.1-32.8°C (lower) and 28.4-31.9°C (upper stretch) in summer and during monsoon the values were 28.8-31.3°C (lower) and 29.5-31.7°C (upper) respectively. Nandy *et al.* (1983) recorded annual variation of water temperature ranging between 17.5 and 32.0°C at Kalna, 19.5 and 31.2°C at Barrackpore, 19.5 and 31.0°C at Uluberia and 18.5 and 33.5°C at Geokhali during 1975-77 after Farakka was commissioned & (Table 41) Das Gupta (1984) documented in "the Ganga Basin Part II" the water temperature of 1981-82 ranging between 26 and 32°C with an average value 30.2°C during May and 13-22°C with an average value 18.9°C) during July in the entire stretch of Ganga from Haridwar to Diamond Harbour.

Turbidity

Turbidity was generally negligible (below 100 JTU) or even low in the Ganga during the period January to June and high value was recorded from July to September/October. The maximum turbidity ranged from 1100 at Ballia and 2170 JTU at Patna. Generally, July or August recorded the highest turbidity values, nevertheless the same was observed during September. (Chakraborty *et al.* 1959; Pahwa and Mehrotra, 1966 and Ray *et al.*, (1966). Bhargava (1981 a) recorded turbidity during summer of 1977 in the regions of upper Ganga plain between Rishikesh and Varanasi, above and below of 9 important cities. Depth of light penetration (Turbidity) ranged between 16 and 78 of which Allahabad and Varanasi evidenced comparatively clearer water (31-78 DPL). Das Gupta (1984) showed turbidity (Jackson Turbidity Unit) value during May and July of 1981-82. The stretch between Hardwar and Farakka indicated much lower values, compared to Nabadwip and Diamond Harbour (196-560) during Summer (May) while during monsoon (July) lower values (1-45) were recorded upto Kanpur and the stretch below Allahabad to Diamond Harbour indicated moderately high values except at Patna and below where the values were highest (1400-1700). Possibly tributaries joining the main river Ganga might have some role in increasing the values, due to sporadic erosion (Table 39).

In the estuarine zone conditions are totally different and the Hooghly waters remain comparatively clear during winter. Thereafter turbidity increases steadily until it attains its maximum value during July and August. Climatic and seasonal variation in the tidal translatory movements effect turbidity significantly. It is also influenced by the lunar phases. The coincidence of the perigee position of the

Table 39 : Physical and Chemical parameters of the Ganga River System

(Chakraborty *et al.*, 1959; Pahwa and Mehrotra, 1966; Ror *et al.* 1966)
(In Jhingra and Ghosh 1978)

	Sampling Centre and Distance from Hardwar (km)	Width of river (m)		Depth of river (m)		Maximum Turb (ppm)	Water Temperature °C	Free CO_2 (ppm)	Bicarbonate (ppm)	Chloride (ppm)	Nitrate (NO_3–N) (ppm)	Phosphate (PO – P)	Silicate (SiO_2) ppm
		Summer	Monsoon	Summer	Monsoon								
R. Ganga													
1.	Kanpur (541)	100 – 150	1500	5 – 6	15 – 20	1283	16.5 – 30.5	0 – 5	81 – 216	2.5 – 16.0	0.08 – 019	0.07 – 0.21	8.2 – 20.3
2.	Allahabad (788)	150 – 200	2000	5 – 6	15 – 20	1670	17 .5 – 31.5	0 – 3.7	72 – 212	2.5 – 22.0	0.11 – 0.22	0.09 – 0.20	6.7 – 17.0
3.	Varanasi (992)	200 – 300	100–1200	4 – 5	13 – 15	1583	18.5 – 31.5	0 – 6.0	84 – 171	4.0 – 15.6	0.08 – 0.14	0.07 – 0.12	4.0 – 12.6
4.	Ballia (1155)	70 – 100	1500	10 – 15	25 – 30	1100	18.0 – 31.0	2.9 – 5.0	85 – 212	6.7 – 35.4	0.08 – 0.17	0.05 – 0.12	9.1 – 18.4
5.	Patna (1274)	70 – 250	800 – 1000	5 – 7	10 – 15	2170	18.5 – 31.0	2.25 – 100	83 – 200	5.0 – 17.0	0.09 – 0.19	0.08 – 0.13	7.2 – 14.0
6.	Bhagalpur (1504)	150 – 200	600 – 800	5 – 6	15 – 20	1530	18.5 – 31.5	1.4 – 10.0	86 – 183	5.0 – 19.0	0.09 – 0.15	0.06 – 0.12	3.5 – 12.3
7.	Rajmahal (1631)	150 – 200	500 – 600	8 – 10	15 – 17	1420	18.5 – 31.5	0 – 6.0	84 – 171	4.0 – 15.7	0.08 – 0.15	0.07 – 0.12	4.0 – 12.6
R. Yamuna													
8.	Allahabad	80	160	5	16	925	19.2 – 32.0	1.0 – 8.0	—	3.0 – 12.0	0.07 – 0.47	0.02 – 0.25	12.0 – 23.0

Source : Jhingran, A. G. and K.K. Ghosh (1978).

Table 40 : Baseline data for Hooghly Estuary expressed by average vlaues for different seasons during the period 1967- 1971

Period	Season	Temp (°C)		turalidity (cm)		pH		DO (ppm)		BOD_5 (ppm)		ALK ($CaCO_3$)		Salimity (%)		Phyto, (u/1)		Zoo (u/1)	
		U	L	U	L	U	L	U	L	U	L	U	L	U	L	U	L	U	L
	W	24.3	23.9	21.5	7.2	8.0	8.1	6.8	5.0	2.1	1.7	259	194	0.07	0.25	78	66	62	27
1967	S	31.9	32.8	–	–	8.2	8.3	4.3	5.1	–		–		0.17	9.75	–		–	
	R	30.6	31.3	4.9	7.4	8.1	8.2	4.8	4.9	–		122	154	0.06	7.07	–		–	
	W	23.8	—	16.8	11.2	8.1	8.2	5.9	6.3	1.9	2.5	218	218	0.06	0.46	352		153	
1968	S	29.2	27.9	5.9	3.3	8.1	8.2	5.2	5.4	3.0	2.9	237	212	0.74	7.66	144	88	49	25
	R	31.7	30.9	3.5	2.6	8.1	8.2	5.5	5.3	1.8	1.9	71	66	0.06	0.06	55	49	21	18
	W	22.6	21.9	19.5	9.4	8.1	8.1	6.3	5.6	2.4	2.3	280	251	0.06	0.45	908	211	209	77
1969	S	28.5	27.1	9.5	3.9	8.3	8.2	6.3	6.4	3.0	3.8	232	216	0.07	3.0	91	300	28	28
	R	30.3	29.6	3.3	3.9	8.1	8.1	6.3	5.5	3.0	2.2	122	144	0.07	0.37	137	103	21	23
	W	22.6	21.8	11.4	9.0	8.1	8.1	6.3	5.4	2.5	2.1	278	252	0.11	0.64	184	70	95	17
1970	S	29.7	28.8	7.0	5.9	8.2	8.2	5.5	5.0	2.0	2.0	229	231	0.18	5.3	107	41	49	7
	R	29.8	29.6	2.5	3.8	8.1	8.1	5.8	5.3	2.1	2.1	105	119	0.06	0.19	103	61	5	3
	W	21.9	21.6	–	6.5	8.1	8.1	8.1	6.3	1.8	1.4	289	267	0.06	0.22	–		–	
1971	S	28.4	28.8	5.7	4.8	8.1	8.2	5.8	5.6	2.8	1.9	190	184	0.165	1.39	–		–	
	R	29.5	28.8	3.0	3.1	8.0	8.1	5.1	5.1	1.5	1.5	32	37	0.33	0.11	–		–	

W — Winter Season.
S — Summer Season.
R — Rainy season or monsoon.

U — Upper Stretch
L — Lower Stretch.

Source : Ray, P., (1981).

Table 41 : Annual Variation of some Physico-Chemical parameters in the Hooghly estuary after Farakka discharge 1975—1977

Parameters	*Zone-I*		*Zone-II*		*Zone-III*
	Kalna	*Barrackpore*	*Uluberia*	*Geokhati*	*Kakdwip*
Water Temp °C	17.5 – 31.0	19.5 – 31.2	19.5 – 31.0	18.5 – 33.5	20.1–34.5
Suspended Solids (ppm)	220 – 2300	120 – 3400	80 – 2240	140 – 3800	50 – 3125
pH.	7.6 – 8.5	7.8 – 8.5	8.0 – 8.5	7.5 – 8.5	8.1 – 8.5
Dissolved Oxygen (ppm)	5.1 – 10.8	5.0 – 10.0	5.0 – 10.2	5.2 – 8.6	4.7 – 10.6
T. alkalinity (ppm)	80 – 175	68 – 195	80 – 180	72 – 180	40 – 158
Sp. Conductivity at 25°C × 10 – 6 (mhos)	194 – 425	143 – 479	199 – 799	211 – 7199	11999 – 18275
Nitrate as $NO_3 - N$ (ppm)	Trace – 0.224	Trace – 0.18	Trace – 0.2	Trace – 0.27	— —
Phosphate as PO_4 (ppm)	0.037 – 0.61	Trace – 0.41	0.03 – 0.31	Trace – 0.265	0.05 – 0.15
Silicate (ppm)	3.0 – 25.0	3.0 – 27.0	1.0 – 25.0	2.0 – 25.0	1.0 – 5.0

Source : Nandy. A.C., M.M. Bagchi and S. K. Majumdar (1983).

moon and the spring tide condition, as observed by Bose (1956) has the meximum mechanical effect in pushing the neritic waters and mixing of column water with the soil particles that had sedimented during the earlier neap phase of the tide. On the other hand, the coincidence of apogee position of the moon and the neap condition of tide is found to have maximum effect on turbidity. Bose further reported increase in turbidity due to daily tidal movements of water in areas where there are curves in the estuary. Further, the turbulence occurred on account of occasional movement of ships also promotes turbidity.

The maximum turbidity of 1000-2500 JTU and the minimum value of 100 JTU were reported by Bose (1956) between Uluberia and Kakdwip. Discharge from the rivers Damodar and Rupnarayan, which join the Hooghly in this area and the river Haldi which joins it further down, also contributes in increasing the turbidity. High turbidity values were observed in the lower stretches during the non-monsoon seasons from 1967 to 1971 (Table 40).

Hydrogen-in-concentration

In the entire stretch of the Ganga irrespective of tidal zone, the water seems to have high buffering capacity, as a result of which only marginal change in pH has been recorded. The pH of water was found to be in the alkaline range 7.45-8.3 in both Ganga and Jamuna at all the centres in and around Allahabad, with the higher values encountered during the summer and winter months and lower values during monsoon months. Bilgrami and Dutta Munshi (1979) recorded pH 6.4-7.4 (av. 6.9) during monsoon (July and August 1976) in a stretch between Barauni and Farakka, while studying the spawn exploration work. Later at 12 centres, spread over the entire Ganga, Haridwar Diamond Harbour, pH recorded in summer (May) was 8.0 while during monsoon (July) 7.0-8.0 (av. 7.5) (Das Gupta, 1984) (Table 39).

The estuary also experienced almost similar trends. Roy (1949) recorded pH 8.3 and 8.4 during 1945-48. Bose (1956) recorded between 7.9 and 8.4. The pH value did not show any yearly variation during 1967-71 (Tables 40). The average value in the upper estuary was 8.1 in winter, 8.1-8.3 in summer and 8.1 in monsoon. In the lower zone however, the pH ranged 8.1-8.2 (Av. 8.1), 8.2-8.3 (Av. 8.2) and 8.1-8.2 (Av. 8.1) during the above seasons (Ray 1981). The above values pertain to pre Farakka release. After Farakka was commissioned a tendency of slight rise in values towards upper level (8.5) was noticed (Nandy *et al.*, 1980). The annual average of 1975-77 recorded at the 5 centres stretched over the estuary, Kalna-Kakwip, was 7.6-8.5 with Kalna 7.6-8.5 Barrackpore, 7.8-8.5, Uluberia, 8.0-8.5, Geokhali, 7.8-8.5 and Kakdwip 8.1-8.5 (Table 41).

Dissolved Oxygen

The pattern and range of DO content were more or less identical at all the centres with the maximum values 8.6-10.4 ppm recorded in the winter months January-February, and minimum during monsoon months (4.9-7.0 ppm) July to September. DO was found to be directly related to turbidity (4.9 ppm at 1125 JTU, Ray *et al.*, 1966). Bilgrami and Duttta Munshi (1979) while conducting the spawn exploration work during monsoon in the stretch of Barauni to Farakka, estimated DO to be moderately high (6.0-9.5 ppm, av. 7.0 ppm). After Farakka was commissioned, the entire hydrological parameters were changed and new Physiographic data of 1980-82 was presented by Das Gupta (1984) for seasonal variation (post monsoon, winter and summer) of DO alongwith other parameters. It is evident that in all the three stretches, upper Ganga plain (Hardwar-Nagbasuki Temple, Allahabad), Middle Ganga plain (Allahabad-Farakka) and lower Ganga plain, except a slight fall in DO value, it had never showed any critical level even in summer. In the upper Ganga plain the DO values ranged between 6.25 and 8.05 ppm (av. 8-02 ppm) during post monsoon, 5.26-98.92 ppm in winter and 4.35-10.10 ppm (av. 5.89 ppm) during summer. In the middle Ganga region, the values were 5.20-8.27 ppm (av. 7.12 ppm); 4.67-11.75 ppm (av. 7.85 ppm) and 6.51-8.84 ppm (av. 6.88 ppm) for the above seasons, while in the lower Ganga such values were 4.53-7.80 ppm (av. 6.26 ppm), 7.47-9.10 ppm (av. 8.27 ppm) and 5.40-6.90 ppm (av. 6.46 ppm) respectively. Although on 2 or 3 occasions in the entire stretch, concentration of DO was slightly below 5 ppm but on the whole it never showed any critical level (Table 39).

In the estuarine stretch DO around Calcutta was more than 4.0 ppm excepting during summer months (Bose, 1956). Between Barrackpore and Kulpi minimum DO values ranged from 8.4 and 5.1 during 1953-55. Subsequent study revealed that the average DO was 5 ppm in all the seasons. At the outfall zones it ranged from 3.8-4.8 ppm (Ray and Ghosh, 1976). Near industrial outfall region the DO was either absent or present in traces. Such low values of DO may be due to combination of several factors *e.g.* high turbidity, great demand of oxygen by decomposing organic material in summer and grater period of effluents retention during the neap tidal phase. The annual variation of DO during the post Farakka release 1975-77, as stated by Nandy *et al.*,(1983) ranged 5.1-10.8 ppm at Kalna, 5.0-10.0 ppm at Barrackpore, 5.0-10.2 pm at Uluberia 5.2-8.6 ppm at Geokhali and 4.7-10.6 ppm at Kakdwip (Table-41) and is certainly an indication of improvement over the years of 1967-71 attributable to release of fresh water.

Biochemical Oxygen Demand (BOD_5), Chemical Oxygen Demand (COD) and Bacterial density (MPN/100 ml.)

Bhargava (1981 a) while assessing the water quality index in the upper stretch of the Ganga above and below the seven cities like Rishikesh, Hardwar, Narora, Kannuj, Kanpur, Allahabad and Varanasi estimated BOD_5 to range between 4.4 and 27.0 ppm with higher values (10-27 ppm) around Kanpur, which indicates the highly polluted area (Table 24). Das Gupta (1984) quoting data for BOD_5 and COD values of the State Water Pollution Control Board, Bihar and West Bengal compared the data both from banks and mid river for the period, January to June for four cities Monghyr, Rajmahal, Kahalgaon and Bhagalpur in the middle zone of the river, has shown that at Monghyr, the mean value for BOD_5 at the bank was 1.7 against 1.2 mg/l at the centre; Likewise, at Rajmahal, the values were 1.3 mg/l against 0.3 mg/l, at Kahalgaon, 1.6 against 1.1 and at Bhagalpur 1.6 against 0.8 mg/l. Similarly, the values for COD were 25.3 and 18.7; 17.0 and 15.7; 16.0 and 15.0 and 20.3 and 12.1 mg/l. Das Gupta (1984) further reported seasonal variation of BOD_5 during 1980-1982 in different stretches of the Ganga as, in the upper plain (Haridwar-Kanpur) the range of BOD_5 was 2.50-17.70 mg/l (average 9.79 mg/l) during post monsoon (October-November); 1.70-18.60 mg/l (average 8.45 mg/l) during winter (December-February); and 2.50-16.30 mg/l (average 10.05 mg/l) during summer (March-June). In the middle Ganga plain (Allahabad-Farakka) such values were 0.47-7.80 mg/l (average 3.55 mg/l), 0.54-6.50 mg/l (average 8.34 mg/l) and 0.72-8.37 mg/l (average 3.4 mg/l) respectively during the above seasons. In the lower Ganga plain upto Diamond Harbour, the values were further lower, 0.53-1.93 mg/l (average, 1.0 mg/l), 0.35-15.23 mg/l (average 2.82 mg/l) and 0.45-15.58 mg/l (average, 3.07 mg/l) during the above seasons (Table 25).

Coliform bacteria are indicator of organic pollution studied by various workers/organisations are presented in Tables 24-26 and 29 & 30. It is estimated in different zones like, upper middle and lower stretches of the Ganga.

Bhargava, (1981a) during 1977 estimated coliform count (MPN/100 ml) above and below seven cities located on the bank of the Ganga in a stretch of about 1000 km from Rishikesh to Varanasi (Table 24). At the upstream of the city the values were mostly below 130 while at the down stream of the cities except Narora and Kannuj the values were slightly higher (1600-2400) Bilgrami and Dutta Munshi (1977) during their study in a small stretch during 1978-79 between Barauni and Farakka presented values which was at par with Bhargava (1981a) (Table 30). Seasonal variation of coliform bacteria in a more wider area viz upper Ganga plain (Haridwar-Allahabad), middle Ganga plain (Allahabad-Farakka) and lower Ganga plain during post monsoon, winter and summer of 1980-82 were also presented by Das Gupta (1984). In this middle Ganga plain and lower Ganga plain and also winter and summer were indicating much higher values (Table 26) although DO was indicating optimum level and BOD remaining reasonably at a low level. In another study for 2 consecutive years, 1981-1982 during summer (May) Das Gupta (1984), showed quite high values between Allahabad and Dakshineswar (24000-2,40,000 average values of MPN. The DO ranged between 5.7 and 8.1 mg/l and BOD mostly below 10mg/l (Table 29). Further Das Gupta (1984) reported that in a short strech study at 6 centres covering Monghyr and Katwa, the coliform count was 80-98 per cent less than the concentration observed in the bank which was indicating quite high value (Table 29). Around Calcutta bank-bacterial density studied by NEERI (1978) during 1975-76 in a stretch, Pathak to Chintamony (about 20 km.) during low and high tide of summer and monsoon, indicated right bank to be more polluted (Left bank, LT: 26.7×10^5, HT: 30.29×10^6 during dry season against the right bank, LT : 37.6×10^5, HT : 11.253×10^6 during monsoon). Some of the values mentioned by Das Gupta, 1984 (Table 26) appears to be quite high (higher values 24,333) (post monsoon) and 9,16,667 (winter). It is clear from all the studies made so far that except the result presented by Bhargava, (1981a), and Bilgrami and Dutta Munshi (1978) all other values shown are abnormally high especially below Farakka and are much above the drinking water standard (5000 MPNV/100 ml) and for this no explanation is offered. Incidentally it may be mentioned that data presented by Das Gupta (1984) are based in the data of WPCB, Uttar Pradesh, Bihar and West Bengal.

The extensive work done by CIFRI (Barrackpore) in the Hooghly estuary in different years prior to Farakka discharge indicate that the organic load of the Hooghly estimated for the first time by Ray during 1967-71 at the lowest dilution level. The average BOD_5 never exceeded 5 ppm during the period in the main body of the estuary (Table 40) except for slight rise in the summer months between Naihati and Calcutta (Ray and Ghosh, 1976). However, around the outfall the values were significantly higher.

Free Carbondioxide and Carbonates

Free carbon dioxide was generally present only during the period July-November, but sometimes also in June and December. It fluctuated within a narrow range of 0.6-6.1 ppm except at Patna and Bhagalpur, when it touched 10.0 ppm. The bicarbonates were present only during the period of total absence of free carbondioxide, *i.e.* from December or January-May or June. This is due to the fact that the presence of free CO_2 facilitates the formation of stable bicarbonate, thereby checking carbonate formation. The carbonate content was found to range 1.0-12.0 ppm (Chakraborty *et al.*, 1959; Pahwa and Mehrotra, 1966 and Ray *et al.* (1966) (Table 39).

Total Alkalinity

Total alkalinity is mainly due to carbonates and bicarbonates of calcium and magnesium. A round Allahabad in the Ganga higher values were recorded during summer months when water level was quite low, the maximum being in June and minimum during monsoon. In general total alkalinity values in the Ganga around Allahabad ranged between 251 and 318 ppm. (Chakraborty *et al.* 1959 and Ray *et. al.* 1966). In a subsequent study made by Pahwa and Ray Mehrotra (1966) at seven centres of the stretch between Kanpur and Rajmahal both carbonate and bicarbonate separately it was revealed that carbonate values ranged between 1.0 and 12 ppm (higher values being between April and June) and that of bicarbonate values 72-84 (lower) and higher values (171.3-216 ppm), Table 39.

In the estuarine region, alkalinity of the Hooghly water remains the same except for some seasonal variation between freshwater zone—Kalna and saline zone Diamond Harbour. September onwards alkalinity increases until it attains maximum value during March and June. It abruptly decreases with the advent of monsoon (Dutta *et al.*, 1954). Bose (1956) observed that from a minimum 2 mg/l (102 ppm) alkalinity increased to 4.5 and 7.0 mg/l (230 to 357 ppm) in January/February. After March it declines but salinity increases. Observations made during 1967-71 by the author shows that the maximum salinity value of alkalinity gradually declined. The highest values were observed during November-February which declined in March-June and were the lowest during the rainy season (Table 40). Subsequent study made by Nandy *et al.* (1983) during 1975-77 after the Farakka discharge shows that the excess freshwater release has brought in considerable dilution to the constituents resulted in the variation 40-195 ppm in the tidal zone. Their split up at 5 centres were—Kalna, 80-175 ppm, 68-195 ppm at Barrackpore, 80-180 ppm at Uluberia, 72-180 ppm at Geokhali and 40-158 ppm at Kakdwip.

Minerals

Chlorinity/Salinity : The chlorides recorded the same trend of fluctuations as alkalinity at all the centres, with high value in summer and lower values in monsoon. During 1960, the chloride content in the Ganga ranged from 4.0 to 35.4 ppm. Ballia records the maximum range and Varanasi and Rajmahal the minimum (Table 39). However, in the earlier years 1954-55 and 1958-59, the chloride values were much higher and ranged from 46-56 ppm and 30-72 ppm respectively.

Later, Bhargava (1981 a) while studying the water quality in the upper plains of Ganga in 1977 between Rishikesh and Varanasi during summer showed that chloride values ranged between 2 and 34 ppm with lower values (2-9 ppm) upto Kanpur. Das Gupta (1984), quoting data from WPCB, Uttar Pradesh, Bihar and West Bengal has shown seasonal variations for the three zones of Ganga. In the upper zone, chlorides ranged from 5-35.3 ppm (av. 18.5) during post monsoon; 14.7-29 ppm (av. 20.2) during winter and 8.3-25.1 ppm (av. 19.0) in summer. In the middle stretch of Ganga plain such values were 8.6-22.7 (av. 15.7); 11.2-38.7 ppm (av. 26.2) and 14.3-37 ppm (av. 22.3) in the above order, while in the lower Ganga plain the values were 7.6-16.1 ppm (av.10.1); 9.5-28.6 ppm (av. 17.2) and 8.8-13.8 ppm (av. 10.6); in the above order. Salinity fluctuations in the estuarine zone is quite significant. In general, observations were made from April to early June due to the intrusion of the sea water. Salinity varies between 8-12 ppm during January-Feb., 10-120 ppm in March, 10-208 ppm in April, 26-180 ppm in May, 18-48 ppm in June and 4-8 ppm in July-December (David 1954).

He also observed wide fluctuations, *i.e.* 72-432 ppm during the course of same day (25th April 1948). During the pre-monsoon period salinity remains high and shows erratic fluctuations. The highest salinity of 4000 ppm was recorded at Palta in 1972. Perhaps there are no vertical gradients of salinity, as it was not recorded.

During summer salinity increases 3 per cent in the brackishwater zone at Kakdwip. It decreases during the monsoon to 1.6 per cent and may become almost like freshwater (Bose, 1956). During April saline water penetrated deep into freshwater zone on account of reduced flow from upstream. Upstream of Barrackpore the water remains fresh throughout the year. Around Barrackpore maximum salinity was 2 per cent during May/June (Bose, 1956). In Brackishwater zone near Diamond Harbour, difference of 2 per cent between surface (20 per cent) and bottom waters (22 per cent) at a depth of 18 m during maximum high tide was recorded. The difference was marginal during the lowest water level. Variations in salinity between surface and bottom waters around Calcutta may be due to local vertical eddy currents at the time of high tide (Bose, 1956).

The Hooghly estuary is free from salinity zones such as marine, tidal and gradient zones (Bose, 1956). Between Kakdwip and Palta lies the brackishwater zone and upstream of Calcutta is freshwater zone. The area between these two zones possesses transitional characteristics. During 1956-58 shetty *et al.* (1961) recorded salinity in traces (less than 0.1per cent) between Medgachi and Khusigali, the upper zone; trace-15 per cent, in the middle or gradient zone between Konnagar and Diamond Harbour, and 6.18-32.85 per cent in the lower zone upto Freserganj. Observations made during 1967-1971 indicated almost similar trend (Table 40). Seasonal fluctuations in salinity before and after Farakka discharges both in fresh and brackishwater zones around Barrackpore and Kakdwip are shown in Fig. 22. A sizable drop in the salinity occurred all over the estuary. At Kakdwip, the Salinity dropped from the range 1.5-32.8 per cent in 1960-61 to 3.2-29.2 per cent in the first year of release (May 1975-April 1976) and further to 2.5-18 per cent. in the second year of release (May 1976-April 1977) Nandy *et al.* (1983). At Barrackpore the values came down to traces from 0.2 per cent.

Specific conductivity : Data on conductivity, a measure of electrolytes indicated lower values 147-180 m mhos from Rishikesh to Narora. From Kannauj to Varanasi the range was 287-515 m mhos (Bhargava, 1981 a). Das Gupta (1984) quoting the data of Water Pollution Control Board of U.P., Bihar and West Bengal for 1981-82 indicated to range 325-2026 micromhos/cm during summer while in July the value ranged 211-809 micro mhos. The drop in value is obviously due to dilution. Nandy *et al.* (1983) estimated conductivity to range 143-18275 mhos (at $25^0C \times 10^{-6}$ mhos) between Kalna and Diamond Harbour. At Kalna, the upper end of estuary, the range was 194-425, further down at Barrackpore, 143-479, Uluberia, 199-799, Geokhali, 211-71289 and Kakdwip 11999-18275 mhos thereby indicating that the water was more of brackish in character. (Table 41.)

Nutrient status : (Nitrate, Phosphate and Silicate)

Not much information is available with regard to nutrient level in the Ganga water except the one indicated here. The concentration of these nutrients were similar in the two rivers Ganga and

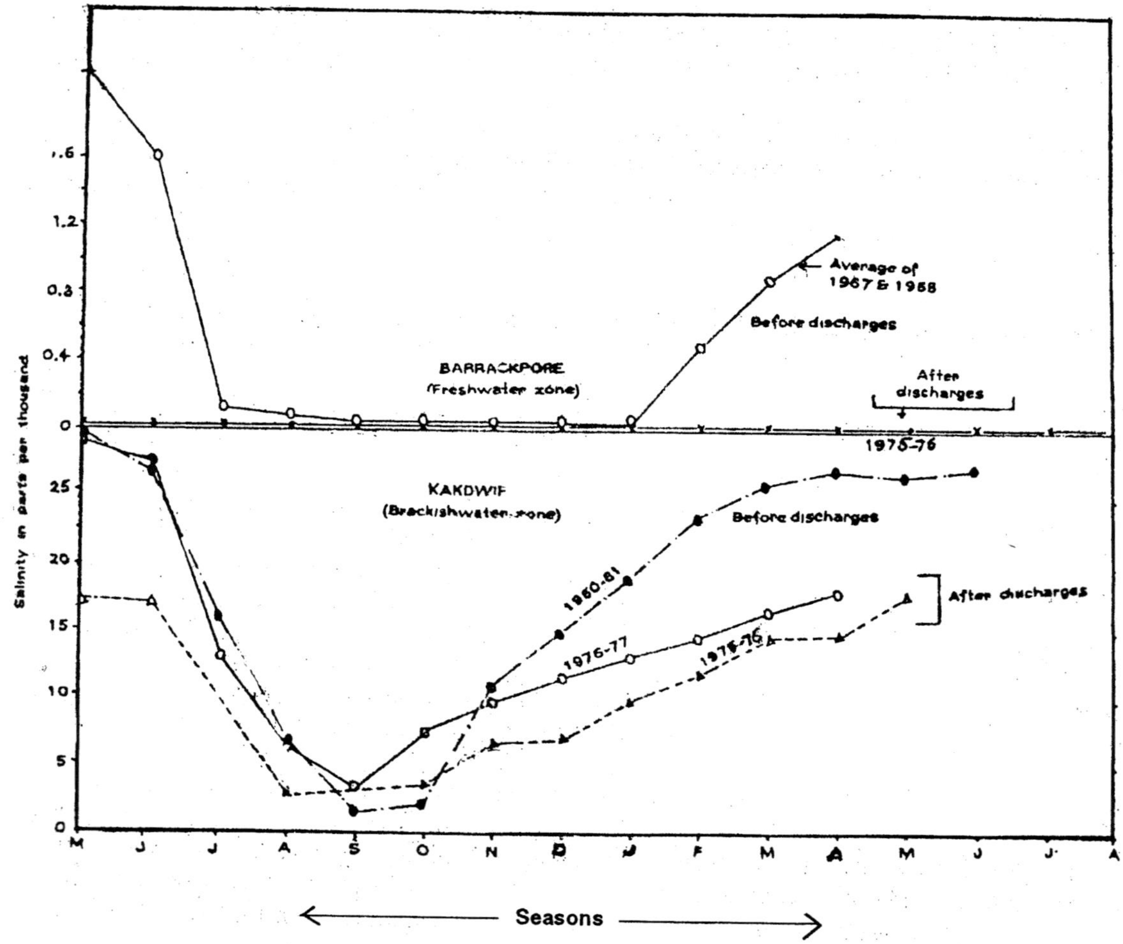

Fig. 22 : Seasonal fluctuations in salinity before and after Farakka discharges.
Source : Ray, P. (1981).

Yamuna at Allahabad. Both nitrates and phosphates were present only in small concentrations generally within 0.1 ppm except during monsoon months when they at times registered concentrations upto 0.2 ppm and 0.21 ppm respectively. On the other hand, the silicates were minimum in monsoon and maximum in summer, their range being 3.5 to 22.0 ppm (Chakraborty *et al.* 1959; Pahwa and Mehrotra, 1966 and Ray *et al.*, 1966). Ray and David (1966) estimated phosphate value between 0.52-2.2 mg/l in and around Kanpur.

The main sources of nitrogen and phosphorus are from the run off of the catchment area, industrial and sewage effluents and decomposition of organic waste which settles on the substratum. Around Palta 0.001-0.0035 ppm phosphate (PO_4)has been reported by Roy (1955). It is the highest in spring and summer and the lowest in winter. The phosphate (PO_4) content in the estuary is low during monsoon but relatively high in summer possible due to its regeneration from the bottom sediments (Bose 1956). He has also found that the phosphate content between Barrackpore and

Uluberia were relatively high, varying from 3.0-4.0 mg atom p/m^3 (0.9201-1.2268 ppm PO_4 compared with 1.5-3.0 mg atom p/m^3 (0.46005-0.9201 ppm PO_4) at other centres. The lower values varied between 0.1 and 0.4 mg atom p/m^3 (0.3067 and 0.12268 ppm PO_4). Silicates, however, varied between 5.25 and 22.2 ppm (Si O_2)(Roy, 1955). The maximum concentration is reached during the winter season and is maintained till late spring or extended upto early summer. Bose (1956) recorded 400-540 mg atom Si/m^3 (0.856-1.556 ppm Si O_2) towards freshwater zone (Uluberia and above); and 180-300 mg atom Si/m^3 (0.3852-0.428ppm Si O_2) in the lower zones. However, the values of Silica gradually increased from 40-160 mg atom Si/m^3 (0.0856-0.3424 ppm Si O_2) from brackish to freshwater zones. In general, there is a dearth of data on the nutrient status of the estuary on a time specified basis. The nutrient status during the post Farakka release as reported by Nandy *et al.*, (1983) ranged between trace and 0.94 ppm for nitrate-N, trace and 0.75 ppm for phosphate P (PO_4) and for silicate it ranged between trace and 31 ppm, between Kalna and Kakdwip. Centrewise the values of nitrate were trace-0.224 ppm at Kalna; Trace to 0.18 ppm at Barrackpore; Trace-0.2 ppm at Uluberia and further down trace-0.27 ppm at Geokhali. In the case of Phosphates the values were 0.037-0.61 ppm, trace-0.41 ppm, 0.03–0.31 ppm trace-0.265 ppm and 0.05-0.15 ppm for the centre in the above order, and finally, silicate ranges 3-25 ppm, 3-27 ppm, 1-25 ppm, 2-25 ppm and 1-5 ppm at the above five centres (Table 41).

***Sulphates** : It ranged between 2.4-35.5 ppm (av. 18.6) during post monsoon, 2.1-22.5 ppm (av. 13.7) in winter and 2.0-18.0 ppm (av. 12.8) during summer of the upper Ganga plain. In the middle Ganga plain such values were almost similar *viz.* 2.4-33.5 ppm (av. 19.5) in post monsoon, 12.8-21.3 ppm (av. 15.8) and 19.8-21.2 ppm (av. 17.1) during summer. In the lower Ganga plain such values were 17.7–27.6 ppm (av. 21.4), 21.5-28.7 ppm (av. 23.3) and 16.9-28.5 ppm (av. 24.3) for the above seasons.

***Sodium** : In the upper Ganga Plain (Haridwar-Nagasuki temple, Allahabad) the values ranged 0.5-20.7 ppm (av. 11.8) during post monsoon, 15-28 ppm (av. 22.1) in winter, and 5.8-10.0 ppm (av. 6.9) during summer. In the middle Ganga plain (Allahabad-Farakka) ranges were 4.8-20.2 ppm (av. 13.3), 15-25.7 ppm (av. 21.2) and 11.1-30.7 ppm (av. 17.3) during the above seasons. In the lower Ganga plain (below Farakka) the values were 4.2-17.8 ppm (av. 11.8), 14-27.4 ppm (av. 21.1) and 10.3-27.4 ppm (av. 14.6) in the order of seasons indicated.

***Calcium** : Increased concentration of calcium was indicated both in the upper and lower Ganga plains. The average values during post monsoon, winter and summer in the upper Ganga plain were respectively 104.2 ppm (59.0-135.5), 84.9 ppm (44.2-119.9) and 77.3 ppm (43.2-111.7). In the middle Ganga plain, 59.1 ppm (19.7-101.3), 64.1 ppm (18.8-122.4) and 51.5 ppm (26.2-88) while in the lower Ganga plain, the values were 83.6 ppm (70-100), 95.6 ppm (62.5-113.3) and 80.1 ppm (63.6-145.7) for the seasons.

Metals and Pesticides

In the fresh water zone of the Ganga, heavy metals were estimated by Joshi (1988) at different centres like Haridwar, Kanpur, Allahabad and Patna for Zinc, Copper, Chromium, Cadmium, Lead and Mercury (Fig. 23). In his document he has demarkated three points like city bank side, mid river and opposite to the city bank side. The result indicated that city side of Kanpur was always showing higher values of Zinc (285 ppb), Copper (175 ppb) and Chromium (200 ppb) which appears to be high. This was followed by Allahabad and was minimum at Haridwar. Cadmium, lead and mercury indicated low values. Neither any specific study on bioaccumulation of metals in fish or any other biotic life has been conducted, nor the effect of such metals on survival, growth and reproduction of biomass specially on fish was studied so far.

**Source* : Modified from Das Gupta, S.P. (1984).

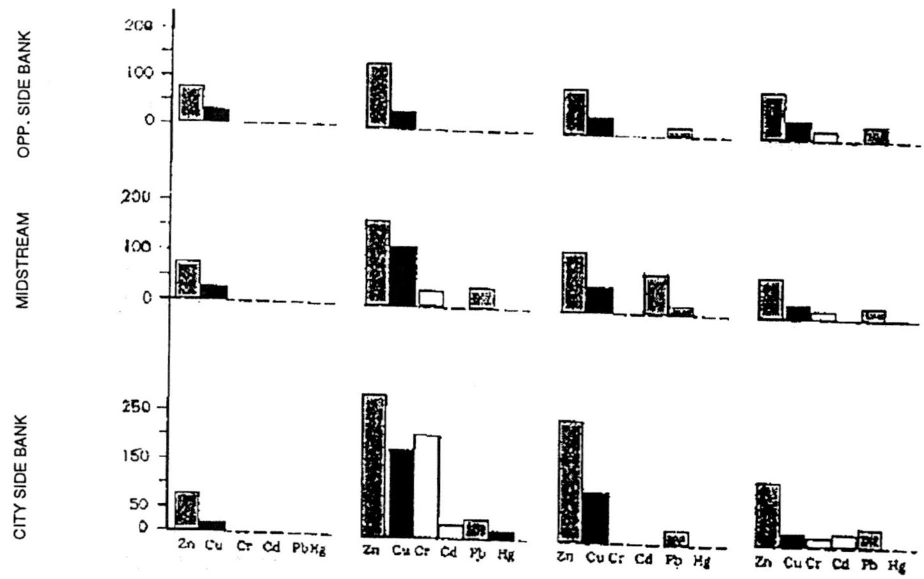

(----------- Broken lines represent below detection levels)

Fig. 23 : Heavy Metals in the River Ganga.
Source : Joohi, H.C. (1988).

Ghosh *et al.* (1983) while studying the pollution status of the Hooghly with reference to heavy metals estimated upper level of Zinc and Chromium above the outfall range as Zinc (0.4 ppm) and Chromium (0.4 ppm). Near the outfall the upper level Zinc ranged between 0.05 and 7.2 ppm, and Chromium 0.2 and 20 ppm, Copper and lead could not be detected. Interestingly concentration of metals in sediment registered much higher values than the wastes Zinc was estimated to be 381.6 ppm (outfall of Rayon industry) followed by outfall of Paints and varnishes (160 ppm) and oil refinery (80 ppm). Chromium level was recorded at 9.6 ppm. Zinc contamination in sediments was found to extend beyond 500 m.

Metals were found to accumulate in fish, shrimp crab and molluscs. The upper level of zinc was found to accumulate in fish (24-120 ppm); in shrimp. (24-112.2 ppm); in crab, (40.ppm) and in molluscs (83.0 ppm). Similarly, the upper level of copper was 20 ppm; Chromium 0.26-4.5 ppm and lead could not be detected.

Fish tissues are also found to accumulate metals appreciably. In gill zinc was found to accumulate in the range 11-60 ppm, in muscle 7.5-22.5 ppm, in kidney 17.5-40 ppm, in gut (intestine) 16-65 ppm and in gonad, 5-146.8 ppm. Copper accumulation was comparatively less and its accumulation in muscle is 2-11 ppm., in liver, 2-8 ppm., in kidney, 5-15 ppm, and in gut, 2-30 ppm respectively. Chromium was found to be present in gill of only one fish (15-19.2 ppm) and lead was absent.

Thus it is clear that zinc is present in highest level and copper although present is in a much lower concentration. Incidence (51 per cent.) and magnitude of DDT and BHC-V residues in fishes of tidal stretch of the Ganga were of higher order in the industrial zone compared to riverine (13 per cent) and estuarine agricultural zone (9 per cent). Greater accumulation of DDT was recorded (Joshi 1966b) in molluses (65-963 ppb), followed by fish (31-460 ppb), plankton (15-150 ppb) and sediment (17-80 ppb) compared to that of BHC-Y in molluses (40-86 ppb), fish (46-210 ppb) and sediments (21-70 ppb). Juveniles of *Hilsa ilisha* (67-104 mm) registered greater accumulation of DDT than larger specimens (122-198 mm), while in case of other estuarine fishes like *Apocryptes bato* it was reverse (Joshi, 1986b), muscle is 2-11 ppm; in liver, 2-8 ppm; in kidney, 5-15 ppm; and in gut, 2-30 ppm; respectively. Chromium was found to be present in gill of only one fish (15-19.2 ppm) and lead was absent.

Thus it is clear that zinc is present in highest level and copper, although present is in a much lower concentration.

Energy Transformation From Light to Chemical in the Ganga

The rate of Energy transformation and photosynthetic efficiencies at different places (Ganga and its tributaries and River Jamuna) are presented in Table 42.

The production efficiencies and flow of energy in the river Ganga from solar radiation to fish from upper stretch (Kanpur) through the middle stretch (Allahabad, Varanasi, Patna) upto Bhagalpur have been shown in Table 43. It is obvious from these observations that the productive potential of the river Ganga has declined sharply at Kanpur due to deterioration in water quality by the discharge of large amount of pollutants in the Ganga around Kanpur. A comparison of productive potential at the above places and the actual output in terms of fish clearly indicates that the extent of energy utilisation is only 11.50 per cent, 27.25 per cent and 25.74 per cent of the potential at Allahabad, Bhagalpur and Patna respectively.

NB : In the energy dynamics of the river and productivity efficiencies at different stations (Location not indicated) only *single value* for each parameter is shown. How can such data be taken as representative in a fluviatile water and based on that conclusion drawn on such a vital study leads to energy utilization ?

B. Biological Characteristics

I. Freshwater Zone

Plankton form an important food of small and large fishes. The principal plankton of the Ganga consists of two types freshwater form in freshwater zone and for the estuarine zone they are mixture of freshwater, brackishwater and marine forms. Results of analyses of plankton (Tables 43-47, 49-50 and Fig. 24-25) and benthos at different centres during different periods and seasons of the Ganga (freshwater zone) and brackishwater zone are discussed in Tables 44-51 and Figs. 24-27. Shannon-Weaver index are also shown (Table 52).

Phytoplankton

Monthly numerical fluctuations of total plankton during 1960 at various Centres of the Ganga indicates that rarely plankton showed in identical pattern at all the centres. It showed two peaks, one during summer (except at Ballia where it has indicated in July) and the other during winter or spring (March). The summer peak, which was invariably obtained during June constituted the primary peak of the year except at Rajmahal. Monsoon and Autumn months (July-October) generally recorded poor concentrations of plankton in August, September/October recorded the year's minimum. The recorded minimum between the two peaks was observed during April except at Ballia. The upper stretches from Kanpur to Varanasi showed a distinctly higher abundance of plankton.

Table 42 : Rate of Energy transformation (Cal m-2/day-1) and photopchetic efficiency at various Centres of the Ganga and its tributary during 1988-89 and 1989-90.

1988—1989	*R Ganga*	*1989—1990*	*Remarks*
1918 - 8733 (5326)	Kanpur	2482 - 12,170 (8563)	Samplings were done at various points of Kanpur, Allahabad, Varanasi and Patna from Fresh water Zone Figures in the Parenthesis are the average values.
135 - 3873 (2004)		1290 - 9,785 (6112)	
118 - 11828 (3359)	Allahabad	299 - 13,004 (5158)	
260 - 15243 (4566)		758 - 19,712 (9498)	
44 - 6281 (2904)		295 - 8,553 (3360)	
2821 - 6010 (4349)	Varanasi	346 - 8,468 (3679)	
		606 - 4,235 (2384)	
119 - 5378 (2872)	Patna	90 - 3,714 (1807)	
324 - 2934 (1306)		35 - 5,024 (2093)	
94 - 5267 (1984)		125 - 3,349 (1454)	
166 - 5758 (2750)		95 - 3,638 (1201)	
699 - 7132 (3142)	Average :	583 - 8,359 (4110)	

	Above outfall	*Below outfall*	*Above out fal*	*Below outfall*	Remarks
	1918 - 8733 (5326)	760 - 3300 (2030)	2482 - 12,170 (8563)	1763 - 14,913 (6905)	Values give are from above and below the sewage out fall of Kanpur and Varanasi.
	135 - 3873 (2004)	663 - 723 (693)	1290 - 9,785 (412)	524 - 9,972 (5128)	
	2821 - 6010 (4349)	711 - 2011 (1361)	346 - 8,464 (3679)	423 - 1,554 (939)	
Av.	778 - 6202 (3893)		606 - 4,235 (2384)	134 - 3,209 (1752)	
Average :			1181 - 5914 (5184)	711 - 7412 (3681)	

	1988—1989		1989—1990	Remarks
R. Jumna		*R. Jumna*		Values are from fresh water Zone from tur Centre.
	58 - 4360 (2023)		509 - 3422 (1746)	
	40 - 3085 (1292)		687 - 3754 (2186)	
Average :	44 - 3722 (1657)	Average :	598 - 3588 (1966)	
R. Pun Pun	891 - 11,690 (4946)	....	403 - 4696 (1489)	
R. Gandak :	937 - 6868 (3989)	....	33 - 2143 (966)	

Source : Annual Report, 1988-89 and 1989-90. CICFRI, Barrackpore.

Table 43 : Energy Dynamics of the River Ganga—Productivity efficiencies at different stretches (1988—1989)

Parameters	*Kanpur*	*Allahabad*	*Varanasi*	*Patna*	*Bhagalpur*
1. Energy on the water face (K cal ha^{-1} yr^{-1})	67,05,050 $\times 10^3$	68,14,550$\times 10^3$	68,43,750$\times 10^3$	68,07,250$\times 10^3$	68,40,100$\times 10^3$
2. Chemical energy fixed by producers (K cal ha^{-1} yr^{-1})	84,05,950	2,61,85,100	2,16,70,050	1,04,82,800	1,50,52,600
3. Potential Chemical energy as fish (K cal ha^{-1} yr^{-1})	96,216	2,98000	2,46,480	1,35,792	1,71,348
4. Actual energy harvest as fish (K cal ha^{-1} yr^{-1})	—	34,428	—	37,0003	44,100
5. *Efficiencies :*					
(i) Light / Chemical	0.124	0.384	0.317	0.154	0.220
(ii) Light / Potential	0.00143	0.0044	0.0036	0.0020	0.0025
(iii) Light / Actual fish	—	0.0005	—	0.0054	0.00064
(iv) Chemical / Fish	—	0.132	—	0.353	0.293
(v) Potential / Actual	—	11.55	—	27.25	25.74

Source : Annual Report, 1988-89, CICFRI, Barrackpore, West Bengal.

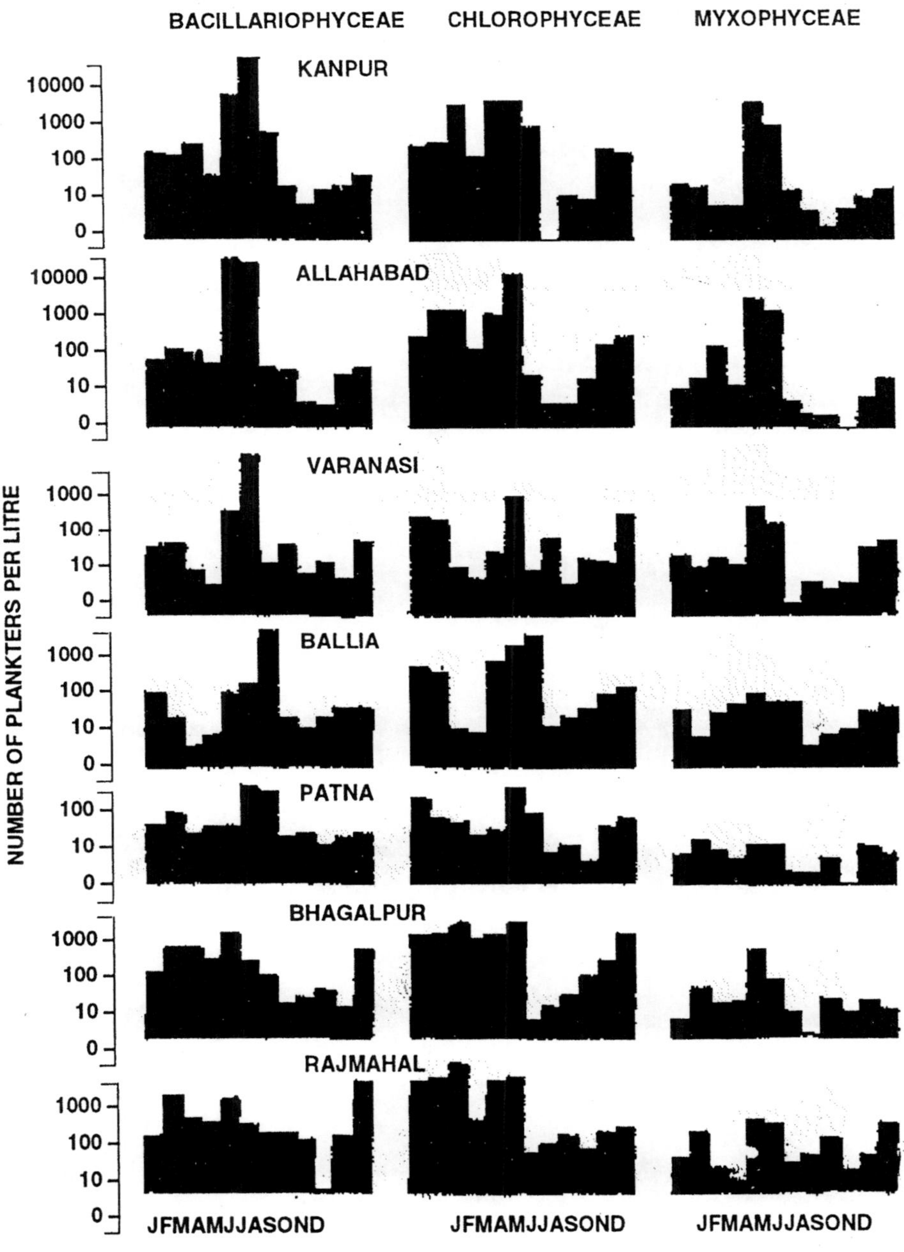

Fig. 24 : Centrewise monthly fluctuations of phytoplanktonic groups (on logarithmic scale).
Source : Pahwa, D.V. and S.N. Mehrotra, (1966).

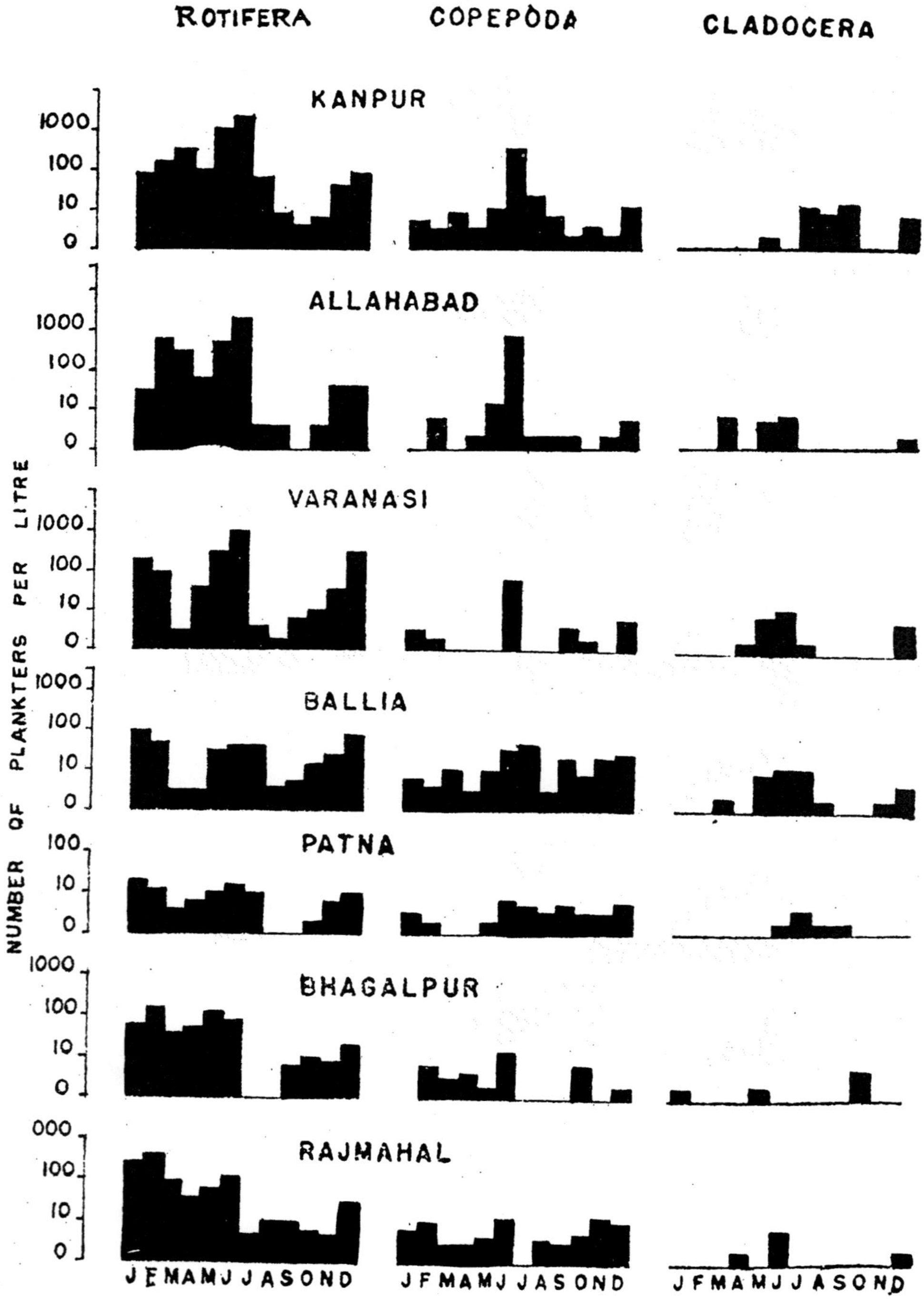

Fig. 25 : Centrewise monthly fluctuations of Zooplankton groups (on Logarithmic scale)
Source : Pahwa, D.V. and S.N. Mehrotra (1966).

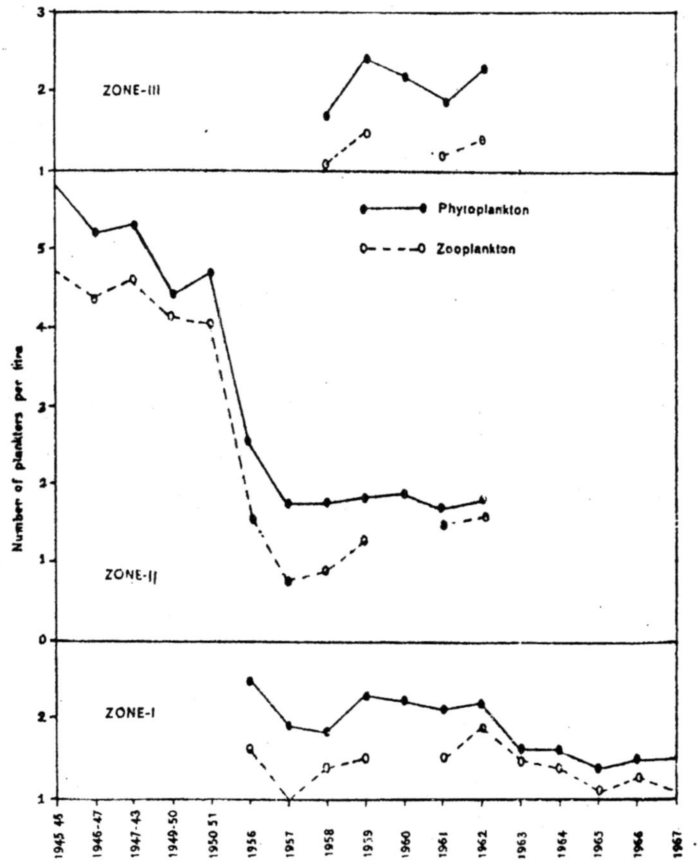

Fig. 26 : Average yearly fluctuations of phyto and zooplankton population on logarithmic scale in three broad zones in the Hooghly estuary.
Source : Ray, P. (1981).

Table 44 : Showing concentration ranges of phyto plankton together with annual average of phyto and Zooplankton at seven Centre spread over the Ganga and also at Allahabad on Yamuna

	Phyto plankton			*Annual average*		
	Maximum	*Season*	*Minimum*	*Season*	*Phytoplankton*	*Zooplankton*
1.	80291	June	33	September	8521	363
2.	45613	June	14	September	6606	372
3.	25125	June	35	September	2323	180
4.	7685	July	39	August	944	51
5.	765	June	32	August	143	13
6.	1444	June	17	August	557	53
7.	1296	June	42	October	593	97
8.	808	June	28	October	245	84

Chakraborty, R.D., P. Ray and S.B. Singh (1959); Pahwa D.V. and S.N. Mehrotra, 1966; and Ray, P., S.B. Singh and K. L. Sehgal (1966) In Jhingran, A.G. and K.K. Ghosh (1978).

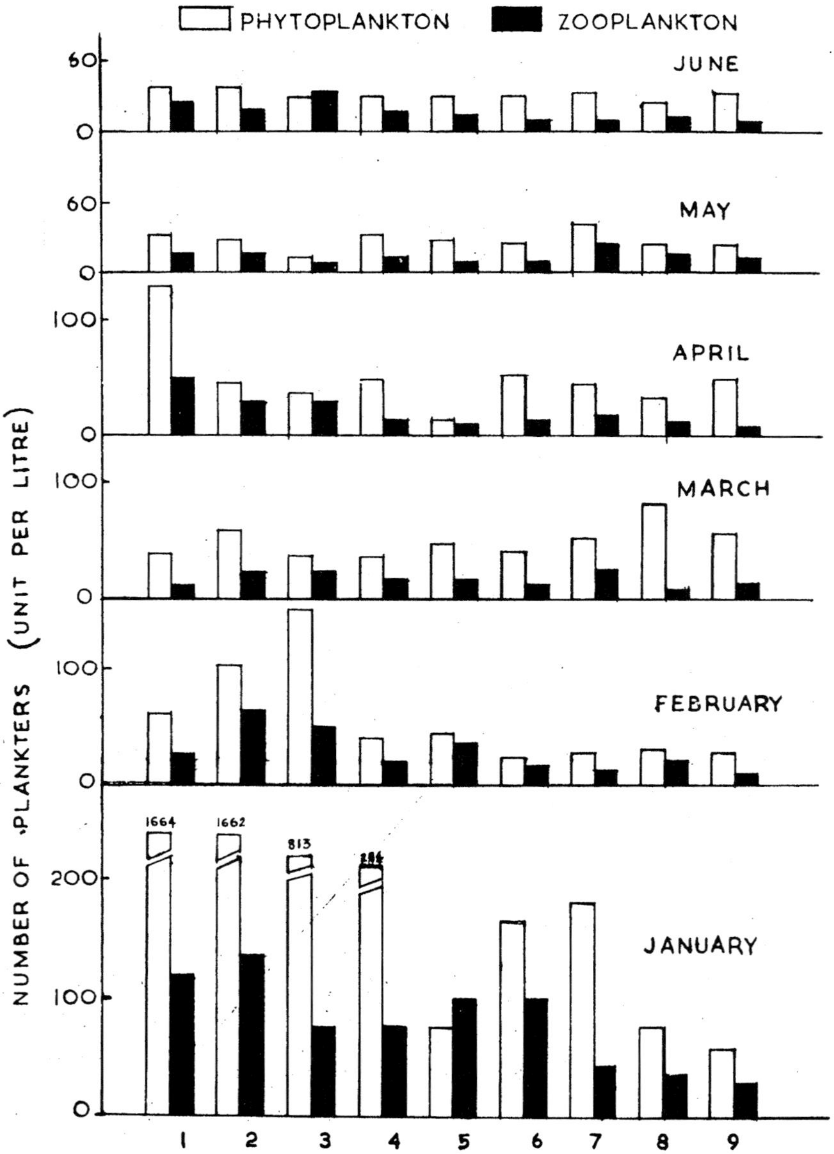

Fig. 27 : Pollution status of the Hooghly with reference to plankton population (monthly average of 4 years) during dry weather period.
1. Dumurda, 2. Tribeni, 3. Naihati, 4. Nawabganj, 5. Titagarh, 6. Baghbazar, 7. Garden Reach, 8. Batnagar, 9. Birlapur.

Source : Ray, P. (1981).

A gradual quantitative decline could be noticed from Kanpur to Patna after which a slight improvement was noticed.

Bacillariophyceae, Chlorophyceae and Myxophycear constituted the dominant groups of phytoplankton. The diatoms dominated the phytoplankton only in the upper stretches and that too essentially during the summer months only. Most of the other centres the green algae distinctly dominated among phytoplankters. The blue green algae formed a sizeable portion of phytoplankton only in the upper stretches.

The diatoms showed two peaks, the major one in summer and the other in winter. *Synedra* spp. dominated at all the centres throughout the year and these were responsible for the peaks. *Nitzschia* spp. was the only other diatom to be encountered in sizable numbers. The green algae followed more or less a trend similar to that of the diatoms with summer and winter peaks. *Mougeotia, Spirogyra, Pediastrum* and *Actinostrum* were the most common green algae encountered all along the stretch. *Mougaotia* was essentially the most abundant, forming as much as 66.1-88.6 per cent of the annual green algae crops at the various centres, and was responsible for the green algal peaks. The blue green algae occurred in abundance only during the summer months May and June, and from August to October constituting their lean period. The contribution of blue green algae which were also more abundant in the upper stretches to total phytoplankton at various centres ranged from 3.2-6.3 per cent only. *Anabaena, Nostoc* and *microcystis* were more common forms encountered. Though the phytoplankton population in the Jamuna at Allahabad showed more or less similar trend of fluctuations as that of Ganga, it was seen that their abundance was very much less than that of the Ganga, and above the confluence. The poorer plankton population of the Ganga, below the confluence than the above stretch it could be attributed to the mixture of Jamuna waters with relatively poor plankton population. This is clear from the data presented in Table 45.

Table 45 : Percentage composition of Phytoplankton in Ganga in and around Allahabad during 1958-1959

Sampling Centres	*Bacillario-phyceae %*	*Chlorophycease %*	*Myxophyceae %*	*Dinophyceae %*
River Ganga (Above confluence)	85.0	3.6	11.4	—
River Ganga (Below confluence)	76.7	9.7	13.6	—

Source: CIFRI Report 1970.

Zooplankton : As in the case of phytoplankton, the Zooplankton also was found to be in greater abundance in the Ganga than in the Jamuna and the Ganga above the confluence was richer than below. The studies on the Ganga in 1960, showed the pattern of Zooplankton fluctuations closely related with that of phytoplankton, thereby showing direct correlation. Similar direction relation was found in subsequent years also in both the Ganga and Jamuna (Devessay *op cit*).

The Zooplankton showed a bimodal pattern of fluctuations with one peak during May-July and the other during December-March, which was identical to phytoplankton. The summer peak constituted the primary peak only in the upper stretches and also at Patna. Monsoon and autumn months (July-October) yielded minimum quantities of Zooplankton. The numerical contribution of Zooplankton to total plankton along the Ganga ranged from 4.8-14.0 per cent only (Pahwa and Mehrotra, 1966).

Table 46 : Yearly average (no/1) of main phyto and zooplanktonic groups and thier respective percentage composition in total phyto and zooplankton.

Plankton Groups	A		B		C		D		E		F		G	
	Yearly average	*% Composition in total phyto/zoo*	*Yrly Average*	*% Comp. in total phyto/ zoo*	*Yrly average*	*% Comp. in total phyto/ zoo*	*Yrly average*	*% Comp. in total phyto/ zoo*	*Yrly average*	*% Comp. in total phyto/ zoo*	*Yrly average*	*%Comp. in total phyto/ zoo*	*Yrly average*	*% Comp. in total phyto/ zoo*
Bacillariyophycae	6642	77.9	4774	72.3	1940	88.5	405	42.9	69	48.3	148	25.7	83	14.2
Chlorophyceae	1347	15.8	1515	22.9	265	11.4	509	53.9	67	46.8	385	70.8	483	81.2
Myxophyceae	532	6.3	317	4.8	118	5.1	30	3.2	7	4.9	24	4.3	27	4.6
Total Phytoplankton	8521	—	6606	—	2323	—	944	—	143	—	557	—	593	—
Rotifers	328	90.4	313	84.2	170	94.4	32	62.8	8	61.5	49	92.4	87	89.7
Copepoda	31	8.5	57	15.3	7	3.9	15	29.4	4	30.8	3	5.7	8	8.2
Cladocera	4	1.1	2	0.5	3	1.7	4	7.8	1	7.7	1	1.9	2	2.1
Total Zooplankton	363	—	372	—	180	—	51	—	18	—	53	—	97	—
Total Plankton	8884	—	6978	—	2503	—	995	—	156	—	610	—	690	—

Source : Pahwa, D.V. and S.N. Mehrotra. (1966) ; A = Kanpur; B = Allahabad; C = Varanasi; D = Ballia; E = Patna ; F = Bhagalpur; G = Rajmahal.

The Zooplankton was dominated by the rotifer and copepods throughout the year all along the Ganga. Cladocerons were encountered in moderate numbers only during a few months, while protozoans, were practically absent at all the centres except around Kanpur. However, in the Jamuna, the protozoans were more abundant and were present almost throughout the year. They even dominated the Zooplankton during the monsoon months. The most common protozoans were *Arcella, Difflugia, Actinoblus, Nebalia, Actinophrys* and *Styloninchia.* (Pahwa and Mehrotra, 1966, Ray *et al.* 1966).

In the Ganga, the rotifers contributed the most prominent group all along the stretch almost throughout the year, with their contribution to total zooplankton ranging from 61.5-94.4 per cent only at Ballia and Patna, copepods were found in abundance. The rotifers showed two peaks, one during May-June and the other during December-February. *Brachionus* spp. and *Keratella* spp. dominated the rotifer crop and were responsible for the rotifer peaks. *Filinea, polyarthra* and *Pedalion* are also encountered in appreciable numbers during few months of a year.

The copepods showed only one peak, mostly in June, in both the Ganga and the Jamuna, the former being richer than the latter in copepod abundance. The nauplii also showed a pattern of fluctuation similar to that of adults. The copepods were dominated by Cyclopoids *Cyclops* spp., were the most common forms encountered and the others were present in lesser magnitude and these were *Diaptomus, Cauthocamptus, Helidiaptomas cinetus* and *Mesocyclops rylove* (Ray *et al.* 1966), Pahwa and Mehrota (1966).

The cladocera were available in large numbers only during March-July. The major genera being *ceriodaphnia, Bosmina, Moina, Alona, Diaphanosoma, chydorus* and *Sida Ostracods,* represented by the genus cypris, were encountered only in the Jamuna. Centrewise monthly fluctuation of Zooplankton and phytoplankton group are indicated in Figs. 24 & 25, while yearly averages of main phyto and zooplanktonic groups and their respective percentage composition in total phyto and zooplankton are indicated in Table 46. A list of identified plankters in the observed stretch of the Ganga is also indicated in Table 47.

Table 47 : List of identified Plankters encountered in the observed stretch of the river Ganga. (Pahwa and S.N. Mehrotra, 1966).

A. Phytoplankton :	
(a) Bacillariophyceae	(b) Chlorophyceae :
Amphora sp.	*Actinastrum spp.*
Asterionella sp.	*Ankistrodesmus* sp.
Cocconeis sp.	*Chlorella* sp.
Cyclotella sp.	*Closterium* sp.
Cymbella sp.	*Cosmarium* spp.
Diatoma sp.	*Denticula* sp.
Fragilaria spp.	*Desmidium* sp.
Gomphonema sp.	*Eudorina* sp.
Gyrosigma sp.	*Hydrodictyon* sp.
Melosira spp.	*Mougeotia* spp.
Navicula spp.	*Paudorina* sp.
Nitzschia spp.	*Pediastrum* spp.
Pleurosigma sp.	*Scenedesmus* spp.
Pinnularia sp.	*Sphaerozosma* sp.
Synedra sp.	*Spirogyra* spp.
Synedra ulna	

Contd....

(c) Myxophyceae
Anabaena spp.
Cylindrospermum sp.
Lyngbya spp.
Merismopoedia spp.
Microcystis spp.
Nostoc spp.
Oscillatoria spp.
Phormidium sp.
Rivularia spp.
Surirella sp.

(d) Chrysophyceae
Botryaccoccus sp.
Dinobryon sp.

(e) Dinophyceae
Ceratium sp.

(f) Euglenophyceae
Euglena sp.
Phacus sp.

B. Zooplankton :

(a) Rotifera
Asplanchna sp.
Brachionus spp.
Conochilus sp.
Filinia spp.
Keratella spp.
Monostyla sp.
Notops sp.
Notholca sp.
Pedalion sp.
Polyarthru sp.
Rathulus spp.
Rotatoria sp.
Salpina sp.
Synchaeta sp.
Tetramastix sp.
Trochosphaera sp.

(b) Copepoda
Diaptomus spp.
Cyclops spp.

(c) Cladocera
Alona sp.
Bosmina spp.
Bosminopsis sp.
Ceriodaphnia sp.
Chydorus sp.
Diapharrosoma sp.
Moinodaphnia sp.
Sida sp.

(d) Protozoa
Arcella sp.
Difflugia sp.
Stentor sp.
Vorticella sp.

II. Brackishwater Zone : In the brackishwater zone, prior to Farakka discharges, Diatoms were prominent among phytoplankton and copepods among zooplankton. Roy (1949) studied the plankton species of the Hooghly estuary around Palta and had recorded 38 genera of algae, 4 genera each of protozoa, rotifera and schizomycetes as given below.

Chlorophyceae	:	*Pediastrum, Micractinum, Tetraedron, Ankistrodesmus, Actinastrum, Crucigenia Scenedesmus, Geminella, Penium, Closterium Cosmarium*
Bascillariophyceae	:	*Melosira, Cyclotella, Stephanodiscus, Coscinodiscus, Fragillaria, Synedra, Navicula, Stauroneis, Pleurosigma, Gamphonema, Cymbella, Encyonema, Rhopalodia Bacillaria, Nitzschia, Surirella.*
Dinophyceae	:	*Glenodinium* and *Peridinium*
Englenophyceae	:	*Euglena, Phacus, Trachelomonass* and *Ascoglena*
Myxophyceae	:	*Oscillatoria, Phormidium, Lyngbia, Anabaena, Aphanizomenon.*
Protozoa	:	*Sarcodina - Difflugia* *Mastigophora - Monas* *Infusoria - Codonella*
Rotifera	:	*Bdelloida - Rotifera*

Schizomycetes or higher bacteria

Chlamydobacteriales or Iron Bacteria—*Leptothrix, Didymohelix* and *Crenothrix*

Thiobacteriales or Sulphurbacteria-*Beggiatoa*

Table 48 : Distribution in the plankton population of the Hooghly during pre and post Farakka discharges

	Kalna 225 km. from sea face		*Barrackpore* 171 km. from sea face		*Uluberia* 126 km. from sea face		*Nurpur* 86 km. from sea face		*Kakdwip* 40 km. from sea face	
	Pre *1960–61*	*Post* *1975–77*	*Pre* *1960–61*	*Post* *1975–77*	*Pre* *1960–61*	*Post* *1975–77*	*Pre* *1960–61*	*Post* *1975–77*	*Pre* *1960–61*	*Post* *1975–77*
Phyto plankton No./1.	68.04	219.38 154.8 (FW) 2.20 (BW) 62.29 (C)	32.18	162.16 73.88 (FW) 24.45 (BW) 63.83 (C)	48.70	91.24 7.02 (FW) 56.31 (BW) 27.91 (C)	18.16	64.16 20.63 (FW) 24.0 (BW) 19.52 (C)	39.81	48.70 Nil (FW) 44.34 (BW) 4.36 (C)
Zooplankton No/1.	26.88	84.99 55.32 (FW) 8.31 (BW) 21.36 (C)	7.97	60.62 30.42 (FW) 19.13 (BW) 11.07 (C)	11.51	53.55 1.90 (FW) 46.10 (BW) 5.55 (C)	7.88	89.17 24.80 (FW) 55.32 (BW) 9.05 (C)	6.90	7.97 Nil (FW) 5.33 (BW) 2.64 (C)

FW—Freshwater ; BW — Brackishwater ; C — Common

Source : Nandy, AC., M.M. Bagchi and S. K. Mazumdar (1983).

Table 49 : Zonewise distribution of the main Plankton Group (No/l) during pre and post Farakka discharge.

	Zone — I			Zone—II			Zone—III			Zone—IV		
A. Phytoplankton Organisms	*Pre 1960-61*		*Post 1975-77*	*Pre 1960-61*		*Post 1975-77*	*Pre 1960-61*		*Post 1975-77*	*Pre 1960-61*		*Post 1975 - 77*
Diatom	55.48		121.08	30.54		133.33	42.78		89.75	17.16		40.83
		FW	71.08		FW	53.23		FW	6.61		FW	10.13
		BW	2.00		BW	24.10		BW	56.29		BW	23.41
		C	48.00		C	56.00		C	26.85		C	7.29
Green algae	10.05		93.25	0.92		23.83	4.51		1.41	0.26		16.50
		FW	80.76		FW	19.27		FW	0.41		FW	7.90
		C	12.49		C	4.06		C	1.00		C	9.00
Bluegreen algae	2.42		5.01	0.69		5.50	0.17		0.08	0.75		6.83
		FW	3.01		FW	1.38		FW	Nil		FW	3.00
		BW	0.20		BW	0.35		BW	0.02		BW	0.60
		C	1.80		C	3.77		C	0.06		C	3.23
Englenoids	0.09		0.04	0.03		—	1.24		—	0.01		—
TOTAL...........	63.04		219.38	92.18		162.16	48.70		91.24	18.16		64.16
		FW	154.89		FW	73.88		FW	7.02		FW	20.63
		BW	2.20		BW	24.45		BW	56.31		BW	24.01
		C	62.29		C	63.83		C	27.91		C	19.25
FW — Freshwater; BW - Brackishwater; C— Common.												
B. Zooplankton Organism												
Copepod	3.49		46.17	2.52		48.11	5.67		46.50	4.32		64.25
		FW	28.4		FW	22.04		FW	1.40		FW	10.15
		BW	5.03		BW	18.00		BW	43.10		BW	48.10
		C	13.10		C	8.07		C	2.0		C	6.00
Cladocerans	11.48		24.33	0.42		7.00	0.04		1.50	0.47		13.00
		FW	23.28		FW	6.00		FW	0.50		FW	10.00
		BW	1.05		BW	1.00		BW	1.00		BW	3.00
		C	NIl		C	Nil		C	Nil		C	Nil
Rotifers	0.43		4.66	0.24		5.51	0.17		0.07	0.26		7.25
		FW	3.0		FW	2.38		FW	Nil		FW	3.05
		BW	0.10		BW	0.13		BW	Nil		BW	2.15
		C	1.56		C	3.00		C	0.07		C	2.05
Others	11.48		9.83	4.79		—	5.58		5.68	2.93		4.67
		FW	1.00		FW	Nil		FW	NIl		FW	1.16
		BW	2.13		BW	NIl		BW	Nil		BW	2.51
		C	6.70		C	Nil		C	3.58		C	1.00
Total :	26.88		84.99	7.97		60.62	11.51		53.55	7.88		89.17
		FW	55.32		FW	30.42		FW	1.90		FW	24.80
		BW	8.31		BW	19.13		BW	46.10		BW	55.32
		C	21.36		C	11.07		C	5.55		C	9.05

Source : Nandy, A.C., M.. M. Bagchi and S. K. Majumdar (1983).

Table 50 : A list of Phyto-plankton identified at the four Stations on the Hooghly during 1949-1951

I. Diatoms

1. *Melosira granulata* (Ehr.) Ralfs
2. *M. granulata var. angustissima* (Ehr.) Ralfs. Mull.
3. *Skeletomema costatum* (Grev.) Cleve
4. *Thalassiosira marginata* Cleve
5. *Cyclotella Meneghiniana* Kütz
6. *C. Kutzingiana* Thwaires
7. *C. striata* (Kütz.) *Grunow*
8. *Coscinodiscus Gramu* Gough
9. *C. radiatus*
10. *C. commutatis*
11. *C. gigas* Ehr.
12. *Coscinodiscus* 5 spp.
13. *Asteromphalus cleaveanius* Grunow
14. *Planktoniella Sold* (Wallich) Schutt
15. *Bacillaria paradoxa* Gmelin
16. *Actinocyclus Ehrenbergu* Ralfs
17. *Corethron hystrix* Hensen
18. *Lauderia* sp.
19. *Leptocylindrus danicus* Cleve
20. *L. minimus* Gran
21. *Guinardia* sp.
22. *Rhizosolenia robusta* Normau
23. *R. Stolterfothii* H. Peragallo
24. *R. imbricata* Brightwell
25. *R. styliformis* Brightwell
26. *R. alata Brightwell* forma *indica* ((Peragallo) Ostenfeld
27. *R. alata* Brightwell forma *inermis* (Castracane) Hustedt
28. *Bacteriastrum hyalinum* Lauder
29. *B. varians* Lauder
30. *B. elegans* Lauder
31. *Chaetozeros orientalis* Schiller
32. *Ch. perwvianus* Brightwell
33. *Ch. affinis* Lauder
34. *Ch. curvisetus* cleve
35. *Eucampia zoodiacus* Enr.
36. *E. cornuta (*Cleve) Grunow
37. *Climacodium* sp.
38. *Ditylum Brightwellii* (West) Grunow
39. *Triceratium* sp.
40. *Biddulphia mobiliensis* Bailey
41. *B. sinensis* Greville
42. *B. rhombus* (Enr.) W. Smith

(Contd...)

43. *Cerataulina* sp.
44. *Hemiaulus sinensis* Greville
45. *Hemidiscus Harmannianus* (Greville) Mann.
46. *Navicula pygmaeea* Kütz
47. *Navicula* sp.
48. *Synedra ulna* (Nitzsch) Ehr.
49. *Lithodesmium* sp.
50. *Rhopalodia gibba* (Ehr.) O. Mull.
51. *Thallassionema Nitzschiodes* Grunow.
52. *Thallassiothrix longissima* Cleve and Grunow
53. *Asterionell japonica* Cleve
54. *Gyrosigma* sp.
55. *Pleurosigma angulatum* (Quekett) W. Smith
56. *Pleurosigma* sp.
57. *Streptotheca* sp.
58. *Diploneis* sp.
59. *Pinnularia* sp.
60. *Amphora* sp.
61. *Nitzschia vitrea* Norman
62. *N. seriata* Cleve
63. *N. closterium* (Ehr.) W. Smith
64. *N. longissima* (Brebisson) Ralfs
65. *N. sigma* (Kütz) W. Smith
66. *Nitzschia* sp.
67. *Fragillaria intermedia* (Grun.)
68. *Surirella elegans* Ehr.
69. *S. tenera* Gregory
70. *Cymbella cistula* (Hemp.)
71. *Gomphonema* sp.
72. *Achanathes* sp.

II. Green Algae

1. *Volvox* sp.
2. *Eudorina elegans* Ehr.
3. *Tetraedron* sp.
4. *Ankistrodesmus* sp.
5. *Pandorina* sp.
6. *Pediastrum clathratum* (Schroeter) Lemmerman
7. *P. duplex* Meyen
8. *Crucigenia* sp.
9. *Scenedesmus quadricauda* (Turpin) Brebisson
10. *Spirogyra* (3 spp.)
11. *Tribonema* sp.
12. *Closterium* sp.
13. *Cosmarium* sp.
14. *Coelastrum* sp.

(Contd...)

15. *Mougeota* sp.
16. *Oocystis* sp.
17. *Zygnema* sp.
18. *Selenastrum* sp.

III. BLUE-GREEN ALGAE

1. *Microcystis aerugionosa* Kütz
2. *M. flos-aquae* Kütz
3. *Merismopedia* sp.
4. *Lyngbya* sp.
5. *Oscillatoria* (3 spp.)
6. *Phormidium* sp.
7. *Spirulina* sp.
8. *Anabaena* sp.
9. *Aphanocapsa* sp.

IV. EUGLININEAE

1. *Phacus* sp.
2. *Euglena acus* Ehr.
3. *E. oxyuris* Ehr.

V. DINOFLAGELLATES

1. *Peridinium* sp.
2. *Ceratium hirundinella* O.F. Müller
3. *C. tripos* O.F. Müller

Source : Dutta, N., J.C. Malhotra and b.B. Bose (1954)

4 Stations are : I—Palta, II—Calcutta, III—Falta & IV—Diamond Harbour.

Subsequently Dutta *et al.*, (1954) also found that diatoms are dominant plankton followed by green algae especially between Palta and Barrackpore. These freshwater forms were conspicuously absent between Palta and Diamond Harbour. The blue green algae although present throughout this section were not so significant as diatoms and green algae. They were found in the following sequence : diatoms 72, green algae 18, blue green algae 9; euglenoids 3 and dinoflagellets 3. zooplankton mostly consisted of freshwater and brackishwater species. A few marine forms were also recorded from Palta and Diamond Harbour (lower stretch) during March-June, when salinity was high. Freshwater species were predominant in the stretch between Palta and Diamond Harbour which had high turbidity. At Palta and Calcutta, zooplankton were maximum during November-March when turbidity was the lowest. The same was true at Palta and Diamond Harbour during October-January.

During 1956-1958 in zones* I and II diatoms were found to dominate throughout the year and they were followed by green and blue green algae. Dominance of diatoms in 1959-1962 has been reported by Saha *et al.*, (1975). Shetty *et al.*, (1961) and Saha *et al.*, (1975) observed that protozoa, rotifera, copepoda, cladocera were dominant zooplankton in different zones during 1956-1967. Shetty *et al.*, (1961) observed that there was a gradual fall in plankton from 1956 to 1968 in freshwater, intermediate and brackishwater zones. The yearly plankton fluctuations in the three zones are given in Fig. 26.

* Zone I : Medgachi-Konnagar; Zone-II : Konnagar Diamond Harbour.

Effect of freshwater release from Farakka Barrage on the ecological condition of the Hooghly shows that plankton population density has increased by 3-5 times in different zones of the stretch during 1975-1977 (Nandy *et al.*,1983) (cf. Table 51) as compared to those reported by Shetty *et al.*, (1961). It now consisted mainly of freshwater forms. Marine forms were encountered from the lower zone. The upper zone from Kalana to Barrackpore, the chlorophyceae members were encountered in large numbers. The zooplankton consisted mainly of freshwater and brackishwater species in the upper and the middle stretch of the Hooghly and Roopnarayan estuary.

Table 51 : List of identified plankters and their zonewise distribution ('x' = Present, .. = Absent)

	Zone		
	I	*II*	*III*
(a) ***Diatoms***			
1. *Synedra ulna*	x	x	x
2. *Coscinodiscus granii*	x	x	x
3. *C. radiatus*	..	x	..
4. *C. commutatis*	..	..	x
5. *C. spp.*	x	x	x
6. *Surirella elegans*	x	x	x
7. *Melosira granulata*	x	x	x
8. *M. spp.*	x	x	x
9. *Nitzschia vitrea*	x	x	x
10. *Navicula spp.*	x	x	x
11. *Pleurosigma spp.*	x	x	x
12. *P. elongata*	..	..	..
13. *Amphora sp.*	x	x	..
14. *Climacodium sp.*	x	x	x
15. *Asterionella japonica*	x	x	x
16. *Stephanodiscus sp.*	x	x	x
17. *Cyclotella spp.*	x	x	x
18. *Cymbella cistula*	x	x	x
19. *Chactoceros spp.*	x	x	x
20. *Diploneis sp.*	x	x	x
21. *Bidhulphia mobiliensis*	x	x	x
22. *Bacillaria paradoxa*	x	x	x
23. *Mastogloea sp.*	x	..	..
24. *Meridion sp.*	x	x	..
25. *Planktoniella sol*	x	..	x
26. *Lithodesmium spp.*	x	x	x
27. *Thalassiothrix longissima*	x	x	x
28. *Ditylum brightwellii*	x	x	x
29. *Rhopalodia sp.*	x	x	..
30. *Gomphonema sp.*	x	..	..
31. *Diatoma sp.*	x	..	x
32. *Eucampia sp.*	..	x	x
33. *Actinocyclus sp.*	x	x	..

(Contd...)

		Zone	
	I	*II*	*III*
34. *Hemidiscus hardmannianus* ..	..	x	x
35. *Hemiaulus sinensis* ..	..	x	x
36. *Sceletonema costatum* ..	..	x	..
37. *Triceratium sp.* ..	..	x	x
38. *Epithemia sp.* ..	..	x	..
39. *Rhaphoneis sp.* ..	..	x	..
40. *Gyrosigma spp.* ..	..	x	x
41. *Rhizosolenia spp.* ..	..	..	x
42. *Bacteriastrum spp.* ..	..	..	x
43. *Lauderia sp.* ..	..	..	x
44. *Cocconeis sp.* ..	..	..	x
45. *Cerataulina sp.* ..	..	..	x
46. *Thalassionema sp.* ..	..	..	..
47. *Thallasiosita marginata* ..	..	..	..
48. *Aulacodiscus sp.* ..	x	x	x
49. *Fragilaria sp.* ..	x	x	..
50. *Pinnularia gibba* ..	x	x	..
(b) Green Algae			
1. *Pediastrum simplex* ..	x	x	..
2. *P. spp.* ..	x	x	x
3. *Mougeotia sp.* ..	x	x	..
4. *Closterium sp.* ..	x	x	..
5. *Ankistrodesmus sp.* ..	x	x	..
6. *Cosmarium sp.* ..	x	..	..
7. *Spirogyra spp.* ..	x	x	x
8. *Eulorina elegans* ..	x	x	..
9. *Pleodorina sp.* ..	x	..	x
10. *Actinastrum sp.* ..	x	..	..
11. *Tribonema sp.* ..	x	..	x
12. *Euastrum sp.* ..	x	..	..
13. *Chlorella sp.* ..	x	x	..
14. *Scenedesmus quadricauda* ..	x	x	..
15. *Zygnema sp.* ..	x	..	..
16. *Desmidium sp.* ..	x	x	x
17. *Pandorina morum* ..	x	x	..
18. *Platydorina caudatum* ..	x	..	..
19. *Borgea planktonica* ..	x	x	x
20. *Treubaria sp.* ..	x	..	..
21. *Chlorococcus sp.* ..	x	x	..
22. *Closteriopsis sp.* ..	..	x	..

(Contd...)

		Zone		
		I	II	III
23. *Staurastrum sp.*	..	x	..	..
24. *Oedogonium sp.*	..	..	x	..
25. *Ulothrix sp.*	..	x	..	..
26. *Micrasterias sp.*	..	..	..	..
27. *Microspora sp.*	..	..	x	..
28. *Volvox sp.*	..	x	x	..
29. *Coelastrum sp.*	..	x	..	..
30. *Kirchneriella sp.*	..	x	..	..
(c) Blue-green Algae				
1. *Microcystis spp.*	..	x	x	..
2. *Phormidium spp.*	..	x	x	x
3. *Nostoc spp.*	..	x	x	x
4. *Oscillatoria spp.*	..	x	x	x
5. *Anabaena spp.*	..	x	x	x
6. *Nostochopsis sp.*	..	x	..	..
7. *Gloeotrichia sp.*	..	x	x	..
8. *Aphanizomenon sp.*	..	x	x	..
9. *Cylindrospermum sp.*	..	x	..	..
10. *Merismopedia sp.*	..	x	..	x
11. *Spirulina major*	..	x	x	..
12. *Lyngbya sp.*	..	x	x	x
13. *Aphanocapsa sp.*	..	x	..	..
14. *Coelosphaerium sp.*	..	x	..	..
15. *Plectonema sp.*	..	..	x	..
16. *Trichodesmium sp.*	..	..	x	x
17. *Rhizoclonium sp.*	..	x	..	..
18. *Wollea sp.*	..	x	..	..
(d) Flagellates				
1. *Euglena sp.*	..	x	x	x
2. *Dinobryon sp.*	..	x	x	..
3. *Ceratium hirudinella*	..	x	x	x
4. *C. tripos*	..	..	x	x
5. *Phacus sp.*	..	x	x	x
6. *Peridinium spp.*	..	x	x	x
7. *Noctiluca miliaris*	..	x	..	..
8. *Trachelomonas sp.*	..	x	..	..
(e) Protozoa				
1. *Difflugia spp.*	..	x	x	x
2. *Arcella spp.*	..	x	x	x
3. *Bursaria sp.*	..	x	x	..

(Contd...)

		Zone		
		I	II	III
4. *Vorticella sp.*	..	X	X	..
5. *Nuclearia sp.*	..	X	..	..
6. *Centropyxis sp.*	..	..	X	..
7. *Proteomyxa sp.*	..	..	X	..
8. *Tintinnidium sp.*	..	..	..	X
9. *Globigerina sp.*	..	..	..	..
(f) Rotifera				
1. *Brachionus spp.*	..	X	X	..
2. *Keratella spp.*	..	X	X	..
3. *Trochosphaera sp.*	..	X	..	..
4. *Tetramastix sp.*	..	X	..	..
5. *Trichocerca sp.*	..	X	..	..
6. *Asplanchna sp.*	..	X	..	..
7. *Filinia sp.*	..	X	..	..
8. *Hexarthra sp.*	..	X	X	..
9. *Notholca*	..	..	..	X
10. *Ploesoma sp.*	..	..	..	..
(g) Copepoda				
1. *Pseudodiaptomus spp.*	..	X	X	X
2. *Diaptomus spp.*	..	X	X	X
3. *Microsetella sp.*	..	..	X	X
4. *Acartiella torianiformis*	..	X	X	..
5. *A. major*	..	..	X	..
6. *Paracalanus sp.*	..	..	..	X
7. *Cyclops sp.*	..	X	..	..
(h) Cladocera				
1. *Bosmina sp.*	..	X	X	..
2. *Bosminopsis sp.*	..	X	X	X
3. *Ceriodaphnia sp.*	..	X	X	..
4. *Daphnia sp.*	..	X	X	..
5. *Holopedium sp.*	..	X	..	..
6. *Monia sp.*	..	X	X	..
7. *Diaphanosoma sp.*	..	..	X	..
(i) Miscellaneous				
1. *Mesopodopsis orientalis*	..	X	..	..
2. *Sagitta sp.*	..	..	X	X

Zone I : Medgachi, Khasbati and Khusigoli (upper zone).

Zone II : Konnagar, Fuleswar and Diamond Harbour (Middle zone).

Zone III : Kakdwip, Namkhana and Fragerganj (Lower zone).

N B. *Source :* Shetty, H.P.C; S.B. Saha and B.B. Ghosh (1961).

Diatoms and copepods were found to dominate the phyto and zooplankton respectively. Phytoplankton mainly belong to two major classes with Bascillariophyceae having the maximum representation in terms of species (18) followed by chlorophyceae (6). The fauna belong to two major phyla of which copepoda was the most abundant and having the maximum representation. Copepod of calanoid and cyclopid forms with various copepodid and nauplii stages. Plankton population during pre and post Farakka discharges at various distances in the Hooghly from seaface are indicated in Table 48. The zonewise distribution of main plankton groups during pre and post Farakka discharges are also shown in Table 49. Besides a list of phytoplankton identified earlier during 1949-51 and also their Zonewise distribution are indicated in Tables 50 and 51.

Jhingran and Babulal (1988) have made some interesting observations on the potential resources over the entire region of the Hooghly estuary (600 km. from the estuary month to Farakka) and also Rupnarayan river in terms of primary production and energy fixation by following C^{14} techniques. Their finding revealed that the mean value of primary production in the lower stretches was generally higher (1.265 g Cm^{-2} d^{-1}) when the intermediate and freshwater zones showed lower values 0.285 and 0.261 g cm^{-2} d^{-1}.

The average annual carbon assimilation in the Hooghly estuary between 1965 and 1986 did not record by significant trend. However, immediately after the Farakka discharge had begun, a spurt in production (120.71 mg $cm^{-3}d^{-1}$) was recorded which ultimately stabilised at 62.50 mg $Cm^{-3}D^{-1}$ by 1986. The observation indicates that overall production potential did not after much due to Farakka discharge (Babu Lal, 1990).

Babu Lal (*op cit*) further observed that the carbon assimilation trend at various centres revealed higher production during post monsoon. (41.51 mg $Cm^{-3}d^{-1}$) at Frezerganj, while minimum was estimated (22.62) mg cm^{-3} d^{-1}) at Barrackpore. On an average the freshwater zone of the Bhagirathi-Hooghly is more productive than saline zone. The primary production of the estuary, studied during different periods are indicated in Tables 33 A, B & C. Impact of pulp and paper mill wastes on the primary productivity are also shown in Tables 34-A and B.

Table 52 : Shannon-Weaver index (H) at various Centres of the river Ganga between Kanpur and Varanasi

	Above outfall	*Outfall*	*Below outfall*
Phytoplankton :	1.36 (0.45 - 1.99)	1.07 (0.43 - 2.12)	1.11 (0.22 - 2.09)
Zooplankton :	0.93 (0 - 1.6)	0.71 (0 - 1.33)	0.88 (0.43 - 1.39)
Benthes :	1.15 (0.16 - 1.89)	0.70 (0.75 - 1.36)	0.98 (0.02 - 1.86)

Values indicated are the average of seven Centres spread over the river Ganga with ranges shown in the parenthesis.
Source : Modified from the Annual Report, 1989-90 CICFRI, Barrackpore.

BENTHIC BIOTA

I. Freshwater Zone

A study of bothem fauna was carried out at seven centres (Kanpur, Allahabad, Varanasi, Ballia, Patna, Bhagalpur, and Rajmahal) along the Ganga in 1960 (Pahwa and Mehrotra, 1966).

River Ganga : In the Ganga, the total benthic organisms were maximum at Kanpur, the yearly average being 3476/m^2 and minimum at Ballia (71/m^2).The quantities at other centres were 1121/m^2 at Allahabad, 880/m^2 at Varanasi, 1593/m^2 at Patna, 218/m^2 at Bhagalpur and 502/m^2 at Rajmahal. (Pahwa and Mehrotra, 1966).

June was the month of peak production of benthic organisms at Kanpur (21143/m^2). Allahabad (3436/m^2), Varanasi (2214/m^2), Ballia (260/m^2), Bhagalpur (473/m^2) and Rajmahal (1859/m^2). However, at Patna, maximum production was observed in the month of September (8415/m^2). Minimum production was observed during August-September in the upper stretches (nil-19 units/m^2), and during January-February in the lower stretches (19-33 units/m^2). Insects were observed to be maximum at Kanpur (98.0 per cent) and minimum at Patna (0.6 per cent) while the molluscs were maximum at Patna (99.4 per cent) and minimum at Kanpur (1.9 per cent). The Annelids were maximum at Rajmahal (29.3 percent), but were completely absent at Ballia and Patna. Among insects, the dipteran larvae (chironomid larvae) contributed maximum at Kanpur and Allahabad, but at Varanasi (Caddis flies) Trichoptera were dominating. At Ballia and Patna, the insects occurred in very few numbers. At Bhagalpur the entire insect population consisted of hemipterans. At Rajmahal though hemipterans were observed in quite good numbers the may flies dominated the insect population.

Among molluscs, the gastropods dominated in the lower stretches, with their maximum production at Patna (94.3 per cent). On the other hand, bivalves formed the bulk of the molluscs in the upper stretches with the gastropods either absent or present only in minimal quantities. *Melanoides lineatus* was the most dominating gastropod at Patna the maximum quantity being (7638/m^2) in September. *Melania tuberculata*, the next dominating form had its maximum number (1197/m^2) in August. The other species, in order of dominance at Patna, were *vivipaus bengalensis, V. variatus, Indoplanorbis exustus* and *Digonistoma pulchella*. At Bhagalpur *Pila globosa* formed the bulk of the gastropoda. Other forms that occurred at Bhagalpur were *V. bengalensis, O. garulia* and *Indoplanorbis exusters*. At Rajmahal, only two species of gastropods were observed, namely, *V. bengalensis* and *I. exustus*.

The bivalves occurred in maximum quantity at Varanasi (155/m^2) and minimum at Ballia (27/m^2). *Corbicula etriatella* dominated at the centres except at Ballia, when the dominating form was *parrysia* sp. Other forms of occurrence were *Novaculina insectica* and *Indocaria Oliveria*.

Among annelids, the polychaetes were dominant at Varanasi, Rajmahal, Bhagalpur and Allahabad with Varanasi recording the maximum quantity (592/m^2). The polychaetes were completely absent at Kanpur, Ballia and Patna. Oligochactes were observed only at Bhagalpur during the monsoon months with a maximum concentration of 57/m^2.

II. Estuarine Zone

Studies on Benthic fauna in the estuary, revealed that prior to Farakka discharge the fauna was contributing poorly towards productivity. Benthic organisms are mostly confined to the freshwater zone of the estuary which is comparatively shallow and mostly free from tidal influence.

The most common forms were Chironomid and oligochaete larvae, gastropods : *Melanoides tarebia* (Müller), *M. lineatus* (Grey), *M. Luteola f. ovalis* (Grey), *M. granifera* (Lam), *M. tuberculata* (Müller), *Thiara scabra* (Müller), *Vivipara bengalensis* (lam), *V. dissimalis* (Müller), *Assiminea francissii* (wood); megalapa larvae; Stenothyra deltae (Benson); *Macrobrachium* Spp. *Palaemon rude*, etc. (Ray, 1980). Their density was low except at Hazinagar which represents the recovery zone of pulp and paper mill effluents. Benthic biota remains absent around the outfall areas.

13

Fish and Fisheries

Natarajan (1989) has given an account of wide variety of species of fish representing in the catch in landings of the Ganga Rivers system over the past few years.

The Ganga river system includes upland streams, extensive warmwater stretches, flood plains, sloughs, oxbow lakes, and back water which present a mix of ecologically distinct habitats that are reflected in the rich range of ichthyofauna of the country over the past 100 years. Menon (1974) listed 141 species occurring in the Ganga river system belonging to 72 genera, 30 families, and 11 orders (Table 53). There is a wide variety of upland stream fish: 9 species of Rasborinae, 11 species of cyforinidae, including tur species of snow trout, *Schizothorax Kumzonensis* (Menon) and *Schizothoraichthys rogastus* (Mc Clelland) and two species of mahseers,*Tor putitora* and *Tor tor:* 3 species of Psilorhynchidae; 2 species of Homalopteridae; 13 species of Cobitidae (loaches); and 20 species of Sisoridae. The warmwater; lotic species of economic importance are: the major Carps *Catla catla, Cirrhina mrigala, Labeo rohita, Labeo calbasu* and the large cat fishes, *Mystus aor, M. seenghala, Rita rita, wallago attu, Solonia silondia, Pangasius pangasius* and *Bagarius bagarius.* The featherback *Notopterus chitala* and murrel *Channa marulius* also belong to this large species category. Species found in marshy habitats include *clarius batrachus, Heteropneustes fossilis, Channa* spp, *Amphipnous cuchia, Anabas testudineus, Mastacembelus* spp., *Notopterus notopterus,* and *Gadusia Chapra* and represent a mixture of families and orders. The anadromous hilsa, *Hilsa ilisha* is of great economic importance. The rest include minor carps, small Catfishes, minnows small clupcids and other. The distribution of species by habitat is in Table 54 excluding exotic carps and salmonids.

I. Commercial Catch

The Ganga (Freshwater zone) Jhingran and Ghosh (1978) and subsequently Jhingran (1989) while analysing the market arrival of fishes from 1958–1969 and subsequently extending upto 1986 at some important centres like, Kanpur and Allahabad, both on the Ganga and in the upper stretch; Varanasi, Buxar and Ballia in the middle stretch and at Patna and Bhagalpur in the lower stretches of the river Ganga recorded the fluctuations and abundance of various fishes (Fig. 28–31). The fluctuations in the annual landings of major carps, selected catfishes *(Wallago attu, Mystus aor Mystus seenghala,* hilsa and other species comprising the rest are also presented by Jhingran Table 55; Fig. 32–35. The stretches feeding these markets being of varying lengths, a comparison of productivity of major carps have been attempted based on the yield per kg/ha stretch area (Table 56.) The average group-wise landing positions of fishes indicated the ratio between major carps and cat fishes, and major carps and others have been shown in Table 57. (Jhingran and Ghosh 1978).

Table 53 : Number of orders, families, genera, and species represented in the catch and the fish species encountered in landings in the Ganga River System

	Class Osteichthyes			
Order	*Family*	*Genera*	*Species*	*Species occurring in landings*
Cluperiformes	2	3	4	*Hilsa ilisha* (Ham.)
				Setipinna phasa (Ham.)
				Gadusia chapra (Ham.)
Osteoglossiformes	1	1	2	*Notopterus chitala* (Ham.)
				Notopterus notopterus (Pallas)
Cypriniformes	4	27	67	*Catla catla* (Ham.)
				Cirrhina mrigala (Ham.)
				Cirrhina reba (Ham.)
				Labeo rohita (Ham.)
				Labeo calbasu (Ham.)
				Labeo bata (Ham.)
				Labeo boga (Ham.)
				Labeo gonius (Ham.)
				Labeo dero (Ham.)
				Labeo dyocheilus (Ham.)
				Tor putitora (Ham.)
				Tor tor (Ham.)
				Oxygaster baciaila (Ham.)
				Chela laubuca (Ham.)
				Amblypharyangodon mola (Ham.)
				Aspidoparia morar (Ham.)
				Puntius sarana sarana (Ham.)
Siluriformes	10	22	41	*Mystus aor* (Ham.)
				Mystus seenghala (Sukes)
				Mystus cavasius (Ham.)
				Rita rita (Ham.)
				Ompok bimaculatus (Bloch)
				Ompok pabda (Ham.)
				Wallago attu (Schneider)
				Ailia coila (Ham.)
				Clupisoma garua (Ham.)
				Eutropiichthys vacha (Ham)
				Silonia silondia (Ham.)
				Pangasius pangasius (Ham.)
				Bagarius bagarius (Ham.)
				Gagata cenia (Ham.)
				Clarias batrachus (Linnaeuss)
				Heteropneustes fossillis (Bloch)
Atheriniformes	2	3	3	
Gasterosteiformes	1	1	1	
Channiformes	1	1	4	*Channa marulius* (Ham.)
Symbaranchiformes	1	1	1	
Perciformes	6	10	14	*Rhinomugil corsula* (Ham.)
				Glossogobius gutum (Ham.)
				Anabas testudineus (Bloch)
Mastacembeliformes	1	2	3	*Mastacembelus armatus* (Lacepede)
				Mast acembelus pancalus (Ham)
				Macrognathus aculeatus (Bloch)
Tetraodontiformes	1	1	1	
	30	72	141	

Source : Natarajan, A.V. (1989)

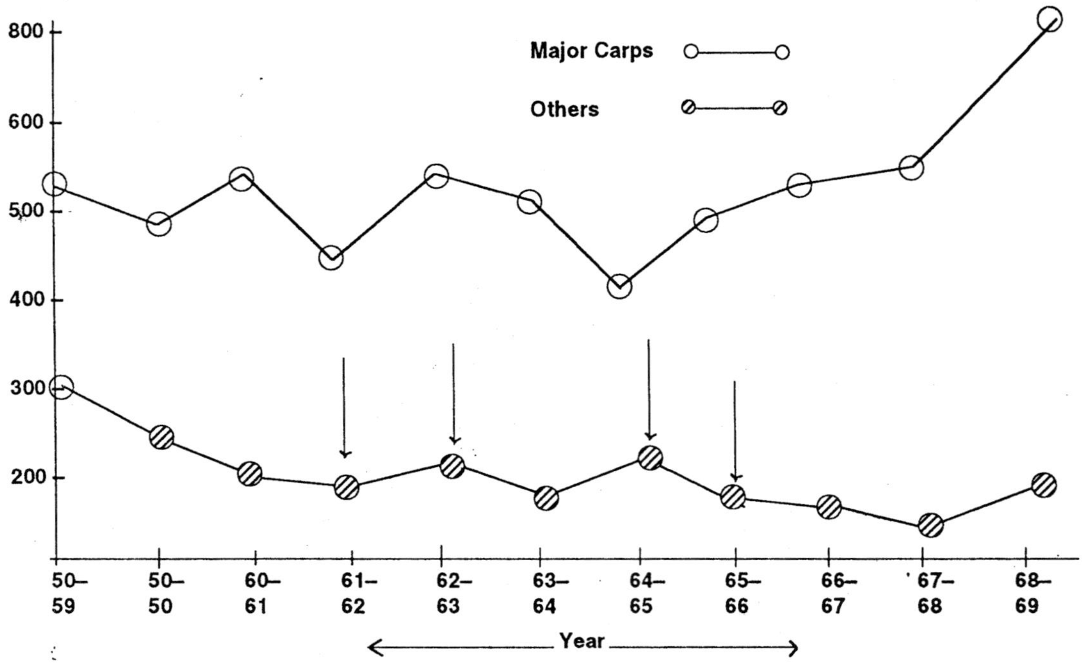

Fig. 28 : Total annual market arrivals for all river stretches considered. Arrows indicate low values probably due to lack of data for one station.
Source : Jhingran, A.G. and K.K. Ghosh, (1978).

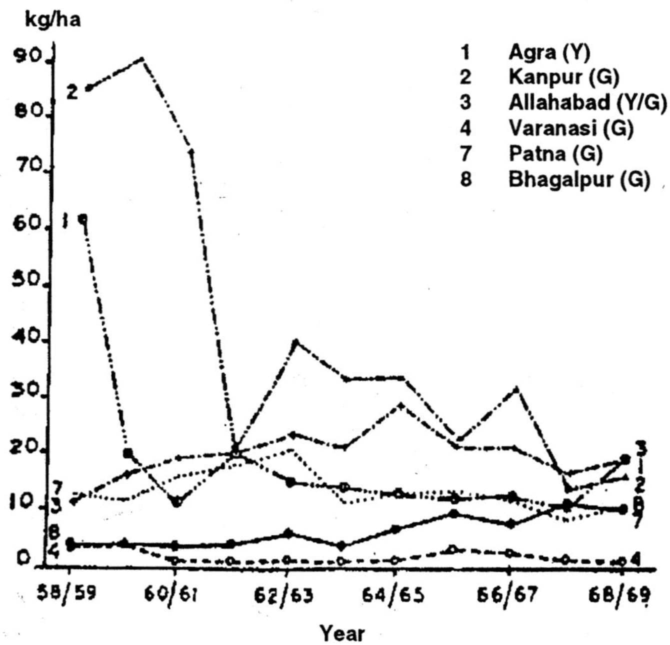

Fig. 29 : Yield development for major carps in different river stretches.
Source : Jhingran, A.G. and K.K. Ghosh, (1978).

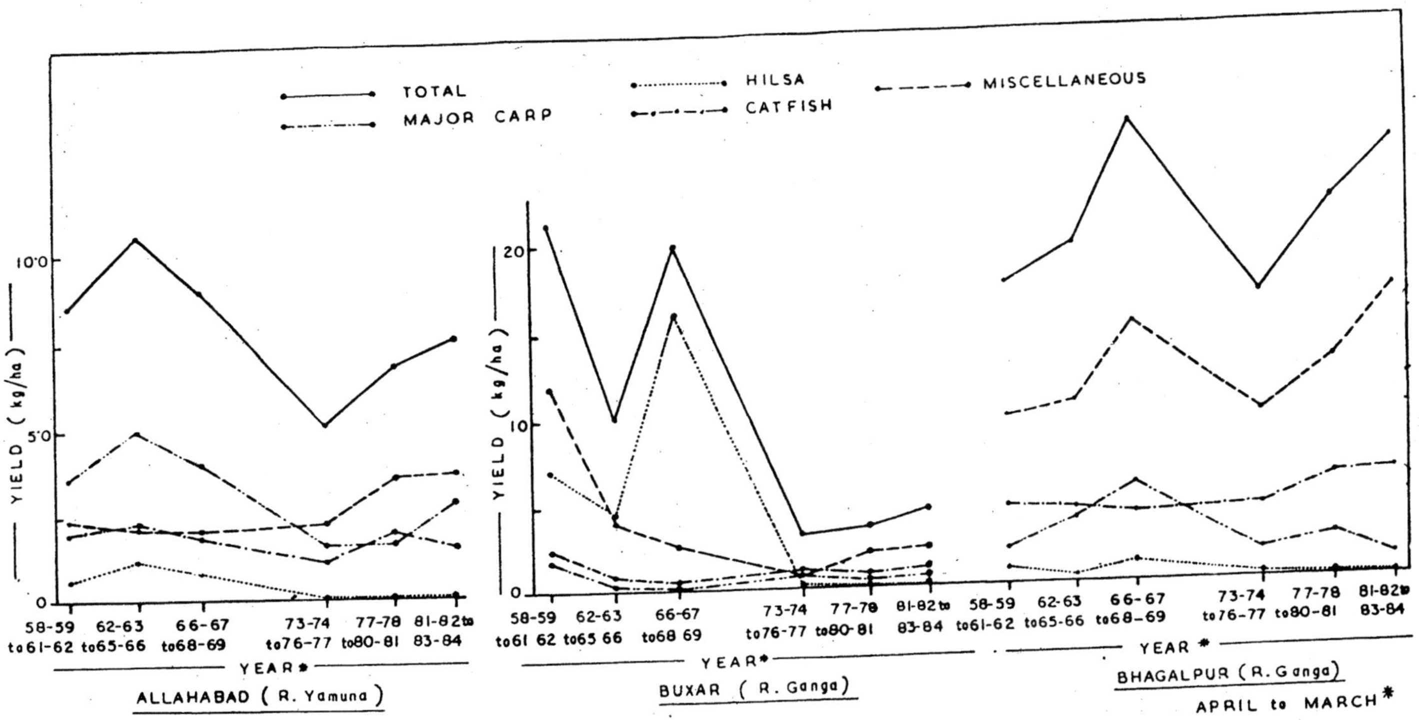

Fig. 30 : Estimated average annual yield rate (kg/ha) of different groups of fishes.
Source : Jhingran, A.G. (1986).

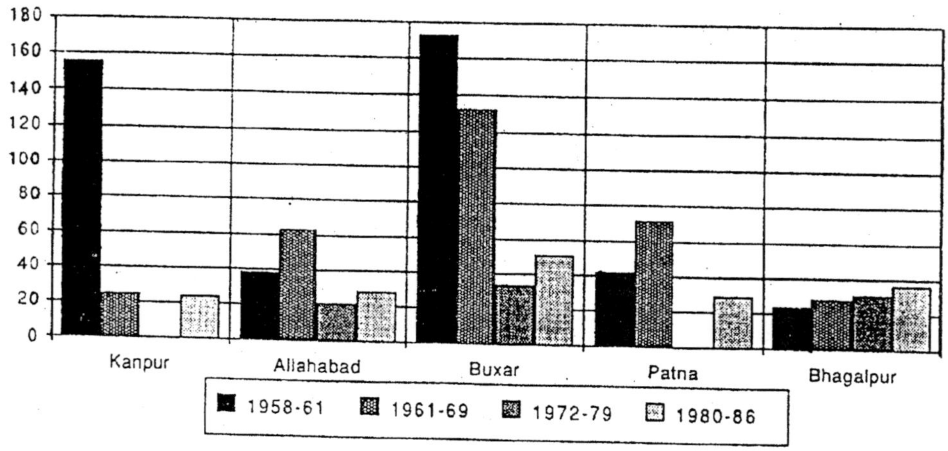

Fig. 31 : The trend In fish yield (KG/HA) from different stretches of the river Ganga during 1958-86.
Source : Jhingran, Arun G. (1989).

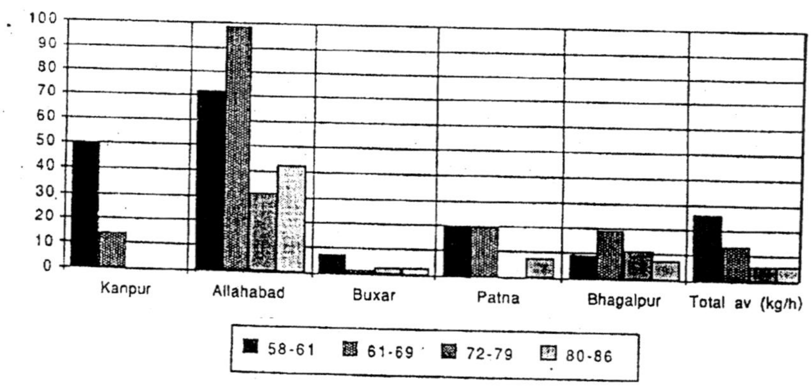

Fig. 32 : Average annual yield (Tonnes) of major carps from various stretches of river Ganga during 1958-1986.
Source : Jhingran, Arun G. (1989).

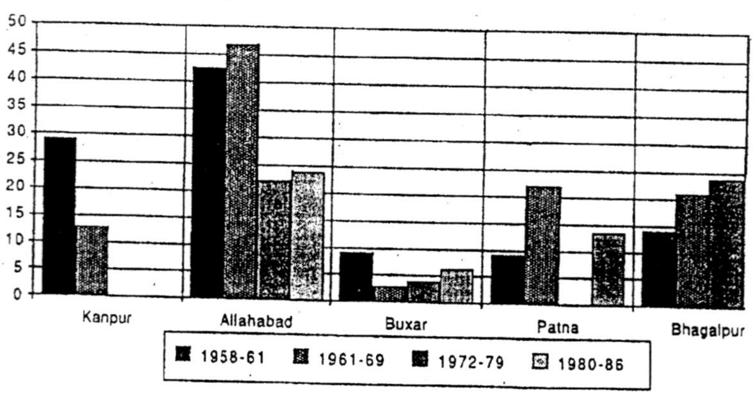

Fig. 33 : Average annual yield (Tonnes) of catfishes from various stretches of river Ganga during 1958-86.
Source : Jhingran, Arun G. (1989).

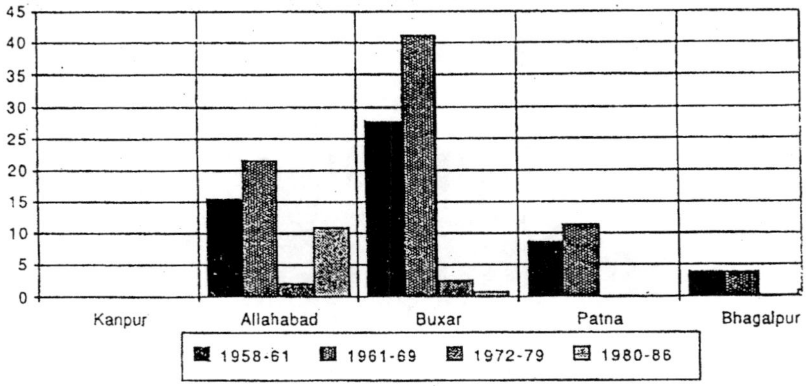

Fig. 34 : Average annual yield (Tonnes) of Hilsa from various stretches of river Ganga during 1958-86.
Source : Jhingran, Arun G. (1989).

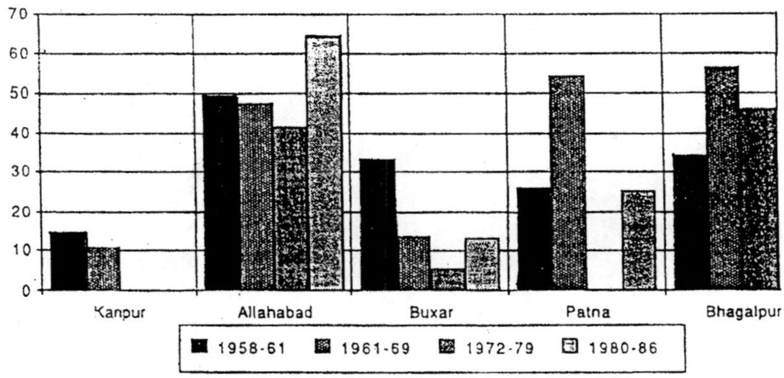

Fig. 35 : Annual average yield (Tonnes) of miscellaneous fishes from various stretches of river Ganga during 1958-86.
Source : Jhingran, Arun G. (1989).

Table 54 : Ecological grouping of fishes and prawns of the Ganga river system

Species	*Species number*	*Economically important fishes*	*Others*
Fish			
Hill stream	55	7	51
Swamp	13	7	6
Plain	72	13	36
Anadromous	1	1	–
	141	28	93
Prawn			
Fresh Water	1	1	–
Deltaic (fresh water and tidal)	6	2	4
	7	3	4

Table 55 : Centrewise average annual landings (in tonnes) of major carps, selected Catfishes, Hilsa and other fishes of R. Ganga during 1958-1969.

Zones with river stretch (Km) and av. width (m) (Non flood period) feeding the Centre		1958–59	1959–60	1960–61	1961–62	1962–63	1963–64	1964–65	1965–66	1966–67	1967–68	1968–69
Agra 160Km. 150m.	MC	148.6 (61.9)	48.5 (20.2)	26.3 (11.0)	48.4 (20.2)	36.4 (15.2)	33.7 (14.0)	30.9 (12.9)	30.0 (12.5)	29.8 (12.4)	22.8 (9.5)	45.9 (19.1)
	CF	123.1	103.6	24.3	22.2	20.7	24.1	24.5	27.4	19.1	10.7	26.9
	H	0.0	0.1	—	0.1	0.0	—	0.0	—	0.0	—	0.5
	O	35.1	13.9	7.4	4.7	5.5	6.1	2.6	2.7	6.0	3.5	10.7
Kanpur 40 Km 150m.	MC	51.0 (85.0)	54.8 (91.3)	44.5 (74.2)	12.3 (20.5)	23.9 (39.8)	20.2 (33.7)	20.3 (33.8)	12.9 (21.5)	19.2 (32.0)	8.0 (13.3)	9.8 (16.3)
	CF	30.4	36.9	20.4	8.3	14.2	13.4	8.6	11.7	15.6	12.1	18.0
	H	—	—	—	—	—	—	—	—	—	—	—
	O	12.0	13.9	16.9	9.1	14.3	9.5	14.9	8.0	13.8	5.10	12.1
Allaha-bad 230Km 200m.	MC	54.0 (11.7)	74.9 (16.3)	86.9 (18.9)	91.7 (19.9)	108.3 (23.5)	95.9 (20.8)	131.3 (28.5)	98.3 (21.4)	98.6 (21.4)	77.2 (16.8)	87.7 (19.1)
	CF	42.4	42.5	42.9	50.0	68.7	40.0	51.0	37.1	37.6	40.4	50.6
	H	7.9	10.7	24.0	9.9	22.3	19.9	30.0	32.5	12.5	15.1	30.6
	O	42.1	61.5	46.3	51.2	53.7	44.6	44.1	44.7	37.5	36.5	60.0
Varanasi 65 Km 300m	MC	5.6 (2.9)	7.0 (3.6)	2.3 (1.2)	2.3 (1.2)	1.7 (0.9)	1.7 (0.9)	2.4 (1.2)	6.6 (3.4)	4.9 (2.5)	2.4 (1.2)	2.3 (1.2)
	CF	7.9	7.7	9.6	14.1	18.2	21.0	20.5	28.3	19.6	24.8	27.0
	H	7.4	12.4	27.9	14.4	36.5	14.9	17.3	18.8	24.0	32.8	71.8
	O	25.8	20.1	29.5	4.5	48.9	37.6	30.4	59.1	49.5	73.9	52.3
Buxar 45 Km 100m.	MC	13.7 (30.4)	7.5 (16.7)	1.8 (4.0)	NA	NA	1.4 (3.1)	1.3 (2.9)	3.0 (6.7)	2.1 (4.7)	0.6 (1.3)	1.5 (3.3)
	CF	14.4	10.4	3.2	NA	NA	3.8	5.8	2.2	2.4	2.0	2.3
	H	29.4	19.9	33.8	NA	NA	8.0	7.4	36.8	48.6	34.1	112.7
	O	51.1	34.0	55.4	NA	NA	19.3	17.1	14.2	11.4	11.6	8.9
Ballia 70 Km 100m.	MC	9.4 (13.4)	8.4 (12.0)	4.1 (5.9)	5.2 (7.4)	2.3 (3.3)	4.9 (7.0)	NA	NA	4.9 (7.0)	4.2 (6.0)	4.5 (6.4)
	CF	6.9	3.8	4.6	15.9	4.7	14.1	NA	NA	4.1	7.2	6.0
	H	10.9	17.3	15.7	4.5	54.1	73.1	NA	NA	32.7	15.5	24.0
	O	18.2	20.8	60.0	25.6	11.2	22.2	NA	NA	9.3	12.9	6.8
Patna 60Km. 250m.	MC	19.0 (12.7)	17.2 (11.5)	24.0 (16.0)	26.8 (17.9)	17.9 (20.6)	17.9 (11.9)	18.9 (12.6)	20.2 (13.5)	16.8 (11.2)	12.8 (8.5)	14.6 (9.2)
	CF	9.0	3.5	15.5	22.1	22.9	34.1	17.8	14.3	10.7	23.0	24.9
	H	2.8	1.7	21.8	14.4	26.5	9.9	8.2	11.4	13.6	4.8	4.2
	O	33.8	17.7	35.2	42.0	49.4-	53.5	32.2	67.0	67.2	58.2	66.5
Bhagal-pur 130Km 200 m	MC	9.0 (3.5)	9.6 (3.7)	9.2 (3.5)	10.6 (4.1)	15.6 (6.0)	10.9 (4.2)	17.9 (6.9)	26.1 (10.0)	19.8 (7.6)	26.6 (11.0)	25.6 (9.8)
	CF	8.2	10.9	24.6	36.9	39.0	16.9	7.9	14.1	8.8	19.6	26.8
	H	4.8	2.4	5.1	4.0	2.1	2.2	2.4	3.9	3.1	5.3	9.6
	O	19.5	37.1	47.1	69.0	41.8	47.7	48.1	50.8	51.7	68.3	77.9

MC = Major Carp; CF = Catfish; H = Hilsa; O = Others; NA = Not available. Values in parentheses indicate yield of Major Carp in Kg/Ha.

Source : Jhingran, A. G. and K.K. Ghosh, (1978).

Table 56 : Trend of major carp catch (Kg/ha) at various Centres spread over the Ganga from 1958-59 to 1966-1967

	1958-59	*1959-60*	*1960-61*	*1961-62*	*1962-63*	*1963-64*	*1964-65*	*1965–66*	*1966-67*	*1967-68*	*1968-69*
Agra	61.9	20.2	11.0	20.2	15.2	14.1	12.9	12.5	12.4	9.5	19.1
Kanpur	85.0	91.3	74.2	20.5	39.8	33.7	33.8	21.5	32.0	13.3	16.3
Allahabad	11.7	16.3	18.9	19.9	23.5	20.8	28.5	21.4	21.4	16.8	19.1
Varanasi	2.9	3.6	1.2	1.2	0.9	0.9	1.2	3.4	2.5	1.2	1.2
Baxar	30.4	16.7	4.0	NA	NA	3.1	2.9	6.7	4.7	1.3	3.3
Ballia	13.4	12.0	5.9	7.4	3.3	7.0	NA	NA	7.0	6.0	6.4
Patna	12.7	11.5	16.0	17.9	20.6	11.9	12.6	13.5	11.2	8.5	9.2
Bhagalpur	3.5	3.7	3.5	4.1	6.0	4.2	6.9	10.0	7.6	11.0	9.8

Note : Modified from Jhingran, A. G. and K. K. Ghosh (1978).

Table 57 : Position of different groups of Fishes in the Ganga indicating the ratio between Major Carp and Cat Fish and also position of other fishes during 1958-59 to 1968-1969

	1958-59	*1959-60*	*1960-61*	*1961-62*	*1962-63*	*1963-64*	*1964-65*	*1965–66*	*1966-67*	*1967-68*	*1968-69*
Major Carp	310.3	227.9	199.1	197.3	206.1	186.6	223.0	197.1	196.1	156.6	191.9
Ratio of MC and CF.	(1.28)	(1.0)	(1.37)	(1.16)	(1.09)	(1.1)	(1.6)	(1.45)	(1.66)	(1.1)	1.51
Cat Fish	242.1	219.3	145.1	169.5	188.4	167.4	136.1	135.1	117.9	139.8	182.5
Others	227.6	219.0	317.8	206.1	224.8	240.5	189.4	246.5	246.4	270.0	295.2

Note : Data presented are in tonnes.

Modified from Jhingran, A.G. and K.K. Ghosh, (1978)

Jhingran and Ghosh (1978) have revealed that all the species were dominating in the upper stretch; Major carp contributed on an average 38.5 per cent at Allahabad and 55.4 per cent at Kanpur to the annual landings, the most dominant carp species was mrigal 50 per cent amongst the total catch of carps. Hilsa was the dominant species at Allahabad. *Mystus* sp. were the next predominant species encountered at Allahabad. The middle stretch landings were dominated by other species. Hilsa and major carp population recorded very low catch and were predominantly juveniles and fingerlings. The lower stretch indicated an abundance of major carps contributing 20.8 per cent at Patna and 18.3 per cent at Bhagalpur. The other fishes also dominated in this stretch. The hilsa fishery had an importance in the Ganga at Allahabad. Hilsa fishery heavily fluctuated in relation to time and space and their occurrence might be due to the existance of different stocks of hilsa. Though Pillay *et al.* (1964) and Ghosh *et al.* (1968) found that atleast three races of hilsa are present in the Ganga river system, but this is yet to be confirmed.

Amongst catfishes *Mystus* sp. were almost evenly abundant throughout with *Mystus aor* the most predominant species available than *Mustus seenghala*. *Wallago attu* was more available in the upper and lower stretches and showed high association with major carp. March to June, the summer, contributed about 50 per cent of the landings in the upper stretches, while year round fishery recorded a low catch during August in the middle stretch. The landings at Ballia during July were higher, while in the lower stretches the landings were higher during winter–November to February (over 40 per cent).

Jhingran and Gupta (1987) have further appraised with the fact that in the middle and lower stretches of the Ganga, the average total yield has declined from 50.3 kg/ha/yr from pre 1961 to 22.0 kg/ha/yr in the post 1972. The average yield rate of major carp dropped from 13.3 kg/ha to only 4.6 kg/ha/yr with lower yields of *C. mrigala, C. Catla and L. rohita* and an upward trend in the yield of *L. calbasu* was recorded for the same period.

Balbir Singh *et al.*, (1987) have reported that at Allahabad the total annual landings was from 120 km. stretch of the Ganga and 60 km. of the Jamuna, for the some period has been reduced to only 49.65 t from that of 91.3 t in the past. Further, decline in total carp catches especially mrigal has been noted. While mrigal used to dominate over the other major carp species until 1972 contributing over 50 per cent has now been reduced to only about 20 per cent. Interestingly, mrigal in recent years is replaced by calbasu which now contributes over 60 per cent of the total major carp catch against earlier contribution of about 13.5 per cent. Calbasu landing which was recorded 16.0 t only in 1973 has gone up to 35.0 t in 1983. The reason of higher catch of calbasu is not however explained.

The drought of 1987 in the country as a whole and the stretches of the Gangetic valley in particular and the scanty rains of 1988 affected the fishery as a whole. An estimated fish landing at Sadiapur (on R. Jamuna) and Daraganj, Patna, Bhagalpur and Lalgola (on the river Ganga) works out to 70.04, 22.12, 65.65 50.23 and 44.72 t respectively between March and December 1988. The specieswise break up of fish landings for the years 1988 and 1989 are given below in Table 58 A, and the impact of drought on mean length of Commercial important fishes in the Ganga & Jamuna at Allahabad are given in Table 58 B.

Table 58 A: Species wise break up of some Commercially important fish landings (tonnes) at different Centres of the Ganga and Jamuna during 1988 and 1989

Species	*R. Jamna*		*R.Ganga*							
	Sadiapur		*Dragang*		*Patna**		*Bhagalpur*		*Lalgola*	
	1988	*1989*	*1988*	*1989*	*1988*	*1989*	*1988*	*1989*	*1988*	*1989*
C.mrigals	3.05	3.82	0.48	0.75	3.69	0.50	0.85	—	0.34	—
C. catla	2.17	1.44	0.09	0.01	3.89	2.34	3.27	—	0.86	—
L. rohita	2.11	3.56	0.09	0.33	1.42	1.65	1.16	—	0.32	—
L. calbasu	7.54	6.88	0.98	0.52	1.04	0.51	0.10	—	0.55	—
Total MC	14.87	15.70	1.64	1.61	10.04	5.00	5.38		2.07	—
M. aor	7.24	11.81	1.09	1.30	9.25	4.16	1.43	—	3.29	—
M. seenghala	7.50	11.47	2.12	2.72	7.06	3.74	1.14	—	0.49	—
W. attu	1.33	1.48	0.83	0.17	3.89	1.22	12.04	—	2.58	—
Selected CF	16.07	24.76	4.04	4.19	20.20	9.12	14.61		6.36	
Hilsha	0.92	0.54	0.18	0.37	0.11	0.03	0.22	—	7.02	—
Others/Misc.	38.18	43.92	16.26	19.27	35.30	25.77	30.02	—	29.27	—
Total :	70.04	84.92	22.12	25.44	65.65	39.92	50.23		44.72	

Source : Modified from Annual Report CICFRI, Barrackpore, 1988-89 and 1989-90.

Data for 1988 (March-December)

* January 1988 – December 1988.

Table 58 B : Impact of drought on the mean length of commercial important fishes in the Ganga and Jamuna at Allahabad

Species	*1987*	*1988*	*Decrease in length (cm)*
Major Carp			
C. marigala	62.3	55.8	6.9
C. catla	73.5	61.8	11.7
L. rohita	65.0	58.0	7.0
Major Cat fish			
M. aor	44.7	43.8	0.9
M. seenghala	54.5	48.8	5.7
W. attu	54.5	47.1	7.4
Others			
Hilsa ilisha	39.6	34.7	4.9

Source : CICFRI, Annual Report 1988-89

In comparison to the landings for the corresponding period of the preceding year, Sadiapur and Daraganj registered a decline catch (19.5 per cent and 21.0 per cent). At Patna, the fishery showed an overall decline of 10.4 per cent than that of previous year, the main composition of the miscellaneous group fishes were recorded in the landing. The fish landings registered an increase of 23.4 per cent at Bhagalpur. Lalgola landings showed a decrease of 66 per cent over the landings of the preceding year. At Bhagalpur, increase in major carps and miscellaneous group of fishes were noticed, where as all the species declined at Lalgola.

The Mean sustained Yield (MSY) of *L. calbasu* (based on the available past data) was estimated at 47.15 t which was far above present level. Among Catfishes, *M. aor* and *M. seenghala* showed an improvement in their MSY levels which stood at 12.09 t. and 12.56 t. Miscellaneous group was the worst affected alongwith *R. rita* and small fishes. The landing of *rita rita* was half the level of 1988 while that of miscellaneous group declined by 30 per cent in 1989. Daraganj fish landing centre did not show much change as compared to 1988 (28.0 t) both qualitatively and quantitatively. At Patna, the landings showed a decline of 33.0 per cent compared to 1986–88. The decline was almost uniform for all the commercially important species (Fig.36).

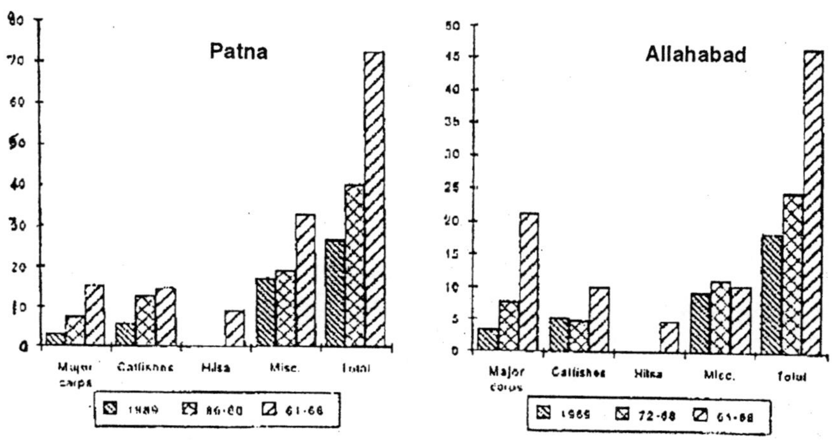

Fig. 36 : Decline in Fish Yield (Kg/ha) at Allahabad and Patna.
Source : Annual Report CICFRI, Barrackpore (1989-90).

Jhingran and Gupta (1987) have recorded that hilsa was abundant till 1972 at all centres in the lower Ganga above Farakka.

Jhingran and Gupta (1987) observed record drop in landing during pre and post 1972 periods viz. a fall from 19 t to 0.9 t from Allahabad centre, 36 t to 0.34 t at Buxar and from 4 t to 0.82 t at Bhagalpur. Similar results were recorded (0.130 to 0.003 t) in 1986 by Dhirendra Kumar *et al.*, (1987) in their study in a stretch of 60 km. around Patna.

Diverse opinion exists about the impact of Farakka barrage on fish migration. Chaphakar and Mhatre (1986) quoting the data of Bilgrami and Dutta–Munshi (1979) has shown that total landing (tonnes) consisting of H. *ilisha*, Major carps, siluroids and miscellaneous were showing lower values at Rajmahal compared to Farakka from 1976–77 to 1978–79.

Location		*1976–77*	*1977–78*	*1978–79*
25°48' N and 87°50' E	Rajmahal	82.2	85.3	101.6
24°49' N and 87°56' (E)	Farakka	433.3	460.4	490.6

It is presumed that adequate care was not taken at the time of construction of fish ladder/Fish pass. As a result the migration of *H. ilisha* to the upstream for breeding is largely affected. While the area below the barrage has become an important point for indiscriminate fishing which never existed prior to the construction of Farakka barrage. More than thousand fishing boats are sometimes engaged in capture operation of hilsa since the barrage acts as a barrier against their migration the fishes get captured before they breed. As a result hilsa has become a vulnerable species at present.

Ravish Chandra *et al.* (1990) more recently observed that commercial landings of hilsa at Allahabad and Buxar fluctuated between 7.87 and 40.15 at Allahabad, and 7.38 and 113.36 at Buxar during the period 1958 to 1972 with an annual average of 19.18 t and 31.97 t at Allahabad and Buxar respectively (Table 58) while during post Farakka period between 1973 and 1989 (cf Table 59). They recorded hilsa landing to range from 0.13 to 3.04 t and 0.07 to 2.60 t with an average 1.01 t and 0.67 t at Allahabad and Buxar.

These authors while studying the hilsa population in the upper and lower stretches of the Ganga between Allahabad and Bhagalpur in two phases *i.e.* initially for 15 years (1958–72) and subsequently between 1973–89 (Table 59) revealed some interesting information. According to them during 1960 a distinct peak recording 24.01 t at Allahabad, 33.80 t at Buxar and 5.10 t at Bhagalpur was observed. A higher magnitude of landing was recorded during 1965 *ie* 40.15 t, 36.15 t and 3.92 t at the respective centres. However the third cycle instead of 1970 was initiated in the 3rd year *ie.* 1968, when 28.28 t, 113.36 t and 9.79 t of hilsa were landed at the above centres. The trend became erratic from 1969 and at all the centres hilsa catch declined considerably, possibly due to constructional work at Farakka which commenced during 1963 and 1964 and with the progress of the work, decline of hilsa landing started. During 1969–71 when the barrage was almost in the final stage of completion, the migration of hilsa was badly affected. They opioned that the concept of 5 years cycle appears to be completely disappeared. Ravish Chandra *et al.* (1990) in a perusal of catch data from three centres of the Ganga for 17 years (1973–1989), Table 59 gave a gloomy picture of the satus of hilsa fishery, after commissioning of the Farakka barrage. The fishery declined drastically at all the centres and no trend was visible till 1983.

Table 59: Landing of *Hilsa ilisha* (t) during pre and past Farakka discharge (1958–1989)

Year	CENTRES			
	Allahabad	*Buxar*	*Bhagalpur*	*Lalgola*
1958	7.87	29.44	4.82	–
1959	14.80	19.19	2.39	–
1960	24.01	33.80	5.10	–
1961	21.06	–	4.58	–
1962	18.06	–	1.93	–
1963	19.71	8.39	2.63	363.24
1964	23.82	7.38	2.39	119.56
1965	40.15	36.15	3.92	66.24
1966	13.64	48.15	3.11	5.75
1967	14.04	34.16	5.28	3.70
1968	28.28	113.36	9.79	6.46
1969	15.06	–	7.56	92.27
1970	16.65	15.69	1.47	72.37
1971	18.40	18.95	3.95	89.27
1972	12.20	18.36	1.84	91.73
1973	3.04	2.00	1.41	227.85
1974	0.30	0.21	0.08	283.21
1975	0.44	0.09	0.01	97.92
1976	0.27	0.48	0.00	44.80
1977	0.21	0.25	0.37	25.25
1978	0.13	0.07	0.12	11.02
1979	1.75	0.07	0.28	9.02
1980	1.65	0.30	2.00	11.64
1981	0.19	0.50	0.25	11.06
1982	0.46	0.09	2.18	15.98
1983	0.24	0.18	2.89	18.82
1984	1.73	1.02	0.80	20.53
1985	1.72	2.60	0.18	29.02
1986	1.87	1.65	0.41	18.31
1987	1.75	–	0.47	17.49
1988	0.92	–	0.22	7.02
1989	0.54	–	–	14.92

Source : Ravish Chandra, R.K. Saxena and R.K. Tyagi (1990)
Date not available.

The hilsa fishery at Lalgola (on R. Padma) during the period of construction of Farakka, reflected wide fluctuations with a maxima of 363.24 t in 1963 to a minima 3.11 t in 1967. During 1973–74 just after commissioning of the Farakka barrage the landing again touched a level 283.21 t after which fishery progressively declined till eighties. Thereafter it maintained almost a steady trend although at a lower catch value.

It is interesting to note that, the year 1975 the starting of decline in hilsa catches at Lalgola coincided with the running of feeder canal originating from the Ganga at Farakka and meeting the Bhagirathi at Jangipur. Heavy exploitation of hilsa has been reported in the lower reaches of the canal as they could not migrate above Farakka. This has resulted in reduction of hilsa fishery in the upper stretch of the Ganga ranging between 82.7 per cent and 98.1 per cent being maximum at Buxar followed by Allahabad and 94.6 per cent and minimum at Bhagalpur.

CICFRI, Barrackpore (1992–93) in its Research Project result has shown that hilsa fishery around down stream of Farakka Barrage, the estimated catch is highest (3505.50 kg) in the month of September followed by 2373 Kg in the month of October coinciding with the peak breeding period of fish. While an estimated catch of 153.25 kg (av. wt. 1.6g) and 413.16 kg (av. wt. 1.5g) of young hilsa were recorded from upstream of Farakka barrage during October and December respectively, reaching its peak in the month of November to the tune of 1003.50 kg comprising three groups of young hilsa of av. wt. 1.4g (60 per cent), 5.5g (30 per cent) and 100g (10 per cent). This supports that hilsa bread at the upper stream of Farakka barrage.

The above statements appear to be reasonably correct. It is clear from the above, that fishery of the Ganga in the upstream, up to Allahabad has considerably dwindled since the time the Farakka barrage was commissioned (Table 59). Simultaneously it is interesting to note that there is a regular heavy congregation of fish below the barrage and they are indiscriminately captured before they could breed. In case these congregated fish had an open passage the set back of fishery and their scanty breeding (in case of hilsa) in the up stream could have been possibly avoided. The above explanation clearly reveals that migration of fish is hardly possible. This statement holds correct especially in the case of hilsa which largely migrates from the sea and are trapped near the barrage.

It is now an established fact that due to faulty construction of Farakka barrage hilsa cannot get a free passage for migration on either side of the barrage, as a result the old stock in the up stream settle down permanently as endemic fauna. Breeding of these fishes occurs during September and October although with descrete concentration of hatchlings in certain sections, some of which are recorded by CICFRI, Barrackpore (1992–93) and these are not from any migrated breeders (except negligible quantity) as they are reported.

As regards other fishery, the reason for their fall in the landing, their intensity of breeding *etc.* are discussed earlier. Thus simultaneously the overall decline in the fishery of major carp and hilsa has threatened the fishery of the Ganga as a whole which has a great economic loss to the fishery industry.

More recently the 'NISADS' (Fisherman community) of Bihar has openly come out to protest against the faulty construction of sluice gate of Farakka Barrage in conference at Gandhi Maidan (Patna) in presence of central welfare minister and state Chief Minister. As a result of which fisherman community of Bihar and Uttar Pradesh are put to severe distress. The organiser of the conference has even openly threatened to damage the barrage if defective sluice gate is not properly mended. Thus it is clear that even the fisherman community has felt the effect of faulty construction of fish pass of the Farakka Barrage for which fish cannot migrate and thus has resulted in the quick fall in the fish population both in Bihar and Uttat Pradesh (Ananda Bazar Patrika, 1993).

Data on fish landings reflects that fishery of Northern India has been greatly affected after the construction of Farakka Barrage. This might be due to defective construction of fish ladder/fish pass at Farakka, as a result, migration of commercially important fish like *Hilsa ilisha* to the upper stretch of the Ganga where it migrates for breeding is reduced considerably (Bilgrami and Dutta Munshi, 1979, of Table 60).

Table 60 : Effect of Farakka Barrage on the Fish landings (tonnes) at two Centres during 1976-1979

Species	1976-77	%	1977–78	%	1978-79	%	Annual average
Rajmahal (25°4'N and 87°50'E)							
Hilsha ilisha	1.2	1.5	1.9	2.2	2.1	2.2	1.7
Major Carps	31.3	38.0	32.4	32.7	43.4	42.9	35.7
Siluroids	42.9	52.0	41.5	48.2	47.5	47.1	43.9
Miscellaneous	6.8	8.5	9.5	11.3	8.6	7.9	8.3
Total Landings	88.2	—	85.3	—	101.6	—	89.6
Farakka (24°49'N and 87°56'E)							
Hilsha ilisha	121.5	28	136.8	29.5	148.6	30	135.6
Major Carps	83.9	19	98.8	21.5	90.8	19	91.1
Siluroids	180.7	42	196.6	42.8	200.5	41	192.6
Miscellaneous	47.2	11	28.2	6.2	50.7	10	42.3

Source : Bilgrami K.S. and J. S. Dutta Munshi, 1979.

Moreover the barrage has become an important point for indiscriminate fishing of Hilsa. Since the barrage acts as a barrier against their migration, the fishes are caught before they migrate to the upper stretches for breeding. As a result, Hilsa has become a threatened species at present, particularly in the upper stretches. The potential (PY) and actual (AY) fish yields from the Jamuna at Allahabad and various stretches of Ganga (freshwater area) are indicated by Jhingran (1989) in (ef. Fig. 37).

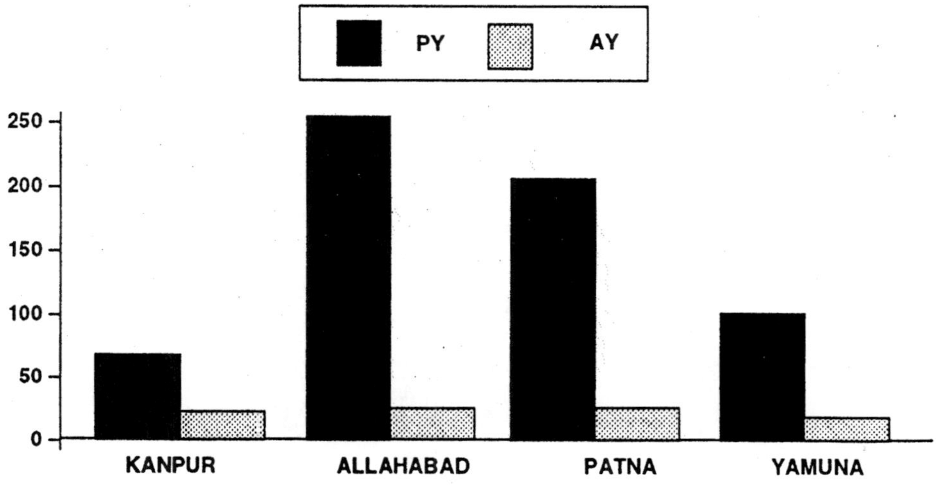

Fig. 37 : Potential (PY) and actual (AY) fish Yields from Yamuna and various stretches of Ganga (kg/ha)
Source : Jhingran, Arun G. (1989)

Estuary (Bhagirathi-Hooghly)

Progressive changes in the fishery composition have been noticed in the estuary in the context of salinity. These may be summerised as :

Hamilton (1982) : Purely freshwater fish except hilsa : Calcutta and in the lower stretches.

Hora (1943) : Considerable changes in the fishery composition was noticed. Estuarine groups in and around Barrackpore

Hora (1953) and

T.V.R. Pilley	:	Mainly brackiswater fishes in an around Barrackpore.
David (1954)	:	Mainly brackishwater fishes. (Salinity range 0.204–1.156 ppt)
Ray (1981)	:	Mainly brackishwater forms in an around Barrackpore (Highest Salinity recorded– 4. Oppt)

"Fishery resources of the Hooghly–Matla Estuarine System, Bull 19, Dec. 1973" a document published by Central Inland Fisheries Research Institute, Barrackpore is perhaps the first of its kind published a comprehensive account on fish fauna covering the period 1960–61 to 1970–71. Although five zones are considered for the Hooghly–Matla estuary and the zone–wise annual landings (specieswise composition) in terms of kilogramme of fish during 1966–67 to 1970–71 have been reported alongwith some other relevant informations specially Gears. From this account it is difficult to pin point the actual species wise landings of the Bhagirathi–Hooghly estuary. Subsequently, Ray (1981) based on the salinity cited by Dutta *et al.*, (1973) delineated the estuary into three zones Fig. 38.

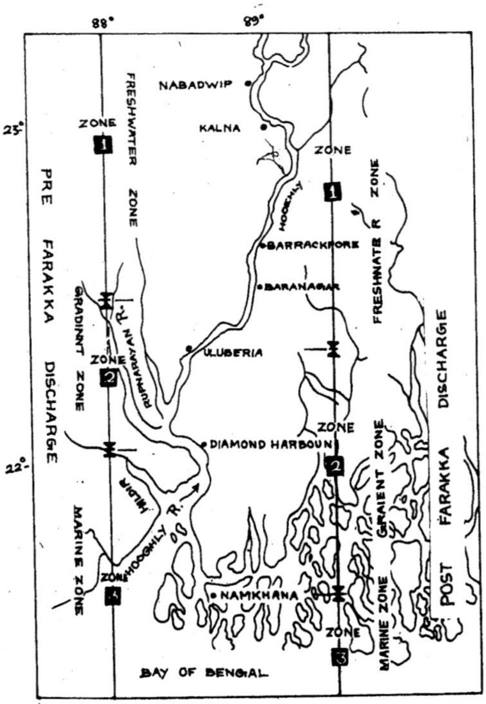

Fig. 38 : Showing different zones in the Hooghly estuary during pre and post Farakka discharges.
Source : Nandy, A.C., M.M. Bagchi, and S.K. Mazumdar, (1983).

Zone–I Nabadwip to Baranagar near Calcutta where traces of salinity (less than 0.1 per cent is found throughout the year.)

Zone–II Middle zone, extends from Barrackpore to Diamond Harbour where salinity ranged between traces and 10 per cent.

Zone–III Lower zone includes the entire Sunderbans and a tract below Diamond Harbour on the main channel having salinity from 10 per cent to nearly neritic values. From the above, Ray (1981) has processed the data to show Yearwise landings in three different zones during 1966–71 in table 61.

Table 61 : Yearwise landing (tonnes) in three different zones during 1966–71[@]

	Zone I		*Zone–II*	*Zone–III*
	Unpolluted	*Polluted* *		
1966–67	720	57	136	4442
1967–68	742	89	415	598
1968–69	735	88	286	5585
1969–70	910	106	289	5751
1970–71	913	120	383	8526

* Polluted zone means where maximum number of industries are located and pollution load is very high.

@ Personal communication with Laha G.C., CIFRI.

In another communication Ray and Laha (1981) divided the estuary into two zones. Unpolluted–Nabadwip to Bandal, a stretch of 86 km. having approximate water spread area, 4509.84 ha and polluted zone, Bandel to Baj Baj, a stretch of 72 Km. having approximate water spread area 4782.24 ha is shown in Table 62 A & B.

The Table 62A shows that total yield of fish in a period of 16 years at the two zones between Nababganj and Baj Baj of the Hooghly estuary system. Average annual fish yield in the respective zones are 719.25 and 124.94 t and catch/ha are 159 and 26 kg which shows that ecological condition in Zone–I is more favourable for biota than Zone–II. The table further shows that there is a year to year upward tendency of fish yield except a year or two which may be due to some non-sampling error during the pre-Farakka period in the unpolluted zone. On the contrary, polluted zone does not follow the same pattern and shows a decreasing trend mostly in 1960s and also in early '70s except a period when it recorded a bumper catch. In late '70s catch shows an increasing tendency.

Considerable changes in the ecology of Bhagirathi–Hooghly has taken place due to release of freshwater from Farakka barrage. Salinity has gone down remarkably. With the increase of freshwater flow, the salinity is lowered, pushed the gradient zone towards sea. The earlier recognised gradient zone largely become freshwater zone, while the marine zone becomes largely gradient zone (Nandy *et al.*, 1983 of Fig 38). In general, biotic profile and their occurrence considerably changed, some of the euryhaline species had shown a declined trend in the upper and gradient zone (Table 63). The total fish landings as well as species composition in the marine zone of the estuary did not show any significant changes.

Table 62 : **A.** Total annual Landings (In Tonnes) Both in the Unpolluted and Polluted Stretches of the Hooghly Estuary (1962–1978)

Year	*Unpolluted* *Zone-I*	*Polluted* *Zone-II*
1962-63	562	297
1963-64	484	142
1964-65	338	103
1965-66	536	68
1966-67	720	57
1967-68	742	89
1968-69	735	88
1969-70	910	105
1970-71	913	120
1971-72	1064	120
1972-73	1064	101
1973-74	1029	106
1974-75	779	154
1975-76	681	170
1976-77	551	161
1977-78	400	118

Note : Zone-I. Unpolluted Zone (Nabadwip to Bandel near Jubilee Bridge), a stretch of about 86 Km. having water spread area 4509.84 ha.

Zone–II. Polluted Zone (Bandel to Baj Baj near Batanagar), a stretch of about 72 Km having water spread area 4782.24 ha.

Source : Ray P. and G. C. Laha (1981).

B. Test of significance for difference in mean values of tubidity, DO and Salinity in Zones I & II

Ecological factors	*Mean differences*	*S. E. of mean differences*	*'t'*	*D.F*	*t 0.05.28*	*Inference*
Tubidity	4.3467	5.55391	2.1209	28	2.048	Significant
DO	0.9540	1.11355	2.2709	28	2.048	Significant
Salinity	1.8392	1.3013	3.8199	28	2.048	Significant

Table 63 : Fish Landing in the Hooghly estuary before and after Farakka Barrage discharge

Species	*Average landing (tons)*		*Landing after the Commission of Farakka Barrage*					
	1972–73	*1974–75*	*1975-76*		*1976-77*		*1977-78*	
	Zone –I	*Zone-II*	*Zone-I*	*Zone-I*	*Zone-I*	*Zone-II*	*Zone-I*	*Zone–II*
Pama pama	7.86	13.50	11.00	21.21	8.57	11.18	1.62	5.73
Liza parsia	0.03	4.61	—	2.71	0.01	2.30	0.01	2.51
Setipina phasa & S. taty	31.53	3.17	38.44	3.29	28.65	0.59	22.75	1.48
Harpodon–nehereus	—	57.83	—	44.27	—	0.07	—	2.17
Trichiurus spp.	—	0.95	—	0.36	—	—	—	0.18
Prawns	403.64	67.81	308.99	77.45	229.31	42.00	160.00	28.52
Total Catch	1077.71	328.71	851.18	421.84	711.78	175.00	518.06	110.35

Source : Nandy, A.C., M.M. Bagchi and S. K. Majumder (1983).

Zone – I – Kalna and Barrackpore.

Zone – II – Uluberia and Nurpur.

Based on C^{14} assimilation technique the maximum and minimum computed values for energy fixation were $2.9{\times}10^4$ and $1.5{\times}10^3$ Cal $m^{-2}d^{-1}$ respectively. The mean values of C^{14} assimilation and energy fixation for the whole estuary works out to about 2.3 $\times10^6$ tonnes of Carbon and $2.25{\times}10^8$ tonnes of calories of energy respectively. Based on this, the potential yield of fish as ratio of carbon production and energy fixation is $5.5{\times}10^4$ tonnes. The mean value of actual fish catch in the estuary is about $2.4{\times}10^4$ tonnes. Efficiencies of fish : Carbon; fish to dry plankton: Fish Carbon produced and fish energy: energy trapped, works out to 1.05; 0.46; 0.11 and 1.04 per cent respectively.

Thus potential resources over the entire Hooghly estuary (including Rupnarayan river) would be of order of the 2.5 to 3 times the present yield of estuarine fishery. (Jhingran & Babulal, 1988).

II. Status of Juvenile Fishery in the Commercial Stock

It is a common feature of the riverine fishery of the Ganga river system that each year during autumn months the riverine catches consist of a sizable percentage of major carp fingerlings. This juvenile fishery is of a very short duration, lasting only for about a fortnight. Observations made on the source of these catches revealed that they comprised small creeks, shallow pools, rivulets, inundated fields during floods and sand bars cut off from the main stream. These isolated water bodies result from the receding of the flooded river during the autumn months and change in the course of the main river and some of its important tributaries like the Ghagra (Uttar Pradesh). The Son and The Kosi (Bihar) (cf. Fig. 6–8).

Surveys were made during the years 1952–54 by Jhingran and Chakraborty (1958) in a stretch of 208 Km. along the Ganga river from Mugalsarai to Patna to assess the extent of damage caused to the major carp fishery by way of large scale destruction of Juvenile fishery.

A total of 3,05,512 specimens of all three major carp species viz. *C. mrigala, C. Catla* and *L. rohita* weighing 12.7 t was estimated to have been captured in a short period of 13 days (September, 1952). Specieswise, the Catch consisted 1,59,783 specimens of mrigal, 11,10,595 of Catla and 35,134 of rohu making 52.3 per cent, 36.2 per cent and 11.5 per cent of the total juvenile catch respectively. By weight Catla being the heaviest of the three contributed 8.9 t (70.5 per cent), mrigal, 2.8 t (22.7 per cent) and rohu 1.0 t (7.4 per cent) of the total weight of the catch respectively.

Similarly in 1953, a total of 3,28,408 specimens of these three carps weighing 14.99 t was estimated to have been captured in12 days (September) from the same stretch. By November, the catch consisted of 1,66,503 (50.7 per cent) specimen of mrigal, 87,028 (26.5 per cent) specimens of Catla and 74.877 (22.80 per cent) specimens of rohu. By weight the percentage contribution of mrigal, Catla and rohu amounted to 4.41 t (29.4 per cent), 8.32 t (55.5 per cent) and 2.26 t (15.1 per cent) respectively.

Natarajan and Jhingran (1963) estimated that catch of juveniles and immature fish catch at Allahabad was 54.6 per cent of the total catch in 1960. A similar picture was also presented by Kamal (1969) .

David (1959), while reporting on mass killing of fingerling and juvenile carps on Kosi 'dhar' estimated the quantity at several tonnes per day, the quantity transported to outside markets exceeds 3.7t/day on some days. This high rate of fingerling exploitation expecially in the middle stretch of Bihar was almost certain to deprive the system of a substantial productive potential. According to Natarajan *et al.*, (1976) 40 t spawners are capable of producing 2000 m spawn.

The relatively lower rates of juvenile exploitation particularly in the upper stretch of the Ganga system correspond much higher abundance of major carps in these regions. A survey made by CIFRI (Survey Report, MSS. 1970) during August 1965 to December 1966 at some important centres like Kanpur, Varanasi, Buxar, Ballia and Patna located on the Ganga and Sadiapur (Allahabad) and Agra on the Jamuna on the status of carp Juveniles in commercial stocks, it has revealed that one year old fish were observed to enter in the commercial fishery from July and these were registered in the catch in appreciable numbers only upto December (cf. Table 64). During 1965 the main juvenile fishery lasted from August to October except at Allahabad centre that lasted upto December and to a lesser extent at Agra where it lasted till November. During 1966 the landings showed appreciable increase during the late monsoon and post–monsoon period of September to December especially during the two months, October and December. The quantum of landings and the extent of increase were especially pronounced in the case of mrigal and catla.

Exploitation of juvenile major carps was more intensive in 1966 than the previous year. A similar increasing trend in regard to Catla had been noticed during 1959 and 1960 at Allahabad (Natarajan and Jhingran, 1963). Recruitment of one year old catla mainly extended from September up to December and the I and II year groups together accounted for about 44 per cent of the two years' pooled total landings of the species. In 1960 alone the I and II year groups respectively consisted about 34 per cent and 20.64 per cent to the annual landings of the species. It is significant to note that combined contribution of the two groups in 1960 was about 54.5 per cent which was distinctly more than their contribution during the previous year *i.e.* 34.09 per cent.

In the estuary adequate data on the recruitment of juvenile fishery and prawn were not available. Karmakar *et al.*, (1987) recorded wanton destruction of early juvenile and young ones of some commercial important fishes like *Gadusia chapra, Hilsa ilisha, Glossogobius giuris, Bacha bacha, Coila sp. Liza corsula Cirrhinus reba, Lepidoce–Phalechthys gunte, Pama pama etc.* captured with the help of small sized nets, particularly bag nets and sine nets in the upper stretch 54–114 km upstream of Calcutta

Table 64 : Landings (in kg.) of one year old major carps at various Centres on rivers Ganga and Jamuna during 1965-1966

	Mrigal	Catla	Rohu	Kalbasu
August 1965				
Sadiapur	169.20	3.56	0.0	54.30
Agra	40.04	—	3.64	7.28
Kanpur	—	—	—	—
Varanasi	82.25	0.0	30.06	0.0
Buxar	0.0	0.0	0.0	0.0
Ballia	...............Data Not Received.................			
Patna	0.0	0.0	0.0	0.0
September 1965				
Sadiapur	65.32	17.16	9.52	5.52
Agra	0.18	—	0.09	0.06
Kanpur	7.31	18.49	4.45	1.13
Varanasi	8.56	0.0	14.47	0.0
Buxar	3.87	0.0	0.0	0.0
Ballia	..................Data Not Received.................			
Patna	0.0	0.0	0.0	0.0
October 1965				
Sadiapur	33.24	0.0	4.42	14.94
Agra	8.08	0.52	2.63	0.0
Kanpur	12.88	19.76	10.23	8.12
Varanasi	0.0	0.0	—	0.0
Buxar	—	0.0	0.0	—
Ballia	.....................Data Not Received.................			
Patna	—	—	—	—
November 1965				
Sadiapur	54.10	4.43	1.28	9.46
Agra	28.59	0.0	0.73	0.0
Kanpur	0.0	0.0	0.0	0.0
Varanasi	0.0	0.0	0.0	0.0
Buxar	—	0.0	0.0	—
Ballia	.....................Data Not Received.................			
Patna	"	"	"	"
December 1965				
Sadiapur	31.97	30.34	1.55	12.61
Agra	0.0	0.0	0.0	0.0
Kanpur	0.0	0.0	0.0	0.0
Varanasi	0.0	0.0	0.0	0.0
Buxar	6.50	0.0	1.93	—
Ballia	.....................Data Not Received.................			
Patna	0.0	0.0	0.0	0.0

(Contd...)

	Mrigal	Catla	Rohu	Kalbasu
January 1966				
Sadiapur	2.26	8.43	0.56	5.13
Agra	—	—	—	—
Kanpur	—	—	—	—
Varanasi	—	—	—	—
Buxar	—	—	—	—
Ballia	—	—	—	—
Patna	—	—	—	—
February 1966				
Sadiapur	20.30	9.40	0.86	8.94
Agra	0.29	—	—	—
Kanpur	—	—	—	—
Varanasi	—	—	—	—
Buxar	—	—	—	—
Ballia	—	—	—	—
Patna	—	—	—	—
March 1966				
Sadiapur	7.62	—	—	8.35
Agra	0.95	—	—	0.39
Kanpur	—	—	—	—
Varanasi	—	—	—	—
Buxar	—	—	—	—
Ballia	—	—	—	—
Patna	—	—	—	—
April 1966				
Sadiapur	12.16	—	1.20	19.76
Agra	1.96	—	—	3.52
Kanpur	—	—	—	—
Varanasi	—	—	—	—
Buxar	—	—	—	—
Ballia	—	—	—	—
Patna	—	—	—	—
May 1966				
Sadiapur	7.02	1.19	—	15.43
Agra	18.10	—	—	18.87
Kanpur	—	—	—	—
Varanasi	—	—	—	—
Buxar	—	—	—	—
Ballia	—	—	—	—
Patna	—	—	—	—
June 1966				
Sadiapur	4.16	—	0.80	7.28
Agra	—	—	—	0.84
Kanpur	—	—	—	—
Varanasi	—	—	—	—

(Contd...)

	Mrigal	Catla	Rohu	Kalbasu
Buxar	—	—	—	—
Ballia	—	—	—	—
Patna	—	—	—	—
July 1966				
Sadiapur	24.83	—	0.78	138.87
Agra	—	—	—	—
Kanpur	—	—	—	—
Varanasi	—	—	—	—
Buxar	2.01	—	—	0.33
Ballia	—	—	—	—
Patna	—	—	—	—
August 1966				
Sadiapur	32.37	8.21	—	12.38
Agra	—	—	—	—
Kanpur	—	—	—	—
Varanasi	0.0	0.0	0.0	5.58
Buxar	0.84	0.0	0.0	0.0
Ballia	0.0	0.0	0.0	0.0
Patna	0.0	0.0	0.0	0.8
September 1966				
Sadiapur	137.86	55.30	31.24	36.44
Agra	22.72	—	—	—
Kanpur	17.62	4.19	13.64	9.31
Varanasi	—	—	—	—
Buxar	5.90	3.74	—	—
Ballia	3.58	—	—	—
Patna	—	—	—	—
October 1966				
Sadiapur	267.23	329.86	39.78	20.40
Agra	4.71	1.35	0.23	0.16
Kanpur	5.92	3.05	4.29	2.00
Varanasi	—	—	—	—
Buxar	—	—	—	—
Ballia	—	—	—	—
Patna	—	—	—	—
November 1966				
Sadiapur	67.71	59.06	1.28	15.30
Agra	1.23	—	—	—
Kanpur	—	—	—	—
Varanasi	—	—	—	—
Buxar	—	—	—	—
Ballia	—	—	—	—
Patna	—	—	—	—

(Contd...)

	Mrigal	*Catla*	*Rohu*	*Kalbasu*
December 1966				
Sadiapur	381.96	—	0.41	40.63
Agra	21.76	2.07	11.17	0.34
Kanpur	48.92	141.89	37.88	2.45
Varanasi	—	—	—	—
Buxar	—	—	—	—
Ballia	—	—	—	—
Patna	—	—	—	—

Source : Fisheries of the Ganga river System. CIFRI Report (MS), 1970.

were very common. An estimated yield of 79.3 t was captured during September 1980 to November 1981. Besides, due to Farakka release, and the indiscriminate exploitation of early juveniles and young ones affected the recruitment considerably.

III. Fishseed Resources

Wide variety of species of fish of the Ganga River system that breed in different habitat are presented by Natarajan (1989).

Major carps of the Ganga breed during the monsoon. Breeding starts earlier in the Bengal basin, where the monsoon extends from April to August (David, 1959). In the Kosi River, the early breeding of major carps in May is caused by snow–melt. With the exception of Gomti, none of the tributaries in the western basin overflow their banks except with exceptionally heavy floods, Spawing is thus delayed as late as August and September in Western Uttar Pradesh. The major carps make migrations of intermediate length during the breeding period and prefer grassy substrates, highlighting the importance of flood plains and similar low lying habitat. *L. dero* and *L. dyocheilus* seek upland reaches for breeding. The mahseer (*Tor tor* and *T.putitora*) more upstream from the plains and breed in shallow areas over gravelly.Bottoms in August–September during the low–flow period. Among the large catfishes, *W. attu* breeds at the same time and in a similar habitat as the major carps. However, most catfishes, like *Mystus, Seenghala, M. aor, Eutropiichthus Vacha, Ompok bimaculatus,* and *Ompak pabda* breed between late summer and monsoon period. *Rita rita* and *Clupisoma garua* breed as early as March (Karamchandani and Motwani, 1955). Karamchandani and Motwani, (1956) report summer breeding for the catfishes *Silonia silondia, Ailia coila, Ailichthys Punctata* and Gangetic cenia. *Pangasius pangasius* which remains in deltaic stretches, seeks freshwater sites for breeding (Day, 1889), *B. bagarious* also migrates up stream to reproduce. Qasim and Qayyum (1862) reported a prolonged breeding season (June–October) for *Channa striatus* and *Channa marulius*. The grey mullet Mugi corsula and spiny eel *Rhynchbdella aculeata* also breed coincidentally with the rainy season. Swamp fishes prefer marshy habitats and breed during July–October. *Hilsa ilisha* ascends as far as 1500 km from the river mouth to Allahabad and is known to move up to Kanpur on the main Ganga and 100 km upstream in the Yamuna River in years of exceptional abundance.

Prawns, *Macrobrachium Choprai* (Tiwari) occur in the middle stretch of the Ganga, while six other species of Macrobrachium remain in the delta, making migrations between non tidal and tidal reaches. *M. rosenbergii* and *M. malcolmsonii* are economically important (Rajyalakshmi, 1980; Kurian and Sebastian, 1982).

Major carps : From time immemorial the Ganga river system is identified as source of carp seed for culture in ponds and tanks. Its tributaries the Kosi and the Son meeting the Ganga in the middle and lower stretches yield very high quality and enormous quantities of fish seed. The exploitation of spawn from this system which has been extended to many centres on many tributaries continues at a very high rate.

The Central Inland Fisheries, Research Institute, Barrackpore, since its inception in 1949 initiated investigations towards exploitation of spawn. This has gradually induced to cover up wide areas and finally in 1964 comprehensive investigations were planned for a systematic survey of selected stretches for assessing the quantity and quality of spawn yield, and to ascertain the availability and flow abundance of spawn to hydrodynamical factors.

Pre–1964 Survey

The survey conducted before 1964 was carried out in a limited scale, yet it yielded valuable basic informations. The pre–1964 survey conducted are summarised in Table 65.

Table 65 : Summerised survey of spawn yield in the Ganga River System

Year	*Stretch/centres surveyed*	*Results*	*River*	*State*
1949	(i) Dhulian, Nimtisa	Developing eggs and spawn of major carps	Ganga	West Bengal
	(ii) Confluence of R. Kosi and R. Ganga	—	Ganga	Bihar
1951	Sultanganj & Mokama Ghat	New spawn collection centre established	Ganga	Bihar
1952	Digwara, Khagaria, Kursella, Karhagola & Buxar	Spawn available	Ganga	Bihar & Uttar Pradesh
1954	Jhansi	Identified as a Natural Spawning ground	Nala of R. Betwa	Uttar Pradesh
1955	*9 Centres on R. Ganga & R. Jamuna*			
	(i) Bagpat (Meerut) and Rajghat, Narora (Bullandsahar)	Productive source of major carp seed excellent quality	Ganga & Jamuna	Uttar Pradesh
	(ii) Minor dhar of Kosi Jheels connected with R. Burhi Gandak	Spawning grounds of major carps	Ganga	Bihar
	(iii) Char lands of R. Ganga (Sakrigoli Ghat)			
1956	10 Centres of Muzaffarpur Saharanpur, Bijnaur, Dehradun	Almost devoid of major carps spawn	Ganga & Jamuna	Bihar/Uttar Pradesh
1958–59	—	Major Carp (24.3) spawn during August	Ganga	Bihar/Uttar Pradesh
	—	– do – (32.9–68.6%) during monsoon (June)	Jamuna	Uttar Pradesh
1962	3 sites of Allahabad	High quality major carp spawn	Ganga, Gomati	Uttar Pradesh
	Jaunpur Hamirpur	Totally devoid of major carp spawn	Betwa	Uttar Pradesh
1963	Jaunpur	Previous results confirmed	Gomati	Uttar Pradesh

Post–1964 Survey

The survey was done round the clock during the entire spawn availability period and was carried out in detail. Simultaneously, an assessment of quality and quantity of spawn yield, and efficiency of different shooting nets and mesh sizes were tested for their efficiency under different sets of hydrological conditions (Anon, 1965, Malhotra *et al.*, 1966 and Shetty 1967). Jhingran (1975) has provided additional information. The results of investigation carried out in Punjab, Haryana, Uttar Pradesh, Bihar and Rajasthan are indicated in Table 66.

Table 66 : Seasonal indices of spawn quality and quantity of the sites investigated during 1964–70 in the Ganga River System

Year	*River*	*Centre*	*State*	*Index of spawn quantity*	*Index of spawn quality*		*Others*
					Major carps	*Minor carps*	
1964	Yamuna	Krishnapur	U.P.	7385.0	85.30	14.60	0.10
	–	Mehewa					
		Jamunapar	"	4404.0	72.30	27.40	0.30
-R	Ramganga	Tejpur	"	3351.2	2.80	96.70	0.50
	–	Sardarnagar	"	965.6	5.90	94.00	0.70
	Tons	Baluha	"	28.4	–	–	–
	Ken	Gouribaba	"	Nil	–	–	–
1965	Yamuna	Anwara	"	3493.0	81.0	19.00	–
	–	Dumanpura	"	2200.0	35.00	24.00	41.00
	Rapti	Bansi	"	4715.0	77.00	21.60	0.70
	Son	Dhundhua	Bihar	637.0	3.5	94.90	1.60
	–	Dangwar	"	2417.0	25.20	74.80	–
	Kosikhana Dhar	Babuaghat	"	Data inconclusive			
1966	Yamuna	Majhawali	Punjab (Haryana)	784.0	18.60	80.84	0.56
	–	Mont	U.P.	17.3	26.30	72.60	1.10
	Ghaghara	Ghagraghat	"	228.4	7.40	92.60	–
	Kosi (Khau–ha Dhar)	Babuaghat	Bihar	664.0	35.1	46.50	18.30
	Badua	Mehdi	"	Data not available	26.1	73.90	–
		Jhajha					
	Burhi Gandak	Khagaria	"	–	–	65.28	34.72
1967	Yamuna	Nethla	U.P.	6006.10	7.00	86.40	6.60
	Gomati	Selempur	"	1373.60	26.40	73.60	–
1968	Banas	Tonk (Sapari)	Rajasthan	1801.00	81.50	17.70	0.80
	Banas	Negria	"	1531.50	59.40	40.40	0.20
	Ganga	Nanamau	U.P.	807.60	76.30	23.10	0.60
	Yamuna	Deolau	"	239.00	83.70	15.20	1.10
	Yamuna	Mahewapatti	"	601.00	18.20	78.80	3.80
1969	Yamuna	Mahewapatti	"	1098.00	52.20	47.40	0.40
	Son	Bahiara	Bihar	252.00	88.50	11.50	–
1970	Ganga	Ahirali	"	552.00	85.49	—	—

Source : Report on Fisheries of Ganga river system, MS. 1970 and Jhingran V.G. (1975).

The fish seed committee has estimated that in 1964 the Ganga river system contributed 89.5 per cent of the total fish seed produced in the country (Govt. of India, 1966). On an average of about 2010 million spawn from Bihar and 1200 million from West Bengal were available with a steady record of production rate, was estimated by the Govt. of India (1966). Tripathi (1976) estimated the production of spawn from of the Ganga river system at 122.2 million in U.P. in 1966 (Table 67).

Table 67 : Crap spawn production from the Ganga river system in Uttar Pradesh during 1966-1975 (Tripathi, 1976)

	Production of spawn (millions)									
River	*1966*	*1967*	*1968*	*1969*	*1970*	*1971*	*1972*	*1973*	*1974*	*1975*
Ganga	42.6	30.8	20.1	25.0	28.4	11.9	7.8	10.2	9.0	13.0
Yamuna	33.7	57.3	29.3	37.2	33.0	41.3	19.8	21.6	18.5	12.0
Gomati	7.5	37.8	28.7	60.1	51.4	30.8	4.1	11.0	70.3	6.0
Son	8.0	6.2	5.9	21.4	33.8	4.2	10.1	8.3	14.0	1.5
Rapti	5.5	4.8	7.4	9.5	7.0	1.0	3.0	1.0	1:3	3.0
Ghagra	10.1	18.6	10.5	18.8	12.9	9.9	8.9	5.0	9.7	14.4
Ramganga	14.3	12.5	19.1	7.3	2.8	0.3	1.0	0.4	1.0	0.0
Total	122.2	165.2	113.6	169.5	162.3	97.4	51.7	57.0	61.6	49.0
	Number of Centres yielding more than 2 million/annum									
Ganga	5	4	6	5	4	2	1	2	1	1
Yamuna	6	4	5	8	5	4	0	2	2	1
Gomati	1	3	2	4	4	3	0	1	3	0
Son	1	1	1	1	1	1	1	1	1	0
Rapti	1	0	1	1	1	0	0	0	0	1
Ghagra	1	4	2	2	1	1	1	1	2	2
Ramganga	2	2	2	2	1	0	0	0	0	0

Source : In Jhingran, A.G. & K.K. Gosh (1978)

Recent analyses show that spawn yield from the rivers in Uttar Pradesh has declined considerably (Table 67) while in Bihar that is of the same order and its recorded yield in 1973 was about 2037 million. The yield from West Bengal registered a declined trend with an average of about 530 million between 1971 and 1975. The best collection in recent past was in 1973 when the production was about 1375 million. Quantity and quality of spawn explored in a stretch of the R. Ganga (Bihar) during July–August are indicated in Table 68.

Hilsa seeds

Ravish Chandra *et al.*, (1990) in his recent document stated that, Hora (1938), Hora and Nair (1940), Jones and Menon (1951), Motawmi *et al.*, (1957), Pilley (1958), Ravish Chandra (1962), Ghosh and Nagpal (1968), Ravish Chandra and Mathur (1979), Ravish Chandra *et al.*, (1984) and more recently by De and Sen (1986) and Sen *et al.* (1990) basing on the observations on the availability of Ova, larvae and Juveniles in different stretches of the Ganga, have earmarked the breeding grounds, differential distribution and period of spawing of *hilsa ilisha*. The delineated stretches are indicated below. A comparative account of embroyonic and larval development of *T. illisha* as observed by different workers in the Ganga is presented in Table 69.

Table 69 : Quantity of Spawn and their quality explored in a stretch of the Ganga in Bihar during July-August 1976 along with some physico-chemical parameters.

Spawn Collection Centres	*Spawn Collection No. of bati (in 24 hrs)*	*Bati Capacity (ml)*	*Seed Composition*		*Currents Speed Km/hr.*	*Sacchi disc Transperancy (cm)*	*Temperature (°C)*	*pH*	*DO (ppm)*
			Major Carp (%)	*Minor Carp, Weed fishes (%)*					
Barauni	1-20	100	26-40	60-74	—	5.2	38.0°C	7.4	6.8
Hatidah	8-115	100	40-70	30-60	2.72	9.1	31.0°C	7.1	6.8
Monghyr	3-40	100	40-70	30-60	1.65	7.8	33.0°C	7.2	6.7
Sultanganj	2-30	100	40-70	30-60	3.89	6.5	29.5°C	6.8	6.6
Akbar Nagar	1-35	123	30-60	40-70	0.8	8.4	31.0°C	6.4	6.0
Sonbarsa	10-124	125	70-80	20-30	3.20	—	30. 0°C	6.8	6.4
Bhagalpur	4-80	100	40-60	40-60	0.70	—	31.0°C	7.0	6.3
Kahalgaon	3-70	125	40-60	40-60	—	7.0	30.0° C	6.7	7.4
Rajmahal	5-100	7.5	60-80	20-40	2.81	9.0	30.0°C	7.1	7.0
Manikchuk	75-115	75-100	60-70	30-40	1.46	10.0	29.0°C	6.8	8.0
Farakka	6-90	100-125	60-70	30-40	—	22.0	29.0°C	6.9	9.5

Source: Bilgrami K.S. & J.S. Dutta Munshi, (1979).

(i) Middle stretches of the Ganga between Allahabad and Patna

(ii) Lower stretches of the Ganga between Sultanganj and Bhagalpur

(iii) The estuary between Medgachi and Baghbazar.

Through extensive work not only on the breeding grounds of hilsa in the river Ganga have been identified but also their breeding period. The observation reveals that the spawning of fish coincides with the run of hilsa from river mouth with the onset of southwest monsoon, the commencement and duration of which varies from region to region. Two spawning runs (post monsoon–August to November with peak in October; and the second peak winter runs which is a half way through between January and March with peak in March) could be identified. Further, from the study of hilsa breeding if reveals that perhaps hilsa breeds almost round the year.

To ascertain the production potentiality of the stretches intensive investigations carried out from 1967–1972 between Allahabad and Varanasi and post Farakka during 1982–1989 the results of which are indicated in Table 70 A & B. (Before and after commission of Farakka).

Estuarine Zone

In the estuary however, available information reveals that some sporadic work on the potentiality of spawning of hilsa and their seed resources were done by various workers. Hora (1938), Hora and Nair (1940) and Jones and Menon (1951) have indicated that the region of Palta water works near Barrackpore as one of the breeding ground of *Hilsa ilisha* which subsequently Pillay and Rosa (1963) have confirmed the earlier findings and they extended the freshwater zone up to Medgachi as breeding Ground. Bhanot (1973) in the subsequent years made more detailed observation on the spawing behaviour of Hilisha (Ham) based on the observations made during 1960–65. The author recorded breeding grounds between Khusigoli and Uchitpur, a tidal freshwater zone, of the 112 km. stretch between Nawabganj and Ghusighata. (The spawning took place about a fortnight earlier to the upper non–tidal region). Among the two main breeding seasons, the major one during monsoon with a long duration, while the other comparatively shorter one, which was during winter. Breeding occurred round the year with high collection obtained from Khusigolia and Medgachi. Abundance of Hilsa larvae during 1960–65 are indicated in Table 71.

Table 69 : Comparative account of embryonic and larval development of T. ilisha as observed by different workers in the Ganga

		Jones & Menon (1951)	*Motwani, Jhingran & Karamchandani (1957)*	*De & Sen (1986)*
1		*2*	*3*	*4*
1.	Locality	Hooghly river, about 190 km from sea within tidal limits	Ganga river over 1200 km from sea	Ganga river near Farakka barrage over 600 km from sea
2.	Season	February, 1950	July, 1954	October, 1985
3.	Material obtained for the study	Developing eggs and larvae from natural collection	Developing eggs and larvae from natural collection	Fertilized eggs by artificial fecundation
4.	Size of mature egg	0.70–0.75 mm	Not reported	0.76–0.87 mm
5.	Nature of egg membrane	Double layered in preserved eggs	Double layered	Single layered
6.	Yolk sac	Segmented	Segmented	Segmented
7.	Optic vesicle	Developed during embryonic stage after 11-13 hrs of fertilization	Not reported	Developed during embryonic stage after about 10 hr of fertilization
8.	Auditory capsules	Developed during embryonic stage, 19 hr after fertilization	Developed 6 to 9 hr before hatching	Developed during embryonic stage after about 13 hrs of fertilization
9.	Twitching movement	After 19-23 hr of fertilization	Not reported	After 16 hr of fertilization
10.	Incubation time	25-28 hr at 23°C	About 24-28 hr.	18-21 hr at 24°-29°C
11.	Heart	Pulsation observed 4-7 hr before hatching	Pulsation observed 3-5 hr before hatching	Pulsation of heart observed in one day old larva
12.	Size of newly hatched larva	2.30 mm	2.50-2.55 mm	2.37-2.44 mm
13.	Mouth	Formed on the 3rd day of hatching, that is, two day old larva.	Not reported	Mouth opening formed when the larva was 3 day old.
14.	Demarcation of vent or anal opening	Observed in newly hatched larva	Observed before hatching	Prominently visible in the 2 day old larva
15.	Pigmentation	Appeared on the yolk sac 12 hr old larva	Appeared just under head and yolk in the newly hatched larva	Few small black melanophores appeared for the first time in the caudal fin of 2 day old larva and subsequently of the ventral side of yolk mass of 3 day old larva
16.	Myotome	More than 40 in newly hatched larva	49 in newly hatched larva	Number of myotomes varied from 40 to 54 in newly hatched larva to five day old larva respectively.

Source : Modified from De, D. K. and P R. Sen (1986).

Table 70 A : Showing Catch/hour of Hilsa larvae during 1967–1972 (Revish Chandra *et al.*, (1990)—Before commission of Farakka

Centre	*1967*	*1968*	*1969*	*1970*	*1971*	*1972*	*Average*
Mahewapatti	507	*NS	13127	96	546	375	2442
Sirsa	5	*NS	NS	21	12	44	14
Vindhyachal	219	NS	959	10	700	8262	1704
Samneyghat	9425	19,855	2717	167	7906	2966	7173
Agiabirghat	NS	4,094	1268	NS	NS	–	894
Sujabad	NS	1,147	4821	51	2637	4279	2156

* NS—Not sampled.

Table 70 B : Showing Catch/hour of Hilsa larvae during 1982–1989 Post Farakka release

State	*Centre*	*1982*	*1983*	*1984*	*1985*	*1986*	*1987*	*1988*	*1989*
Uttar Pradesh	Madhuka	Neg.	Nil	2	1	Nil	250	neg	210
	Sirsa	6	2	2	neg.	20	27	2	1
	Manda	1	neg.	1	nil	53	1	–	–
Bihar	Buxar	–	26	1813	–	780	×	×	–

×– Centre closed

Table 71 : Abundance of *H. ilisha* larvae from the Hooghly estuary (net/hr)

	1960	*1961*	*1962*	*1963*	*1964*	*1965*
Jan.	0	0	0	90.5	54	10.2
Feb	0	0	0	1444.3	127.5	26
March	0	0	0	111	261.6	45.2
April	0	0	0	36	55	2.0
May	0	0	32	27	23.7	34.5
June	7.0	30	92	112	110.7	2
July	155.3	33	197.3	166.3	0	32.8
August	773.0	83	49.3	27.5	0	2
Sept.	15.0	43	108	40	0	3
Oct.	487.0	0	44.5	180	0	10.5
Nov.	103.0	0	26.5	182.5	0	20.5
Dece.	1.3	0	16.0	83.5	0	2

Source : Bhanot, K. K. (1973).

Further, to the above exploitation De and Sehgal (1989) in the recent past (after Farakka release 1975) tried to make some assessment on the recruitment of hilsa seed (*Tenualosa ilisha* (Ham) at 5 centres Nabadwip, Medgachi, Nawabganj, Uluberia and Diamond Harbour which included non-tidal and tidal zone. During 1987 and 1988, and 1989 they covered a stretch of about 225 km. Studies were restricted from August to May only, since the post larvae, frg and fingerlings of hilsa were available in the freshwater stretch of the Hooghly estuary from October to May with a peak during October and December. Diamond Harbour and Nabadwip are covering a distance of 75 and 300 km. from sea face.

Table 72 : Availability of hilsa seed in a 225 Km. stretch of the Hooghly estuary during 1987–1989

Year		*August*	*Sept.*	*Oct.*	*Nov.*	*Dec.*	*Jan.*	*Feb.*	*Mar.*	*Apl.*	*May*
Max. no. of seed per/net/hr.	1987–1988	–	–	37	467	16	36	46	27	71	–
	1988–1991	–	16	130	149	39	45	164	39	93	–
Length range (mm) of seed	1987–1989	–	12–15	9–22	9–25	18–25	8–20	18–20	5–15	10–24	–

Source : De, D.K. and B.N. Sehgal, (1989).

The peak spawning of hilsa deserved by various workers (Sujan Singhani, 1957; Pilley, 1958; Ravish Chandra, 1962) is in conformity with the above findings. The species spawns in the entire freshwater zone of the estuary, while earlier observation inferred that the lower and upper limits of the spawning grounds of hilsa were restricted between Baghbazar (Calcutta) to Medgachi a distance of about 155 and 251 Km. respectively from the sea face. The probable reason for the extension of spawning grounds as observed and opioned by De and Sehgal (1989) might be due to the increased post Farakka discharge of freshwater into the Hooghly estuary. The barrage has changed the entire ecology significantly by reducing the salinity and thus converting the gradient zone into almost freshwater zone which might be conducive for spawning of hilsa.

IV. Prospects of beel/mun fishery

As already stated that shifting of river course from the main river bed due to heavy silting especially around the deltaic region of the rivers result in a number of disconnected water spread areas both seasonal and perennial in character. These are known as mans, chars, beels etc. Such phenomenon are prevalent in Bihar followed by Uttar Pradesh and a few in West Bengal.

These areas are ideal places for natural recruitment. Jhingran and Ghosh (1978) reports that a very high larval mortality in carps is most likely since the spawn freely drifts at the rate of about 100–150 cm/sec. in flooded rivers once they flow out from the shallow impounded breeding ground. The only larval masses that gets locked up in the natural nurseries, develop to recruit to sustain the stocks, rest flow down the river where a high percentage ultimately perish.

Since the inundation of shallow areas are recurring affairs during the monsoon, no systematic, fish culture of commercially important species is possible. The locked up larvae grow fast due to the fertile character of the eco–system into juvenile stage in a few months and from January onwards they are intensively netted. David (1959) reporting on mass killing of fingerling and juvenile carps on Kosi Dhar (Bihar) estimated the quantity at several tonnes/day. The quantity transported outside markets exceeds 3.7 t/d on some days. Such high rate of fingerlings exploitation especially in the middle stretches in Bihar is almost certain to deprive the system of a substantial reproductive potential.

Shetty and Malhotra (1983) made an extensive survey of Champaran, Muzaffarpur and some other districts of Bihar which reveals valuable information on the status of carp Juvenile fishery. Available figures relating to estimated total fish catch from 48 mans of various sizes (4,405 ha); 43, extending over 360 Sq. Km.) of the Champaran district indicate that catches had progressively declined from 1965 to 1968 are as follow :

Year	Production	Remarks
1965–66	2883.30 t	Heavy infestation of water hyacinth
1966–67	2672.30 t	
1967–68	2537.40 t	

The fall in the catch is due to rising of height in recent years of Gandak embankments and the construction of new embankments which has evidently cut down the amount of water flooding over the chors and mans.

Investigations carried out subsequently in some lakes of Muzaffarpur district more scientifically revealed valuable information. The study pertains to morphometry of lake, commercial catch, physical character together with biotic and abiotic character of the ecosystem. The study was further modified in the form of penculture experiment in the same waterbody to achieve maximum benefit of the water and soil. (Bhowmik, 1988; Rai and Singh, 1988). The results are presented in Tables 73 A & B.

Table 73 A shows that soil and water quality are highly fertile. This is evidenced from their biotic and abiotic character. In pen–experiments (Table 73.B) the production on different occasions come to 2.53 t/ha/5 months and 4.0 t/ha/6 months and also 5.5 t/ha in another pen, while in commercial catch (Table 73 A) the annual landings in the first two cases are 1.2 and 6.3t from an area of 60 ha (CICFRI, 1989–90). This is undoubtedly a poor catch. Further, composition of the catch at times consist of some non–commercial species and does not follow the same pattern every year. In some year, the catch is high while in the subsequent year, it registered low catch. Some time the catch is dominated by major carps while in the following year minor carp and non–commercial species dominates (CICFRI 1988–89 and 1989–90). The weed infested water sheets are not conducive for proper growth of spawn/fry/fingerlings. The swampy areas being shallow, fingerlings and early stage juveniles (immature) are largely captured easily by intensive netting operations. Mass killing of juvenile carps in Kosi Dhar (Bihar) estimated the quantity at several tonnes/day (David, 1959).

Thus pen culture in ox–bow lakes (beel/man) are not only most profitable but can also be used as nurseries where seeds of small fish of commercial important species can be reared upto immature juvenile stage safely in 5–6 months under a highly fertilized ecological condition.

V. Fish Culture in fluviatile waters

Apart from drawing water for *mans*, chars or bundhs, the vast network of irrigation Canals may also be made use of for stocking fish fingerlings, for enhancing the fish production in the area. In some countries of the Far East, very high fish production have been obtained in running water fish culture system and cage culture. While most of the canals are not deep enough for cage-culture, it is worthwhile exploring the possibility of establishing running water culture in such areas making use of the constant supply of running water from the canals.

Unfortunately, so far, no such work on fish culture in running waters has been initiated in the country excepting in a very limited way culture of fingerlings of *Acrossochilus hexagonolepis* commonly known as chocolate mahseers in running water canals have been initiated in the Barapani farm of ICAR Research Complex, Shillong (Mandal and Nayak, 1988). Fish grew in 12 months at a density of 5,10,15,20 and 30 Nos/m^3 water area attained weight 425 ± 20 ; 402 ± 25 ; 388 ±18 ; 350 ± 22 and 310 ± 15 g respectively when survival percentage recorded between 91–97. These results appears to be encouraging. Hence, culture at other places in running waters of the Gangetic plains may be taken up.

Table 73 A : Commercial Catch in a lake with some important Biotic and Abiotic Chartacters during 1988-89 and 1989-90

Morphometry of the lake	*Species Caught*	*Total Catch*	*Percentage*	*Physical character*	*Biotic Character*	*Abiotic Character*
Muktapur (Bihar) 60 ha.	Minor Carp and smaller fishes	....	82.71	Infested with macrophytes	Bottom biota Dominated by gastropods (96.7%)	pH 7.8 – 8.15 (7.97) DO. (ppm) : 1.7 – 9.9 (4.82) T.D.S. (ppm) : 87 – 122 (104)
Mean depth : (0.56-5.28m)	Major Carp	1205.46 Kg.	9.96	Mainly Submerged vegetation	Tubex larvae- 2 - 3%	Sp. Cond. (μ. mhos) 181 – 249 (212)
	Large Cat fishes	....	7.33			
	Channa marulius					
	Nandas nandas	....	48.7			*Primary Productivity* : mg C/m^{-2}/d^{-1} (C^{14} method) 134.5 – 675 (387.93) *Detritus load* : 232.8 – 417.22 gm^{-2} (284 gm^{-2})

Source : Annual Report 1988–1989 _ C I C F R I, Barrackpore

—do—	Major and Minor Carps, Catfish, feather backs, Murrels, Miscellaneous including shrimps	6309.75 Kg (From 19th January to 18th July 1989)		—do—		Water : pH 7.5 – 8.1 (7.9) D.O. (ppm) 4.6 – 9.0 (5.9) T.D.S. (ppm) 65 – 88 (81.5) Sp. Conductivity 134 – 183 (168.5) (μ. mhos)
	Major carps	....	52.1			Soil : pH. 6.8 – 7.4 (7.1)
	Minor carps	....	12.3			Org. C (%) 2.8 – 9.0 (6.5)
	Cat fishes	....	5.1			Av. P (ppm) 40 – 185 (112)
	Feather backs	....	3.1			Av. N (ppm) 605 – 985 (780)
	Miscellaneous with Shrimps		23.8			*Primary Productivity* (mg C/m^{-2}/d^{-1}) 57.26 – 384.2 (236.3) *Energy transformation* : 562 – 3773 Cal m^{-2}d^{-1} (2321)

Source : Annual Report 1988-1989 : CICFRI, Barrackpore

Table 73 B : Results of Pen Culture Experiment

Name and location	*Size or area*	*Species stocked*	*Density of stocking*	*Ratio*	*Observation Period*	*Total Production*	*Weight range* Initial	Final
Karti Ox-Bow Lake Muzaffarpur (Bihar)	40m × 20m (0.08 ha)	Catla — 1100 Rohu — 800 Mrigal — 750		C: R : M 7 : 12 : 6	6 months	317 Kg/6 months i.e. 4t/ha/6 months	C : 177g. R : 75g M : 102g.	— 1100 g — 800 g — 750g.
Source : S. R. Rai and R. C. Singh (1988).								
Muktapur (Bihar)	(0.08 ha)	Catla —80 – 165 mm (122.5) (15 – 19 g)m Mrigal —60 – 122 mm (91.0) (10-15 gm) Rohu —90-120 (105) (17 – 20 gm)	250 300 250	C : R : M 1 : 1 : 1.5	19 January-18 June (5 months)	203.5 Kg. i.e. 2.53t/ha/		
Source : CICFRI, 1989 – 1990, Barrackpur.								
Kanti Mann (*Muzaffarpur city*)	100 Sqm.	*Fingerlings* *Catla* *16 Fingerlings /Sq.m.* Rohu Mrigal			*133 days*	*Net Production :* *5545.5 Kg/ja.* i.e. 5.5t./ha.	*Initial Wt.* C : 26.0 R : 8.5 M : 8.0	*Final Wt.* — 90.0 — 37.2 — 26.5

Source : Bhowmik, M. L.(1988)

VI. The Fishing gears and Their Mode of Operation

Considerable effort has been made by various workers highlighting the informations on various types of fishing gears used for exploring the fishery resources in the Ganga river system (Freshwater and estuary) during different seasons. (Anon, 1949); CIFRI Report MS, (1970); De, (1910); Dutta, (1973); Faruqui & Sahay, (1943); Anon, (1969); Jab & Pantulu, (1953); Naidu, (1938); Pillay & Ghosh, (1962); Saxena, (1966); Saxena and Ravish Chandra, (1968) and Saxena, (1991). Most of the workers have described the important gears and their modus operandi. But the constructional design of the commercial fishing gear, their operation, seasons and nature of catch are however discussed by Saxena (1966) (in a section of middle stretch of the Ganga). Such fishing gear are also commonly used in the entire middle stretch of the Ganga. CIFRI (1970) covered the larger section of the Ganga, Bullandshar (Uttar Pradesh) to Murshidabad, (West Bengal) and Agra to Allahabad (Uttar Pradesh) of the river Jamuna while Dutta (1973) discussed the Gill nets used in the Hooghly - Matla estuarine system. Dutta *et al.*, (1973) discussed the principal gears, area and seasons of operation and the main species caught in the estuarine zone. This was followed by De (1980) who enlighted with further information on fishing gear.

Classified fishing gear mainly used in the Freshwater zone of the Ganga river system with a comparative account of major gill net are indicated in Appendix IA, B & C while that for estuarine zone. in Appendix II A, B & III.

Saxena (1991) observed that the steady state of ecosystem in the Ganga river basin (middle stretch) till the 20th century, experienced a serious setback when human interference started exerting their influence of the river courses. The unabated derangement of the ecosystem for the last four decades has exerted tremendous strain on aquatic organisms in general and fish in particular. As a result, the fishing practices followed till yesteryears also witnessed a sudden changes to suit the necessities and requirements in the changed milieu.

The riverine fishing gears being low energy fishing devices under the stresses, experienced by the fishing community, was subsequently modified and accordingly changed. Those craft and tackles which involved lower energy in operation with sufficiently better output have been retained and those which were on the higher side of the scale discarded and abandoned.

From an assessment Saxena (1991) has shown that dragnet maintained the status of principal gear till the end of 1960s, when 66.6 per cent of the catches were drawn from this fishing, were slowly withdrawn. Gill nets which contributed in a minor contribution of 2.6 per cent gradually took over and have taken the position of principal gear within the last three decades. This is undoubtedly an important finding.

Saxena (1991) has further tried to correlate the constructional design of certain nets to meet the needs of a particular environment and as such have a localised use appropriate to the habital.

As for example, Gochail—a gill-net with foot rope which operates efficiently only in fast currents, is also well suited to the Ganga and is therefore, operated mostly above the confluence in the Ganga. Similarly, Tiar—a gill net having no foot rope is operated in Jamuna where water remains usually calm. At places Jamuna provides very large deep pools where a net with a considerable depth is effective and hence the use of Karia jal in Jamuna usually operated. Kamel is operated in Jamuna just above the confluence where *Hilsa* and major carps ascending from below the confluence are caught. Bandal is largely deployed below the confluence as it requires 3–4 feet of water over a considerable stretch with fast current. Some times it is also operated above the confluence where required depth over a wide stretch is available.

The operation of various nets in different seasons has taken advantage of the seasonal behaviour of different commercial species of fish. The main fishery of Allahabad comprises of major and minor carps, catfishes and *H. ilisha,* the yield of such species fluctuating from year to year. Major carps are caught mainly with different sized drag nets. Kuriar and Kamel are operated after winter of till the outset of monsoon and mostly big sized major carps are caught. From October to January Kamel is used to capture *H. ilisha* which ascends in large numbers. When the flood water subsides and water level is considerably lowered, the operation of Bandal starts mainly to catch *H. ilisha*. Its operations continues upto the end of May. *Mystus aor* and *M. seenghala* of big sizes are captured in large numbers by Gopal jal and Chaundhi besides other drag nets. During February to June *M. aor* and *M. seenghala* breed in shallow waters of river pockets and are caught by Gopal jal. The operation of major nets is practically suspended during the flooded months due to excessively strong current. At that time small nets like Dodandi, Jali *etc.* are put into operation and juveniles of *Mystus* and small sized fish of various other species and prawns are caught.

It will be seen from the account given above that a large number of fishing gears are used in this stretch, some selective to size and species and others are somewhat, non–selective. Before the catch data can be used for population studies, it is essential to define the size selection characteristics of different gears as very few gears are used to capture a particular species throughout their life span. Drag nets, for example have a lower limit to the length range of fish caught below which fish has a chance to escape through the mesh, where as gill nets provide an upper limit as well as a lower limit to the probability of capture. The possible bias due to gear selection is reflected on the length measurements of fish taken at assembly centres which should be duly considered and accounted for before studies on population, age and growth are taken up.

On the other hand Dutta (1973) has simply described some important gill nets operated in the Hooghly estuary during different seasons with catch, without mentioning the importance of the environmental condition like different tidal phases, transitional periods between tides, current velocity, depth etc. on the use of gears. Further, impact of Farakka barrage discharge on the gear technology is yet to be highlighted.

14

Tributaries—Their Pollution Problems and Impact on Biomass

R. Jamuna

The Jamuna emerges from the Mussoorie hills into the plains where it flows taking a broad curve near Delhi, Mathura and Agra to join the Ganga at Allahabad (Prayag). Its tributaries in the plains are the Chambal and Sindh which joins it below Etawa, the Betwa at Hamirpur and the Kim above Allahabad. Its total length up to Allahabad (Prayag) where it meets the Ganga is about 1,380 km (Fig. 1).

Major share of water flowing through the Jamuna from Haryana is reduced considerably as it passes through a long dry area from Munakh escape to Delhi and is stored in the Wazirabad Dam which is around 150—200 cusecs. The water is released as per the requirement of greater Delhi. It is thus clear that very little water spills in the main flow of the Jamuna. The river Jamuna is practically kept alive by the discharged water from 17 nallas located around Delhi. Table 10, indicates that in the Jamuna upstream of Delhi, the winters are characterised by low flow, while in the Down stream of Delhi the minimum flow occurs in summer and the river bed gets almost exposed. The river Jamuna is practically fed by metropolitan wastes through 17 drains from Najafgarh nalla to Tuglakabad nalla and as a result more than 20 km has been converted into a septic zone where dilution ratio is 1 : 1.2 only . It is thus evident that such low flow can only help high silt precipitation annually in the river bed from its catchment area 366223 km^2.

Fortunately, most recently an agreement on the distribution of water of the Jamuna has been signed by five river basin states—Haryana, Rajasthan, Uttar Pradesh, Himachal Pradesh and Delhi (The Statesman, Calcutta May, 13,1994) and accordingly Delhi is supposed to get 0.724 bem. of water, but how far such volume of water will help the situation is yet to be seen.

No systematic qualitative and quantitative studies on plankton and physico-chemical parameters were done on the river except some stray work at different stretches were done during 1954-55 (Chakrabarty *et al.* (1959), during 1958-59 by Ray *et al.* (1966) in a few km above the confluence of the Ganga and the Jamuna. This was subsequently followed by the workers of CICFRI, Barrackpore during 1988 on evaluation of fish community structure in the context of environmental modifications in the Jamuna. (CICFRI, Barrackpore Report, 1988-89) in a longer stretch at Agra, Mathura and Etawa.

Pollution studies however, were made by George *et al.* (1965) and Pillai and Agrawal (1979), but was restricted around Delhi only. It was only in the recent past that the pollution study in respect of heavy metal pollution was initiated in the entire river up to Allahabad by Jhingran and Joshi (1987) with a greater details.

Ranges of physico-chemical conditions recorded by Chakrabarty *et al.* (1959) and Ray *et al.* (1966) were 17.7 – 32°C for temp; 7.8-8.5 for pH, 3.5–10 for DO, 116–323 ppm for alkalinity and nitrate, phosphate and silicate generally within 0.1 ppm with exception of nitrate which increased some what during monsoon. CICFRI (1988-89) subsequently recorded 15–31°C for temp, 7.15-8.2 for pH, 10.0-32.5 cm for transparency, 1.2-6.8 mg/1 for DO, 7.3-8.4 mg/1 alkalinity, 124-310 mg/1 chloride, 1.8-4.5 mg/1 hardness, 6.8-14.8 mg/1 BOD, 124–321 Nicro mhos per cm for SB conductivity. Total dissolved solids, 62-161 and silicate, 8-11.5 mg/1. The gross and net oxygen production was respectively 25.94-81.58 and 10.43-29.09 mg C m^{-3}/h^{-1}.

Phytoplankton mainly dominated with Diatoms being maximum at all the three places, 66-100 per cent of the total population at Agra, 33-100 per cent at Mathura and 50-100 per cent at Etawah. Main Diatoms were Navicula and Synedra at Agra and Etawah. Mathura, was represented by Navicula, Cymbella and Eunotia. Zooplankton were encountered only in the month of December at Agra, Mathura and Etawah (42 per cent, 33 per cent and 50 per cent respectively of total plankton population). Dominant forms were *Difflugia* and *Areclla* sp. No benthic fauna were encountered above the out fall. Monthly variations of phytoplankton in three sectors of the Jamuna (Agra, Mathura and Etawah) are presented in Fig. 39.

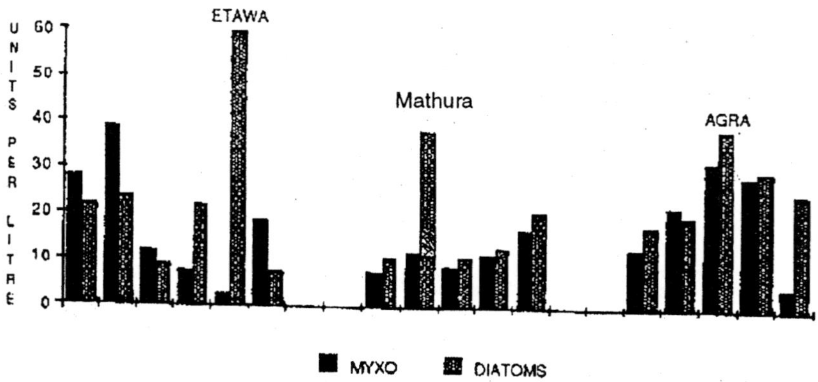

Fig. 39 : Monthly variations of Phytoplankton in three sectors of the River Yamuna.
Source : Annual Report, CICFRI, Barrackpore (1988-89).

Earlier during 1958-59 (Ray *et al.* 1966) recorded percentage composition in the Jamuna as Bacillariophyceae, 52.2, Chlorophyceae 17.09, Myxophyceae, 26.7 and Dinophyceae, 0.2 per cent.

Protozoans were more abundant and were present almost throughout the year. They even dominated the Zooplankton during the monsoon months. The most common protozoans were *Arcella, Difllugia, Actinoblus, Nebalia, Actinophrys* and *styloninechia*. (Ray *et al.* 1966).

Burthic biota Polychaet works, Chironomid larvae, molluscs and amphipods constituted the main components and their percentage abundance is depicted in Table 74.

Table 74 : Percentage abundance of major groups of benthic fauna in the Jamuna (u/m) during 1958-59

Stretch	*Chironomial larnae*	*Molluscs*	*Polychact worms*	*Amphipods*	*Other organisation*
19 km above confluence	48.27-79.20 (63.73)	1.88-20.30 (15.59)	3.35-20.30 (11.82)	2.30-19.59 (10.94)	0.80-8.83 (4.81)

Source : Ray *et al.* (1966).

Table 75 : Characteristics of sewage wastes disposed into the river Jamuna (1988-89) along with the conditions above and below the outfall

Parameters	*Raw Sewage*	*Above Outfall*	*Outfall*	*Below Outfall*
1	*2*	*3*	*4*	*5*
Temperature °C	15.1 – 30.3	15.1 – 30.6	15.0 – 30.5	15.0 – 30.5
	(15.0 – 30.5)	(15.0 – 31.0)	(15.0 – 30.5)	(15.0 – 30.5)
pH	7.0 – 8.48	7.1 – 8.20	7.06 – 7.79	7.05 – 8.4
	(7.0 – 8.85)	(7.0 – 8.4)	(7.0 – 8.0)	(7.0 – 9.2)
Alkalinity	436 – 666	128.6 – 240	246 – 416	236 – 543
(ppm)	(340 – 750)	(124 – 310)	(210 – 416)	(218 – 642)
Dissolved	2.6 - 5.1	6.8 – 8.4	5.66 – 7.86	5.06 – 7.48
Oxygen (ppm)	(0.5 – 6.8)	(6.4 – 8.6)	(5.0 – 9.2)	(4.8 – 8.84)
Sp. conductivity	562 – 1277	187 – 256	395 – 796	390 – 805
(micro-ohms/Cm.)	(453 – 1410)	(142 – 356)	(315 – 918)	(310 – 1052)
Transparency	1.2 – 13.4	11.3 – 32.1	11.8 – 23.0	9.3 – 20
(cm)	(0.5 – 18.5)	(10.0 – 32.5)	(10.5 – 32.5)	(4.5 – 28)
Hardness (mg 1^{-1})	63 – 76	62 – 71	64 – 74	65 – 78
	(60 – 78)	(60 – 78)	(62 – 78)	(62 – 78)
TDS (mg 1^{-1})	281 – 614	94 – 128	198 – 399	195 – 402
	(227 – 708)	(71 – 179)	(158 – 459)	(156 – 527)
CO_2 (mg 1^{-1})	2.5 – 16.3	1.4 – 6.5	10.3 – 16.26	8.1 – 14.8
	(1.5 – 28.5)	(1.2 – 9.5)	(5.5 – 23.5)	(4.8 – 18)
NH_3 (mg 1^{-1})	1.5 – 13.5)	—	0.9 – 13.2	1.26 – 8.8)
	(0.5 – 21.5)		(0.5 – 17.5)	(8.6 – 15.1)
Chloride (mg 1^{-1})	6.3 – 17.8	1.7 – 4.3	8.7 – 15	8.6 – 15.1
	(10.0 – 18.5)	(1.5 – 4.8)	(6.8 – 17)	(7.5 – 16.5)
Silicate (mg 1^{-1})	5.2 – 15	8.0 – 10.6	9.5 – 15.6	9.1 – 14.8
	(1.8 - 18.5)	(8 – 11.5)	(8.5 – 16.5)	(8.0 – 17.5)
BOD_5 (20°C)	44 – 268	7.0 – 12.0	32.6 – 44.6	46.6 – 123
	(52 – 171)	(6.8 – 14.8)	(18 – 196)	(31 – 184)
Gross Carbon	68.68 – 317.22	33.68 – 71.85	—	—
(mg/cm^{-3} hr^{-1})	(64.50 – 416.66)	(25.94 – 81.58)	(61.51 – 219.54)	
Net Carbon	40.44 – 85.2	11.19 – 29.67	—	—
(mg/c m^{-3} hr^{-1})	(28.24 – 91.56)	(10.43 – 30.83)	(19.59 – 67.05)	

NB : Values indicated above are the averages of low and high values of Raw sewage disposed into the river Jamuna. Above outfall, outfall and below outfall values are also shown in the same manner. Values in the parenthesis are the extreme ranges.

Source : Modified from Annual Report CICFRI, Barrackpore 1988-89.

Chironomial larvae and polychaeta worms formed the dominant groups during the major part of the year. *Corbicula, Occidens, Melania tuberculata* and *Viviparus bengalensis* were the main molluscs forms, while the amphipods were represented by a single species *Gammarus pulux.* The maximum abundance was only 84 units/m^2 at 19 km point and 824 units/m^2 at 3 km point above confluence.

Rate of Energy transformation and photosynthetic efficiency of the Jamuna during 1988-89 and 1989-90 are shown in Table 42.

To understand the impact of organic waste released into the Jamuna, raw sewage was also characterised during 1988-89 and are presented in Table. 75.

Pollution Load

George *et al.* (1965) reported that the DDT manufacturing unit discharging untreated waters into the Jamuna through Najafgarh Nalla which is highly toxic and are responsible for fish kills to those migrating during early monsoon. The effluents were characterised by the ranges; pH, 0.1-1.2; chlorides 4-37 mg/1; total solids, 4710-6560; dissolved solids, 4700-6560; total volatiles, 2280-3150; sulphate, 3650-141000; chlorides, 7600-18450; COD, 1350-62510; iron, 14-225; chromium (cr), nd-3.9; Total DDT, 9-290 and chloral hydrate, 440-5230 mg 1^{-1}. Findings after a series of bio-assay tests revealed that the waste required to be diluted 8000 times for the survival of fish.

Pillai and Agrawal (1979), reported different types of isomers of DDT and BHC which were detected in the sediment of the river Jamuna near Delhi, their concentrations being 11-21 ng/g in the former and 21-412 ng/g in the latter. In water DDT have been reported to be 0.60 - 3.42 ppb and in fish, 0.06-36.9ppb and are much higher compared to lidal stretch of the Ganga. The biomagnification in fish in this stretch was of an alarmingly high order (61,500 times).

According to Jhingran (1988) the Jamuna receives total pollution load in terms of 3000 kg/d suspended solids; 1800 kg/d oil and grease; 100kg/d acid; 180 kg/d heavy metals; 0.432 MLD of DDT from DDT factory around Delhi. Besides oil rearing wastes at Mathura and effluent from tannery at Agra are also released.

Jhingran and Joshi (1987) have revealed some valuable information on metals. The values recorded were higher in the unfiltered water (dry wt) and highest accumulation of metals were estimated in the sediment (Fig. 40). The authors have further indicated accumulation of heavy metals in fishes and molluscs collected from various centres (Fig. 41 & 42). Metal contents expressed in µg/1 (wet wt) in the Jamuna were in the ranges stated in Table 76.

Table 76 : Metal content µg/l in the Jamuna

Metal	*Filtered water*	*Unfiltered water*	*Sediments*	
1	*2*	*3*	*4*	
Zinc	22.0 – 54.76	34.02 – 82.46	44.33 – 284.8	: City side
			43.04 – 108.45	: Opp. bank
Copper	2.34 – 18.03	5.38 – 38.13	12.57 – 81.35	: City side
			10.57 – 36.93	: Opp. bank
Chromium	ND – 1.32	ND – 2.32	3.54 – 99.45	: City side
			3.09 – 34.24	: Opp. bank
Cadmium	ND – 0.38	ND – 1.51	ND – 3.85	: City side
			ND – 0.49	: Opp. bank
Lead	0.82 – 6.90	0.96 – 12.85	20.48 – 106.12	: City side
			16.04 – 151.1	: Opp. bank
Mercury	ND – 0.34	ND – 0.34	0.06 – 1.05	: City side
			0.064 – 0.442	: Opp. bank

Source : Modified from Jhingran, Arun G. and H. C. Joshi (1987).

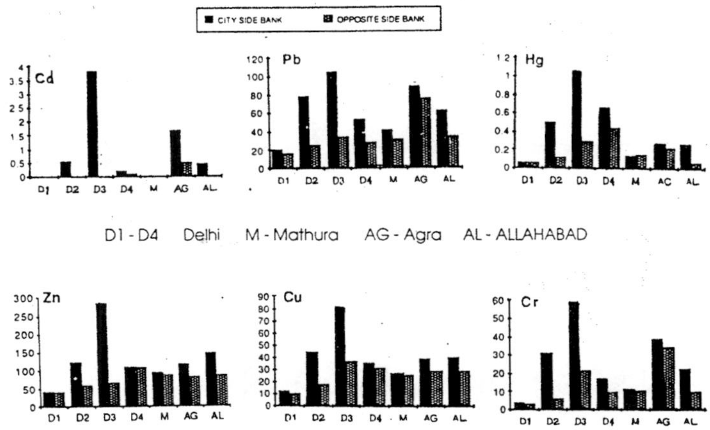

Fig. 40 : Heavy metals in sediments in the River Yamuna (μg g^{-1} dry wt.)
Source : Jhingran, Arun G. and H.C. Joshi (1987).

Maximum metal concentration in the water and sediments were found at Delhi (ITO), followed by Agra, Mathura and Allahabad. The metal levels at all the centres were considerably higher than the unpolluted stretch. Further, metal enrichment in sidement on the city side bank was significantly higher than the opposite bank (Table 76).

Sharma *et al.* (1981) have reported 0.1296 – 0.7512 ng/l^{-1} of copper in one of the drains of Agra. Likewise, high values of zinc, copper and lead have been reported for Najafgarh and Rajghat drains at Delhi (Dakshini and Soni, 1979). The values for zinc and Calcium in unfiltered water at Mathura were considerably high as compared to those reported by Handa *et al.* (1985). Mercury was below the detection levels at most of the stations.

In fish, zinc concentration ranged, 6.16 - 36.3 ; copper, 0.41 - 3.57; chromium, nd - 1.25; Cd, nd - 0.188; lead, 0.46 - 1.56 and mercury, nd - 1.38 (Wet wt) while in molluses the rangcs of concentration were, 28.1 - 193; Copper, 2.3 – 1.88; Chromium, nd – 6.72, Cadmium, nd – 1.8; lead, 7.3 – 38.6 and mercury, nd – 0.62 μg/g^{-1} (Wet wt) (Figs. 41 & 42).

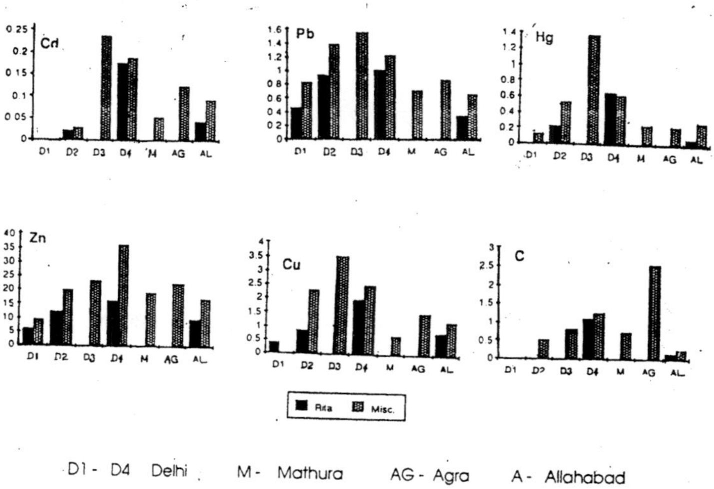

Fig. 41 : Heavy metals in fish in the River Yamuna (μg^{-1} wet wt.)
Source : Jhingran, Arun G. and H.C. Joshi (1987).

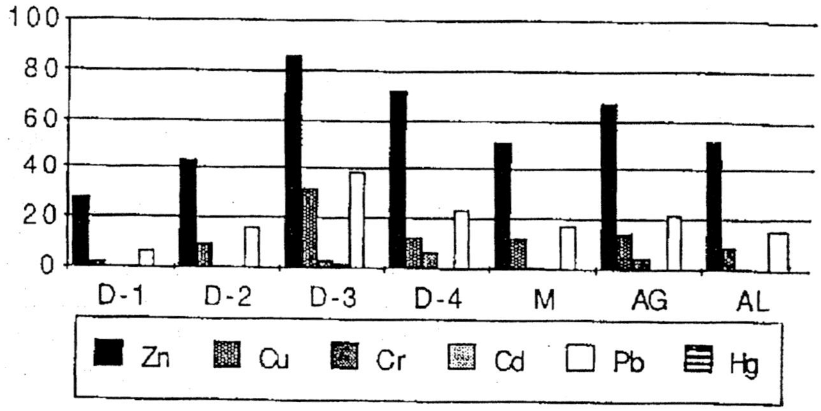

Fig. 42 : Heavy metal concentrations in molluscs at selected centres of river Yamuna (µg/g wet wt). *Source* : Jhingran, Arun G. and H.C. Joshi (1987).

The mercury content in the trash fishes collected around Delhi (Station 3) exceeded the safe limits (FAO, 1983). According to Mc Neill *et al.* (1979) zinc and mercury concentrations in water have exceeded the maximum admissible levels. Again copper and zinc levels in fish (muscles) and molluses exceeded the world base line levels (cu, 0.2-3. $\mu g\ g^{-1}$; Zn, 2-17 $\mu g\ g^{-1}$ wet wt (ICES, 1974).

The overall impact of the pollution load on the entire Jamuna water shows that, the water quality in the river downstream of Wazirabad barrage of Delhi showed significant changes in alkalinity, hardness, chlorides and specific conductivity. There values during summer showed gradual but considerable increase up to Agra (Table 77).

Table 77 : Physico chemical parameters of water in R. Jamuna

	Sampling Zone	*pH*	*ALK*	*Hardness*	*Chlorides*	*Conductivity mm mhos/cm*
	Delhi					
(1)	3 km above Wazirabad water works. (1 Km.)	8.16 ± 0.05	75.3 ± 0.57	59.6 ± 0.49	9.03 ± 0.25	180.3 ± 11.01
(2)	Below Wazirabad barrage at confluence with Nazafgarh drain (1 km)	8.26 ± 0.05	101 ± 19.51	69.8 ± 11.58	36.85 ± 22.91	217.6 ± 18.5
(3)	Near ITO down stream of Rajghat drain and Indraprastha power plant (1 km)	8.16 ± 0.15	98.3 ± 5.68	94.2 ± 40.70	52.03 ± 34.50	267.6 ± 88.0
(4)	Okhla pumping station (1 km.)	8.26 ± 0.11	93.0 ± 3.0	89.5 ± 12.04	45.86 ± 27.15	236 ± 34.7
	Mathura					
(5)	Asteunda Ghat-Visram ghat (1.5 km.)	8.0 ± 0	83.3 ± 1.15	87.8 ± 1.55	49.5 ± 1.80	404.0 ± 28.98
	Agra					
(6)	Jawahar Bridge-Behind Tajmahal (2.0 km)	7.86 ± 0.05	97.0 ± 6.08	85.7 ± 8.05	62.76 ± 10.85	475 ± 98.50
	Allahabad					
(7)	Kankariaghat—Gaughat	8.16 ± 0.20	166.0 ± 7.0	94.4 ± 3.75	37.96 ± 19.65	396.0 ± 78.08

Source : Jhingran, Arun G. and H. C. Joshi (1987)

Fishery status

Average annual landing (t) of major carps, selected cat fishes, Hilsa and other fishes recorded by Jhingran and Ghosh (1978) at Agra of the Jamuna along with other Centres of the Ganga during 1958 - 69 are indicated in Table 55. A sudden fall from 61.9 kg/ha. in the landing from 1959 onward was noticed which subsequently fluctuated within a range 9.5 - 20.2 kg/ha. (Table 56).

Subsequent detailed study shows that carps dominated at all three centres Agra, Mathura and Etawah during August and at Mathura during September. Catfishes during September to December at Agra while others and miscellaneous fishes were abundant at Agra followed by Etawah. However, the estimated fish landings of the river Jamuna during 1988 is shown in Table 78 and Fig. 43.

Table 78 : Estimated fish landings at different centres in tonnes of the river Jamuna during 1988

Centres	*June*	*July*	*August*	*Sept.*	*Oct.*	*Nov.*	*Dec.*	*Total*
MAJOR CARPS[1]								
Mathura	—	—	366.47	897.00	328.59	442.50	471.97	2506.53
Agra	1896.19	7989.57	4092.80	1514.81	852.58	840.14	668.04	17854.13
Etawah	—	—	4017.59	633.75	1003.62	674.25	816.84	7146.05
CAT FISHES[2]								
Mathura	—	—	298.37	722.25	1647.64	1284.00	2672.97	6625.23
Agra	3670.12	2014.29	2539.41	5750.80	7011.57	8180.35	5397.32	34563.83
Etawah	—	—	417.72	3797.30	4336.89	1545.75	3957.15	14054.81
OTHERS AND MISCELLANEOUS[3]								
Mathura	—	—	864.12	892.50	1674.00	480	1433.75	5344.37
Agra	1570.20	1283.12	1909.37	3416.61	1772.96	3873.40	2483.86	14809.52
Etawah	—	—	982.70	1230.00	2449.00	1220.00	1480.25	761.75

1. *Major Carps* : Rohu, Catla, Mrigal and Calbasu.
2. *Cat fish* : Aor, Seenghala and Wallago.
3. *Others* : Miscellaneous : Rita, Silondia, Bata, Bagarius, Murrel Vacha, Garua, Shrimps and Chitala.

Source : Modified from Annual Report, CICFRI, Barrackpore 1988 – 1989.

In a stretch of Jamuna around Allahabad, Natarajan (1989) stated that the catch rate of major carps (Kg./ha^{-1}/yr^{-1}) decreased from 2.85 to 2.02; for catfish, from 2.10 to 1.58 and for hilsa from 0.91 to 0.04, but the catch rate for miscellaneous group increased from 2.17 to 16.0 between the periods of 1958- 69 and 1973 – 84 as stated in Table 79.

Natarajan (1989) has further stated that by species, *C. mrigala* decreased in annual yield by 71.64 per cent; *C. catla* by 69.45 per cent and *L. rohita* by 68.14 per cent between the periods of 1958- 66 and 1974-84. For the same period, large catfishes (*M. aor* and *M. seenghala*) decreased by 24.73 per cent; *W. attu* by 60.12 percent; *Hilsa ilisha* by 95.73 per cent while the yield of *L. Calbasu* increased by 157.88 per cent and other miscellaneous fish increased by 51.66 as shown in Table 80.

Table 79 : Estimated mean annual landings (tonnes) of species groups and yields (Kg./ha^{-1}/yr^{-1}) of the Jamuna at Allahabad for the period, 1958 - 1984

	Major Carp	*Catfish*	*Hilsa*	*Others*	*Total*
1958 – 59 to 1961 – 62	76.88	44.45	14.12	50.28	185.75
	(3.53)	(2.045)	(0.65)	(2.319)	(8.54)
1962– 63 to 1965 – 66	108.45	49.20	26.17	46.78	230.60
	(4.99)	(2.27)	(1.20)	(2.15)	(10.61)
1966 – 67 to 1968 – 69	87.33	42.87	19.40	44.67	194.77
	(0.04)	(1.97)	(0.89)	(2.05)	(8.95)
1973 – 74 to 1976 – 77	35.87	25.69	1.16	48.76	111.48
	(1.65)	(1.18)	(0.055)	(2.25)	(15.13)
1977 – 78 to 1980 – 81	36.32	32.18	1.13	78.15	147.78
	(1.67)	(1.98)	(0.05)	(3.59)	(6.80)
1981 – 82 to 1983 – 84	49.13	34.59	0.34	79.44	163.50
	(2.76)	(1.59)	(0.015)	(3.65)	(7.52)

Source : Natarajan (1989).

Note : Figures in the parenthesis indicate yields in Kg/ha.yr.

Table 80 : Estimated mean annual landings (tonnes) of economic species of three sampling centres in a stretch of 200 Km. on the Jamuna river (1958 - 66 and 1977 - 84)

Species	*1958 – 66*	*1977 – 84*	*Percentage of increase (I)/ decrease (D)*
Major Carps :			
C. mrigala	36.0	10.21	71.64 (D)
C. Catla	16.86	5.15	69.45 (D)
L. rohita	16.67	5.31	68.14 (D)
L.Calbasu	9.57	24.68	157.88 (I)
Cat fish :			
M. aor	34.28	25.80	24.73 (D)
W. attu	12.54	5.00	60.12 (D)
H. ilisha	20.16	0.86	95.73 (D)
Miscellaneous	49.65	75.30	51.66 (I)

The mean annual lengths of major carps in the Jamuna is also indicated by Natarajan (1989) for the years 1967 and 1976– 83 and are indicated below.

	1967	*1976–83*
C. mrigala	48.3 cm.	50.8 cm.
C. catla	59.5 cm.	61.7 cm.
L. rohita	56.7 cm.	66.8 cm.
L. calbasu	40.3 cm.	43.8 cm.

Spawn quality and quantity investigated during 1964-69 in the river Jamuna are also indicated in Table 65 along with the Ganga river systems.

The potential (PY) and actual fish yields (AY) in kg/ha from R. Jamuna are shown in Fig. 37 along with the Ganga.

The status of carps juveniles in commercial stook surveyed by CIFRI, MS. 1970, during August 1985 to 1966 December at the two centres of the Jamuna at Allahabad and Agra along with the Ganga, are indicated in Table 63. It was observed that one year old fishery entered the commercial fishery from July to December at Allahabad and upto November at Agra in appreciable number. (Also refer P 106).

UTTAR PRADESH

R. Chambal

River Chambal also receives substantial volume of wastes from Gwalior Rayon Industry and contribute appreciable pollution problems.

R. Gomati : River Gomati near Lucknow receiving total effluent load R. Gomati 19.84 mgd from one pulp and paper factory, one distillery and 11 sewage outfalls. Although the river maintained a dilution ratio of 1 : 11 during summer months, pollutional effect still persisted as dissolved oxygen value at some stretches below the outfall indicated value below 5 ppm. On the whole about 20 km stretch is affected (Bhaskaran *et al.*, 1965).

R. Kali : A more hazardous picture was revealed when pollutional investigations are made by Gerge *et al.*, (1966) in a 160 km stretch of the R. Kali near Aligarh. The river receives wastes from a number of factories like —Sugar, paints, distillery, soap, glycerine, rayon, silk and yarn etc. which are highly putrescible in character. The flow during the dry months was only 7 mgd (31.5 MLD) an incident of large scale fish mortality was observed which resulted in a loss of 300 quintals of fish and none of the air breathing fishes like *channa* sp.; *Heteropneustes* sp. *clarius* sp. and *Notopterus* sp. were affected. Depletion of oxygen which resulted in from the highly putrescible organic wastes was attributed to be the cause of mortality. Plankton population decreased considerably showing that self purification completely failed. However, recovery was initiated below 160 km when concentration of dissolved oxygen rose to 5.4 ppm.

R. Suvaon : Pollutional hazards in the river Suvaon a small tributary of the Rapli caused by the effluents of Balrampur sugar factory were investigated by Banerjee and Motwani (1960). October to May is the period when R. Suvaon gets disconnected from the parent river and assumes the

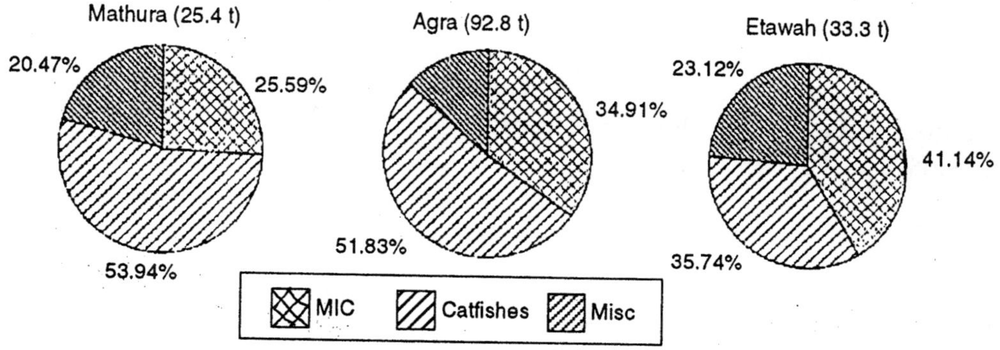

Fig. 43 : Estimated fish landings in Mathura, Agra and Etawah.
Source : Annual Report, CICFRI, Barrackpore, 1989-90.

characteristics of a land-locked channel showing virtually no flow. The liquid wastes, estimated at 780,000 gall/day, create anaerobic conditions in the environment which in combination with other factors like low redox polential and constant supply of sulphate from waste, stimulate activation of sulpher reducing bacteria, producing hydrogen sulphide, a gas highly toxic to fish life. Fish mortality, which recurs in a 3 km stretch below the outfall annually, takes a heavy toll of fish. Bio-assay experiments reveal that the effluent do not contain any material directly toxic to fish.

BIHAR

R. Daha

Wastes from sugar and distilleries, are the sources of pollution of river Daha in the Ganga river system of Bihar.

A small perennial river with an average depth of 0.5 m during the dry months touching up to 1.5 m at few points. The flow during major parts of the year as reported ranged between 2-3 cusecs except during monsoon. The river bed in general is full of submerged and emergent vegetation like *hydrilla, vallisnaria, ceretophylum, Ipomia, Nitella, Ottellia, Eichhornia* and varieties of algae especially *spirogyra* covering the river bed extensively.

The investigations carried out in the river Daha (North Bihar) by David and Ray (1966) revealed that although a small river with tremendous self purifying capacity (Ray 1961 b) the hazardous conditions existed (1.6 km severe and 34.4 km moderately) particularly below the outfall of Siwan sugar mill in Uttar Pradesh all the year round except during rains. Distillery wastes of low pH and high oxygen demand created a chain of reduction and oxidation processes involving iron compounds. On one occasion flocculation of iron salts affected the gill of fishes causing asphyxiation resulting large scale fish mortality in a stretch of 275 – 300 m (Ray and David, 1962).

Below the outfall of sugar mill, soil conditions indicated a high C/N ratio and oligochaets and chironomids were found in abundance with a sharp decline of sensitive molluscs *e.g.* vivipara, *Melanoides* and *Garulus* sp. Conditions improved 16 km. below the outfall of suger mill where fish catches were also maximum with high feeding intensity. The high organic load during the process of decomposition and black humus played an important role in maintaining the fertility of water. The thick growth of aquatic marginal and bottom vegetation, plankter and insect life is an evidence of it. The continuous supply of high dissolved oxygen produced from aquatic weeds above the outfall of sugar mill had been substantially contributing in oxidising the highly reduced organic matter of the waste below the outfall (Ray, 1961 b). Otherwise the polluted zone perhaps would have been more hazardous and extensive. Number of fish species recorded was 32 against only 14 species in the unpolluted zone of upstream.

R. Son

River Son although a mighty river, most of its water is diverted to the canals by constructing an anicut across the river above the factory complex thereby leaving only 50-75 mgd of fresh water to flow during the dry season. Wastes from pulp and paper mill, soda recovery plant, chemical factory, sulphuric acid plant, cement factory and sugar factory waste containing dilute molasses, lime cake slurry and domestic sewage from Dalmia nagar industrial complex are disposed into the river. The total pollution load maintain a dilution ratio of 1 : 12-18 with the river water. The high organic content of the waste in the absence of adequate dilution drew heavily on the oxygen budget of the stream and changed the water quality and bottom condition of the river up to a stretch of 24 km (Motwani *et al.*, 1956). Such changes adversely affected the environment both for fish and fish-food organisms. It acted as a barrier both for upward and downward migration of fish. Fishes are found to be off-flavoured.

A case of heavy fish mortality was recorded by Banerjee *et al.*, (1956) and was attributed to chlorine released along with pulp and paper mill waste in low oxygen level. Besides, colloidal particles like cellulose, lignin etc. of paper and pulp mill clogged the fish gill and resulted in mortality. (Ray, 1961 a).

WEST BENGAL

R. Damodar

The 538 km long Damodar river originating from Chotanagpur hills, Bihar, joins the Hooghly river near Geokhali in West Bengal. The catchment area of the Damodar and the Barakar rivers are 8500 sq. m. out of which 7000 sq. m is occupied by hilly dry uneven areas. Heavy silt load brought down into the Damodar river system during monsoon. An assessment of silt load settled on the bed of the Damodar river system during different periods reveals valuable information on the extent of reduction of water holding capacity of the river. The enormous silt carried by the Damodar to the middle stretch of the river is due to the nature of silt in the upper area and such silt will have positive effect on Calcutta post and its part thereof was carrectly anticipated and predicted by Sir William Wilson Hunter long back in his valuable document "Statistical accounts of Bengal vol. I, 24-Parganas Sundarbans, Director General of Statistics Govt. of India. Low water carrying capacity is prone to floods and inundation of large areas even by low rainfall. Some estimate on the rate of silting in different stretches of the river bed made by different workers/Institutions during different periods are indicated in Table 81. The continuous silting of mighty Damodar has reduced its flow to a narrow channel above the barrage.

Various industries and a number of coal washeries located in the upper Damodar contributing wastes containing various toxicants are surveyed by Basu (MS, 1966). In the lower stretch however Dhaneshwar *et al.* (1972) have done a commendable work by studying various aspects of pollution like survey of industries, total pollution load, identification of pollutants, assessment of extent of pollution etc. Various agencies drawing water from rivers and its tributaries and the list of establishments discharging effluents in the river system are indicated in Fig. 44 and Table 82.

Impact on fish breeding and their production were studied by Job *et al.*, (1952); Gopalakrishnan *et al.*, (1966); Natarajan *et al.*, (1968) and Jhingran and Natarajan (1969). Their studies were restricted only to the upper Damodar.

Unfortunately, except breeding success and fish production, no other biological studies have been made so far in the entire stretch of Damodar.

Table 81 : Silting of the Damodar river system during different periods

Reference	*Year*	*River with stretch*	*Silt load*
1. From a Survey	1938	*Damodar River* From Bhajudi to 13 km down stream at Katrijore. (Of the stretch from Bhula-barari Coal Co. Ltd. to Jharia coalfield area)	1 crore 20 lakh tonnes.
2. N.K. Bose (Expert in river Science)	1945-48	*Barakar river* Near Giridih and Kulti	2.08 acre ft./sqm 1.47 acre ft./sqm
3. Oden		*Damodar river* at Randia near Panagarh	1.3 acre ft./sqm.
4. D.M.Sen (Geologist)	1946–48	*Damodar river* (i) From Decergarh to Serampur. (ii) From Asansol to Adra Rail (3.2 km above and 9.6 km below Rekhaplu)	5 crore 30 lakh tonnes 10 crore tonnes.
5. A.B.Dutta (Geologist)		*Damodar river* (i) Near Jharia mine area (ii) Near Raniganj mine area Barakar river Damodar river Ajoy river.	14 crore tonnes 99 crore tonnes Dutta in subsequent estimate states, silting at the rate of 1 crore tonnes/ year
6. Committee formed by Central Govt. (Ministry of Production)	1956	*Damodar river* at Jharia	12 crore tonnes.
7. Government of India (A Report)	1970	*Damodar river* (i) Near Jharia coal mine area. (ii) After construction of Tenu Ghat dam. *Barakar river* After construction of Maithan Dam (1955) according to DVC estimate in a stretch 16-24 km.	15 crore tonnes 15 lakh tonnes/yr. 4 crore tonnes
8. Dhanbad coal mines planning and dredging organisation.	1980	(i) From Panchet to opposite Ondal (46 km stretch) (ii) Up stream of Durgapur Barrage in 12 km stretch (iii) Maithan to 9 km stretch below. (iv) In a 17 km stretch in the upstream of Damodar river up to the confluence of Jamunia river. (Katrijore)	55 crore tonnes (accumulated) 34 crore tonnes (accumulated) 7 crore tonnes 1 crore 40 lakh tonnes
9. Coal Board		(i) In a 16.5 km stretch up to Sudamdi in the river bed (ii) 10 km below Sudamadi Rly. bridge, Bolpur to 16 km stretch below (iii) Also in a 50 km stretch in that area.	3 crore tonnes 6 crore tonnes 15 crore tonnes.

Source : Sakti Dhar (1984) : In Damodarer Banya—3 (The Flood of Damodar River—3); The daily "Aajkal", January 17, 1984.

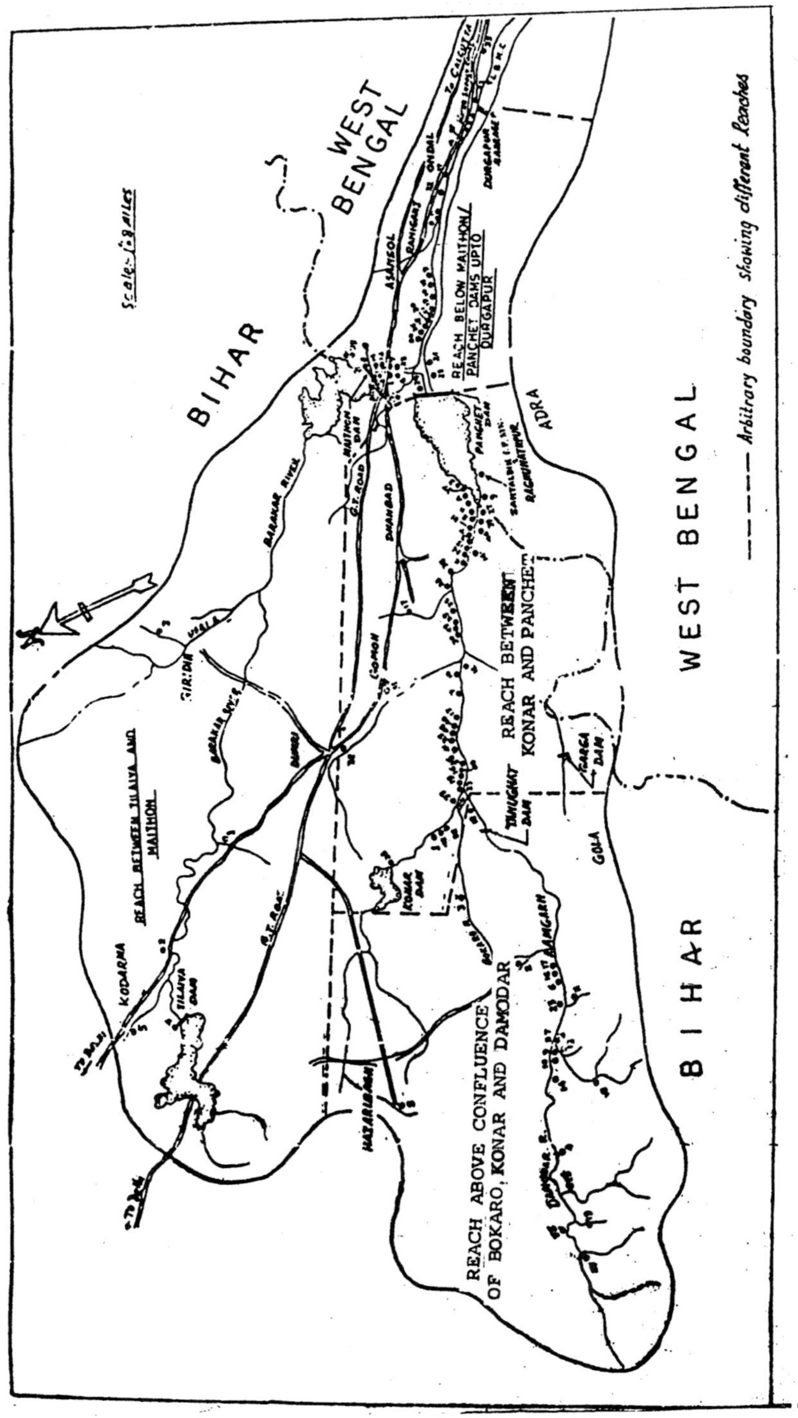

Fig. 44 : Upper and Middle Damodar Valley Location of agencies drawing water from the rivers and its tributaries
Source : Banerjee, R.K. (1988)

Table 82 : List of establishments discharge effluent in the riverine system

Sl. No.	*Name of Agency*
A.	**Reach Between Konar to Panchet**
1.	Indian Explosives Ltd.
2.	Kathara Colliery and Washeries
3.	Sindri Fertiliser and Chemicals
4.	Superphosphate Factory
5.	Fuel Research Institute
6.	Jharia Water Board
7.	South Eastern Railway, Bhaga
8.	Eastern Railway, Patherdih
9.	Bhojudih Coal Washery
10.	Dugdha Coal Washery
11.	Patherdih Coal Washery
12.	Garga dam for part requirement of Bokaro Steel Project
13.	Asian Refractories Ltd.
14.	Amlabad Colliery
15.	Bhowrah Kankanee Colliery
16.	Eastern Railway, Bermo
17.	Eastern Railway, Katrasgarh
18.	Kargali Colliery
19.	Bokaro Colliery
20.	Kargali Washery
21.	Sudamdih Colliery, NCDC
22.	Swang Colliery, NCDC
23.	Jarangdih Colliery
24.	Jamadoba Colliery (Tata's)
25.	Khasdhari Colliery
26.	South Eastern Railway, Mahuda
27.	South Eastern Railway, Bhojudih
28.	Moonidih Project, NCDC
29.	Mahuda Power House
30.	Bhatdhee Colliery
31.	Eastern Railway, GOMO
32.	Eastern Railway, Pareshnath
33.	Bokaro Thermal Power Station
34.	Chandrapura Thermal Power Station
35.	Konar Colony
36.	Harry Refractory and Ceramics Works
37.	Pyrites and Chemicals Development Ltd.
38.	Kapuria Colliery
39.	South Eastern Railway, Sudamdih
40.	Pichri Colliery
41.	Bokaro Ramgarh Ltd. (Dhori Colliery)

Contd....

Sl. No.	Name of Agency
42.	Gomia Water Supply
43.	Murulldih Colliery
44.	S.T.P.S. (S.B.)
45.	Phusro Rly. Stn.
B.	**Reach Between Tilaiya and Maithon**
1.	Eastern Railway, Hazaribagh
2.	Eastern Railway, Sarmatand
3.	Public Health Engineering Department (Khandoli Dam)
4.	Tilaiya Colony
5.	Jhumri Tilaiya (Public Health Engineering Department)
C.	**Reach Below Maithon/Panchet Dams Upto Durgapur**
1.	Kumardubi Engineering Works
2.	Chittaranjan Locomotive Works
3.	Durgapur Projects Ltd.
4.	Durgapur Steel Project
5.	100-Beded Hospital, Maithon
6.	Housing Department, Kumardubi
7.	Barakar Water Supply Company
8.	Raniganj Coal Field Area
9.	Asansol Municipality
10.	Asansol Satellite Township
11.	Raniganj Municipality
12.	Aluminium Corporation of India
13.	Burn and Co., Gulfarbari
14.	Indian Iron and Steel Co., Kulti
15.	Indian Iron and Steel Co., Burnpur
16.	Eastern Railway, Barakar
17.	Eastern Railway, Ondal
18.	South Eastern Railway, Radhanagar
19.	South Eastern Railway, Damodar.
20.	Dishergarh Power Supply Co.
21.	Dhakeswari Cotton Mills
22.	Bengal Paper Mills
23.	Begunia Colliery
24.	Saltora Colliery
25.	Ranipur Colliery
26.	Dishergarh Packaged Plant
27.	Durgapur Thermal Power Station
28.	Maithon Colony
29.	Panchet Colony
30.	Bhaladih Sand Gathering Station
31.	Maithon Hospital, Ganga

Contd. ...

Sl. No.	Name of Agency
32.	Fertiliser Corporation of India, Durgapur
33.	Hindusthan Cables Ltd.
34.	Kumardubi-Chirkinda Water Supply
35.	Panagarh Rly. Station.
D.	**Reach Above Bokaro**
1.	Indo Asahi Glass Co.
2.	Assam Silimanite Ltd.
3.	Eastern Railway Dansa
4.	Eastern Railway, Hondigir
5.	Public Health Engineering Department (Charwa Dam)
6.	Argadda Colliery, NCDC
7.	Giddi Colliery, NCDC
8.	Bhurkunda Colliery, NCDC
9.	Soundya Colliery, NCDC
10.	Sayal Colliery, NCDC
11.	Barkakana Workshop and Township, NCDC
12.	Kathara Colliery (from Damodar)
13.	Bird's Soundya Colliery
14.	Eastern Railway, Patratu Station
15.	Eastern Railway, Ray Station
16.	Ramgarh Cantonment
17.	Indian Fabricks and Insulation Co. Ltd.
18.	Patratu Thermal Power Station
19.	Associated Cement Co., Khalari
20.	New Bharat Refractories
21.	Eastern Rly. Mac. Clausiganj
22.	Bachra Colliery
23.	Barakakana Rly. Stn.

Source : Banerjee, R. K. (1988).

In the upper stretch between Bokaro and Panchet, the Fertilizer Corporation of India, discharging 4 mgd (18MLD) of wastes containing pollutants like ammonia, cyanide, phenol, naphthalene, traces of tar and oil, sulphates, fly ash, alkalies, chromates etc. Besides huge quantities of ammonium sulphate from cement factory, Bokaro thermal power station with a number of coal washeries contribute large volume of fly ash, the total solids in the effluent ranged between 250-12,000 ppm. (Basu, MS 1966).

In the lower reaches, the river receives wastes from a number of coal washeries, chemical industries, still mill and its alloys, coal based industries, numerous engineering industries, paper mill and distilleries. Some major industries under Asansol and Durgapur agglomerations contributing waste water appreciably together with their pollution load and toxicants released in t/d in the Damodar river are indicated in Table 83.

Table 83 A : Major industries and their pollution load under Asansol agglomeration

Name of industry	*Effluents disposed m^3/d*	*Pollution load*
1. *IISCO*		
(i) Cold rolling mill	29523	5600 kg iron/day 1254 mg/1 Solids, 17per cent in suspended form
(iii) Coke oven effluent	14383	657 kg Phenol/day 26 kg Chanide/day 2300 kg ammonia/day 140 kg iron/day 2150 kg grease /day
2. Sen Raleigh	1180	0.8 mg/d. domestic sewage
3. Carew and Co.	150	Domestic waste/day combies waste containing 6.4 per cent DS. of which 70 perc ent volatile having 25,000 mg/1 BOD 71,000 mg/1 COD.
4. Reckitt Colmans & Co.	160	BOD, COD, SS. and silica (quantity not indicated)
5. Bengal Paper Mill	46350	Paper mill effluent, 11583 kg/d, BOD; 1200 mg/1 SS. paper mill effluent from new and old plant, all these 3 wastes contribute 218 tonnes SS/d, 225 tonnes BOD/d.

B : Pollution load contributed from some major industries under Durgapur agglomeration

Name of industry	*Effluents disposed m^3/d*	*Pollution load*
1. Hindustan Steel	4730	82.5 mg/1 Phenol. 8.9 mg/1 Cyanide 285 mg/1BOD 1220 mg/1 SS *Blast furnace and Rolling mill :* 42.5 mg/1 BOD 1286 mg/1 COD.
2. Durgapur Project Ltd.	14000	*Coke oven and bye-product units :* 180 kg phenol 160 kg cyanide. 155 kg sulphide 2900 kg ammonia and 1000 kg light oil
3. Coal Washery		*Coal Washery effluent :* 4.2 per cent SS 110 mg/d BOD 77800 mg/1 COD.
4. Phillips Carbon Black	1600	3232 mg/1 SS 250 mg/1 BOD 15800 mg/1 COD 250 mg/1 oil/grease

Source : Banerjee , R.K. (1988) based on the data of Dhameswar R.S., C. S. G. Rao, M. K. A. Kutty, D. Ray and R. C. Trivedi (1972).

Thus the total pollution load contributed to the Damodar river from some important industries of Asansol and Durgapur region are : total solids, 22377 t/d; SS, 11132 t/d; BOD, 3141 t/d; cyanide, 24 t/d; phenol 15 t/d and COD, 20311 t/day (Table 84).

Table 84 : Total quantum of Toxicants discharged from Asansol and Durgapur agglomerations

Toxicant	*Asansol*	*Durgapur*	*Asansol + Durgapur*
	tonnes/day		
Total solid	12,317	10060	22377 tonnes/day
Suspended solid	1,782	9350	11132 tonnes/day
B.O.D.	2,984	157	3141 tonnes/day
Cyanide	2.88	21.12	24 tonnes/day
Phenol	4.5	10.5	15 tonnes/day
C.O.D	8,374	11537	20311 tonnes/day

Source : Banerjee, R. K. (1988)

Nunia nullah and Tumbla nullah, the two important canals carry load respectively 1,04,087 m^3 and 1,28,690 m^3 per day. In the recent past fire broke out in a stretch of 1.5 km of the Damodar river near Dhanbad, where more than 11,000 kilo litres of furnace oil leaked (Statesman, 1990).

Only from five major industries under Asansol agglomeration is the total effluent discharge 91,746 m^3/d and three from Durgapur agglomeration is 1,12,076 m^3/d. It is thus clear that 0.73-0.4 or 0.33-0.22 comes to 1.6 *i.e.* the dilution ratio of effluent to river water is approximately 1: 1.6 except during monsoon which is quite alarming (Banerjee, 1988).

R. Churni

River Churni an offshoot of R. Mathabhanga, which joins the Bhagirathi-Hooghly at Gournagar of Nadia district, was earlier known for its commercial fisheries especially, for prawn seed is now highly polluted by wastes from Dharsana sugar mill of Bangladesh and has lost its potential value. The study reveals positive indices of pollution of a noxic condition at bottom (nil-1.9 mg/1) high BOD (74.8 mg/1) with a luxuriant population of *Tubifex* sp, *Chironomus* sp, *sphaerium* sp, *Oscillatoria* sp and *Lyngbya* sp (Ghosh, Mukhopadhyay and Bagchi, (1990).

15

Overall Stressed Environment and Biotic Life

A little background information on the principal river Ganga will enlighten how the fast deterioration of its physiographical condition resulted in low production potentiality.

Around 1869 the depth of the Ganga was maximum between Allahabad and Rajmahal. At Rajmahal the depth ranged between 3.0 and 4.57m and velocity of water during summer season was 1.34m/sec against 3.12-3.57m/sec during other periods, when maximum discharge recorded in September was 1,50,000 cusecs (Sarkar, 1988). The Bhagirathi flowing through Jangipur (West Bengal) recorded 1,40,160 cusecs and further down near Berhampur it was 1,40,762 cusecs with water velocity 1.8-2.31m/sec, is now reduced to less than 20,000 cusecs so much so that the maximum available water depth between Tribeni and Farakka is about 2 meters only (Mukherjee 1993).

As regards major carp fishery (Catla, Rohu and Mrigal both in the Jamuna and the Ganga) the maximum total length of Catla was 12.3 cm., while Dey, (1871; 1878; 1889) and Raj (1939 and Chacko and Kurian (1950) recorded maximum size as 180 cm (Jhingran, V.G.1968) (FAO Fisheries Synopsis, No. 32, Rev. I)

Rohu attained a maximum size of 91.11 cm. (Dey, 1978); 73.7 cm. (Shaw and Shebbeare, 1937) and mean length 450 mm-470 mm from February to July (FAO, Fisheries Synopsis No. 111, SAST-*L. rohita*, 1975) and Mrigal, maximum size 992.0 mm at the time of XII completed annuli (FAO, Fisheries Synopsis No. 120, 1979). Such sizes have been reduced to a mean length during 1987 and 1988 and are as follows :

Species	Size attained (in mm)		Fall in Length (mm)
	During 1987	During 1988	
C. mrigal	623	558	(69)
C. catla	735	618	(117)
L. rohita	650	580	(70)
M. aor	447	438	(9)
M. seenghala	545	488	(57)
W. attu	545	471	(74)
T. hilsha	396	347	(49)

(Annual Report, 1988-89, CICFRI, Barrackpore).

From the above it is interesting to note how major carps are loosing their sizes. Even in one year time (1987 and 1988) the sizes of major carp, catfish and hilsa decreased by 69-117 mm, 9-74 mm and

49 mm respectively. It will be worthwhile to investigate the probable factors responsible for such fall in fishery potential of the major river like the Ganga and the Jamuna. The relevant points discussed are as follows :

1. Continued deforestation in the catchment area brings enormous volume of silt in the river bed every year during monsoon. The accumulation of such silt load gradually raises the river bed, fills the deep pools either completely or to a great extent, reduces the slope (viz. upper stretch : Rishikesh to Allahabad (800 km) 1 in 4100; middle stretch : Allahabad to Farakka, (1005 km) 1 in 13800, deltaic non-tidal plain, Farakka to Nabadwip (230 km) 1 in 23,000 and Nabadwip to outfall (240 km) 1 in 24,000 (Table 4) with water velocity, 0.35-0.9 m^1 sec between Haridwar and Allahabad; 0.35-0.44 m/sec between Varanasi and Farakka and 0.09-0.56 m/sec between Nabadwip and Diamond Harbour during May and July. Reduced velocity again encourages further siltation. Thus the shape factor is increased (Table 4) which means deterioration of the river condition. The river gets widened by loosing depth. In the Hooghly estuary the quantity of silt load brought down with water during monsoon comes to 20 crore f^3 and is more than double the value estimated 12 years before (Bera, 1984). According to the Ministry of Irrigation and Power estimate (Mukherjee, 1993) about 2.8 m/m^3 silt (about 7m tons enter the Bhagirathi-Hooghly channel annually, about two-thirds of the silt is deposited below the Bansberia point, northern limit of Calcutta Metropolitan area during summer. In addition the river receives about 8 million tonnes of sand from the sea every year that chokes the flow. As a result a number of sand bars have also come up in the estuarine bed.

2. Silt even causes the river to shift its normal course. The Rapti, the Ghagra in Uttar Pradesh; the Gandak, The Kosi, the Son and the Ganga itself in Bihar and the river Damodar in West Bengal, have shifted their courses frequently flooding wide area throughout from the historic times, is leaving remnants of their former beds in the form of ox-bow lakes, mender loops, creeks, chars, tals and ruins of settlements *etc*. For example, the Ganga had shifted 35 km between Bhojpur Tal and sarha Tal in Uttar Pradesh. The Koshi has shifted 120 km in the course of 200 years from Purnea to Tiljuga, Darbhanga district in Bihar, and distributaries of the Bhagirathi-Hooghly and also the Damodar river in West Bengal. The shallow inundated low lying areas which are remnants of the river bed formed due to its shifting.

Near the confluence, of the Koshi river, the minor dhars are interlacing channel ecosystem spreading over an area of 250 sq. miles in Saharsa and Darbhanga districts of north Bihar annually become alive by spilling flood waters of the main Koshi over its banks and serve as the foremost areas where the carps are found to spawn. The Jheels connected with the Burhi-Gandak and the slightly elevated alluvial sandy islands and banks of the Ganga are also known to attract fish for spawning. With the commencement of flood, gravid adults congregate for breeding.

3. An examination on how siltation affects feeding and reproductive activities of fish may be more fruitful than comparative taxonomic inventories, because the primary influences of the increased sedimentation on stream fish are believed to be the disruption of normal reproduction and the destruction of food supply (Allen, 1969; Cordone and Kelley, 1961; Ellis, 1936; Karr and Dudlay, 1981; Smith, 1971). Berkman and Rebeni (1987) while studying the effects of sedimentation on stream fishes in some small north-east Missouri streams and to relate the effects of sedimentation to ecological characteristics of the fish fauna revealed some valuable informations and at the same time they have quoted findings of others. These are :

(a) As siltation increased the distinction among riffle, run and pool fish communities decreased, because the abundance of many riffle species was reduced.

(b) Physical attributes that have been related to fish diversity in streams include depth (Sheldon,

1968) or a combination of depth, substrate and flow (Gorman and Karr, 1978) while Berkman and Rabeni, (1987) was of the opinion that substrate was of overriding importance, because, even the maintenance of characteristic current velocity and depth was not sufficient to retain integrity of the typical riffle assemblage if the percentage of fine particle sizes increased.

(c) Berkman and Rabeni (1987) further suggested that fish with similar feeding or reproductive strategies were similarly influenced by siltation.

(d) The feeding guilds most negatively correlated with siltation were those most specialized to feed from the substratum. Benthic insectivores were adversely affected.

(e) Increased sedimentation reduces insect diversity density and species richness. (Brusvan and Prather, 1974; Chutter, 1969; Gammon, 1970; Nuttall and Bielby, 1973).

(f) Schlosoer (1982) demonstrated the possibility that an increase in algae and invertebrate production could possibly influence the growth, spawning success and recruitment of fishes in the mid western stream. The behaviour of insectivore and omnivore fishes never been correlated with siltation and are more opportunistic in their foraging behaviour and might have been modified their feeding habits so as not to be food limited (Starrett, 1950).

(g) There is evidence that fish may be more affected by alterations in their spawning conditions than by alterations in their habit (Balon, 1975); Siltation may degrade spawning grounds, cause behavioral changes in spawning fish, decreases in fecundity and increase in egg mortality, reduction or slow in larval growth and affect development rate and survival of larval fish (Langlois (1941), Cordone and Kelley, 1961; Muncy *et al.* 1979 and Smith, 1971). Remarkable fall in reproductive yield may be correlated with an increase in siltation.

The above findings of various workers are interesting and are important from the point of view that such phenomenon are common occurrence in rivers with biotic life Fresh water fish normally breed in shallow inundated zones rich in reproductive feeding guilds. This means increase in algal and invertebrate production could possibly influence growth reproduetive success and recruitment of fish. But with the onset of monsoon and heavy silt precipitation such natural ecosystem is destroyed. Garvid brood fish when remains under stressed conditions prior to their breeding, result some behavioural changes which compel them to hunt for a safe spawning ground and by doing so partial or unsuccessful breeding is resulted. The breeding centres may thus get shifted which may be far away from original harbouring zones of breeding occurs under physiological pressure in places where the area soon gets exposed and huge quantity of eggs/larvae are thus get perished (Ray *et al*, 1964). According to Jhingran and Ghosh (1978) a very high larval mortality in carps is most likely, since fish larvae are freely drifted by about 100-150 cm/s in flooded rivers or more along with the spilled water. Only the larval masses that get locked up in natural nurseries develop to recruits to sustain stocks, the rest are drifted down and get perished. High turbidity, 800-2000 ppm is not conducive to high survival. Silting, thus not only degrade spawning grounds of major carps, but also increase egg mortality, decrease larval growth and affect development rate and survival of larval fish. For catfish and other miscellaneous groups of fishes, there are no such problems, year to year with the increased silt load reduces the breeding and as a result stock and quality fish seed resources is not encouraging. Such phenomenon is observed by the workers of CICFRI (Tables 65-67), in the records of Jhingran (1975) and Tripathi (1976).

The fish seed committee has estimated in 1964 that the Ganga river system contributed 89.5 per cent of the total fish seed produced in the country (Government of India, 1966). An average of about

2010 million spawn from Bihar and 1200 million from West Bengal were the two major states for seed production while Tripathi (1976) estimated the production of spawn from rivers of Ganga system in Uttar Pradesh to the tune of 122.2 million in 1966.

Recent record shows that the spawn yield from the rivers in Uttar Pradesh has declined, while that in Bihar is of the same order, the yield in 1973 being 2037 million. The yield from West Bengal also appears to have much declined with an average of about 530 million during 1971-1975 (Jhingran and Ghosh 1978). If silting goes on a steady rise, the major carp industry will undoubtedly loose its importance, and spawn of such commercial important fishery will completely change into non-commercial species like catfishes and trash non-commercial species. This fact is gradually appearing in the ecosystem (Table 57). Further, fry of a number of predatory species which abound during the carp breeding season have been observed with stomaches gorged with major carp larvae (Anon, CIFRI Report, 1965).

4. High silt precipitation reduces the water holding capacity of river. With the cessation of rains and heavy obstruction of water by various sources (stated elsewhere) the water level of the river goes down considerably from September onwards. It is during this time the spawn which just attained to juvenile stages (immature) are indiscriminately caught by the fisherman community easily.

As reported earlier a total of 3,05,512 juvenile specimen weighing 12.7 t was estimated to have been captured in 13 days (September, 1952) from a 208 km stretch from Mughalsarai to Patna. In 1953 a total of 3,28,408 juvenile specimen (14.99t) have been captured in 12 days (September) from the same stretch. By November, the catch consisted of 1,66,503 specimen (14.99t) were harvested (Jhingran (1975).

Natrajan and Jhingran (1963) estimated catch of juveniles and immature catla at Allahabad to be 54.6 per cent of the total catch in 1960. A similar report was presented by Kamal (1969).

More alarming picture was reported by David (1959) on mass killing of fingerlings and juvenile carp on 'Kosi dhar' in Bihar, estimated the quantity at several tonnes per day. The quantity transported outside the market exceeds 3.7 tonnes/day. Such high rate of fingerling exploitation especially in the middle stretch of Bihar is almost certain to deprive the system of a substantial reproductive potential. According to Natarajan *et al.* (1976) 40t spawners may produce 2000 m spawn.

The relatively lower rates of juvenile exploitation in the river Jamuna and in the upper stretches of the Ganga system generally correspond to much higher abundance of major carps in these regions.

5. The number of deep pools are very limited where major carp spawners can harbour. The fishermen can thus easily locate such areas and fish them out. This is how large sized mature and spawners are reduced day by day. The same picture holds true in case of the river Jamuna. Although number of mature spawners have not been estimated so far and keeping in view of the above findings the statement of Jhingran and Ghosh (1978). It may be assumed that the total quantity (mature spawners) may not be less than several tonnes in the Ganga river system whose spawning potential may be enormous may not be correct and requires to be substantiated with adequate data.

Further, the remark made by Natarajan *et al.* (1976) that, decline in fisheries of major carps after 1961 as compared to the situation prevailing during 1958-1961 might have been the result of higher exploitation of the resources. This view is based on the fact that major carp fishery is exploited in the form of spawn, fry, fingerling, juveniles as well as adult stages, and more and more effort was put to exploit the younger stages. A similar view (higher exploitation) is also put forward by Gupta and Tyagi (1992) while reviewing the trend of fish production in the middle stretch of the Ganga river system. But all the authors have failed to give adequate emphasis on the important factor, which is silt precipitation that brings enormous physiograpical changes in the ecosystem for which higher exploitation has become possible (Major carp brood fish generally harbour in deep pools of the river. But heavy silt load brought down during floods, a recurring annual phenomenon, has considerable impact in filling up such deep pools).

To get more catches, competition of effort pattern increases with more improvised fishing technique, but this holds good for deep water only. In the prevailing situation where river is shallow, low gradient and low water velocity, intensive fishing effort is also done and for this no improvised gear is required. Such ideal shallow ecosystem that tempted the fisherman to exploit fishes using simple gears. Young juveniles are not allowed to grow to large sizes, those could have bred profusely. R. Jamuna, is also suffering from similar problems. Perhaps it is based on the above findings Saxena (1991), more recently opined that qualitative and quantitative transfer in the effort pattern becomes inevitable due to physiographical changes while discussing on status on riverine gears. Thus ultimately the stock and quality of major carp brood fishes are on the process of extinct and in its place catfish and other miscellaneous non-commercial species are dominating (Table 57). Kamal (1991) opious, conspicuous shifts in population structure have occurred with the commercial species giving way to miscellaneous groups of fishes. Concomitantly the changes in the catch structure are also reflected in spawn quality index with increasing dominance of minor carps and forage species. Further, it revealed from his study that habital destruction, water quality deterioration and changes in hydrological regime have led to such changes.

6. Next to silt load, presence of heavy metal in the water has considerable impact on the healthy growth and reproduction of fishes.

Factors influencing the toxicity of heavy metals in solution have been discussed by Bryan, 1976 in a simple way in the document "*Metal Pollution in aquatic environment*" by U. Förstner & G.T.W. Wittmann - Verlag, Berlin, Heidelberg, New York, 1979.

Form of metal in water	Inorganic	Soluble	Ion Complex ion Chelate ion Molecule
	Organic	Particulate	Collidal Precipitated adsorved
Presence of other Metals or poisons	Joint/action	More than lethal level	
	No interaction	Additive (productive level)	
	Antagonism	Less than additive (productive level)	
Factors influencing physiology of organism and possibly form of metal in water.	Temperature pH Dissolved Oxygen Light Salinity Stage in life history (egg, larva, etc) changes in life cycle (e.g. moulting, reproduction)		
Condition of organisms	Age and size Sex Starvation Activity Additional Protection (e.g. shell) Adaptation of metals		
Behavioural response	Altered behaviour.		

The above facts give a base line information on factors influencing toxicity and are important in pollution studies.

Metals like zinc, lead, copper, cadmium *etc.*, commonly found almost in all polluted wastes behave differently on different organisms on the health, survival, growth and reproduction of aquatic biomass including fish, either directly or through synergistic action by combining with other metals when treated in a medium of water of any kind soft or hard. Some interesting results obtained are reviewed by European Inland Fisheries Advisory Commission (EIFAC, Published by FAO, 1973), water quality criteria (1963) and also by others are presented below. These will certainly help in understanding the complications of pollution problems of the Ganga river system which the biomass are facing for their healthy survival, growth and reproduction. The ninth technical paper on water quality criteria for European Freshwater Fish (Report on Zinc and Fresh water fish) prepared by the European Inland Fisheries Advisory Commission (EIFAC), published by FAO, 1973 opines that although extensive data on the toxicity of zinc to under laboratory conditions supported to some extent by field data on fish kills are available there are virtually no field observations to indicate the concentrations of Zinc that are not inimical to fish populations or fisheries. They observed that information on water hardness perhaps the most important factor affecting the toxicity of zinc to aquatic animals is missing and details of the status of the fish population are not presented as a result of which only tentative criteria can be suggested. Some interesting findings are however indicated below—

Zinc

The toxicity of zinc in solution is mainly attributable to the zinc ion and perhaps also to particulate zinc present as the basic carbonate or the hydroxide held in suspension. It is influenced by water quality reduced in particular by an increase in hardness and also temperature salinity and suspended solids and increased by a decrease in dissolved oxygen. The effect of pH however is uncertain.

The acute toxicity of zinc in presence of other heavy metals and other common pollutants seems to be largely simply additive but there is no evidence that the chronic toxicity of different poison in a mixture is also additive. The effect of zinc is modified, and can be reduced, by acclimation and by the age of the fish

A low but significant mortality has been found among rainbow trout exposed continuously for 4 months to constant concentrations of 0.2 of the 5-d LC_{50} and among rudd exposed for months to 0.3 of the 7-d LC_{50}.

Laboratory studies of avoidance reactions have shown that Atlantic salmon and rainbow trout may avoid concentrations of zinc in soft water which are 0.14-0.01 of the 7-d LC_{50}. Avoidance reactions have also been observed at 0.35-0.43 of the 7-d LC_{50} by migrating Atlantic salmon in a river polluted with copper and zinc. Carp and gold fish show avoidance of 0.3-0.45 of lethal concentrations under laboratory conditions.

Field observations show that brown trout populations were present when the zinc concentration was less than 0.06 of the 2-day LC_{50} rainbow trout or when the annual 50 and 95 percentiles were upto 0.05 and 0.19 of the LC_{50} respectively; coarse fish were present when the corresponding percentiles were 0.02 and 0.11 of the 5-d LC_{50} to roach.

Because concentrations of zinc in freshwater fluctuate, being distributed approximately lognormally over a period of a year, tentative criteria are expressed as annual 50 and 95-percentile values.

Pending the availability of more information it is tentatively recommended that for the maintenance of the thriving populations of fish the annual 95 percentile of concentration of zinc should not be higher 0.1 of the approximate 7-d LC_{50} at 15°C; thus the criteria in terms of concentration of zinc would depend upon water hardness and type of fish as shown in Table 85.

Table 85 : Maximum annual 95-percentile concentrations of zinc (mg Zn/l)

Water hardness mg/1 $CaCO_3$	*Coarse fish except mirrow*	*Salmonids*
10	0.3	0.03
50	0.7	0.2
100	1.0	0.3
500	2.0	0.5

The concentration of 0.03 mg Zn/1 for salmonids in very soft water may be too severe if brown trout only are present, since this species appears to survive sucessfully at higher concentrations; in such cases a 95-percentile concentration of 0.2 of the 7-d LC_{50} (0.06 mg Zn/1) may be more appropriate. For the minnow, it might be more appropriate to set a more stringent standard, but further data would be desirable to support and explain the existing laboratory findings.

The values of the corresponding annual 50 percentiles would be appropriately 0.25 of the proposed 95 percenties unless the distribution was much wider in the range than has hitherto been found or was not Lognormal resulting in a larger ratio between the 95 percentile and higher percentiles, in which cases more stringent standards would be appropriate. Where other poisons are present and dissolved oxygen concentrations are below the air-saturation value, allowance should also be made for their contribution to the toxicity.

Brungs (1969) found that reproduction in fathead minnows was almost totally inhibited at zinc concentrations that had no effect on survival, growth and maturation. The number of eggs produced per female at 0.18 mg Zn/l was only 17 per cent of that produced at 0.03 mg Zn/1. In tests lasting about 4 months during which rainbow trout were exposed to concentrations between 0.6 and 2.0 mg/L in hard water, there was a small but significant mortality among the fish (Edward and Brown, 1966).

Water quality criteria (1963)

Zinc exhibits its greatest toxicity to fish and aquatic organisms. A concentration ranging 0.1-1.0 mg/1 have been reported to be lethal in soft water, but calcium is antagonistic towards such toxicity. For a mature fish lethal limit for zinc in water containing 1 mg Calcium is only 0.3 mg/l but with 50 mg/l of calcium as much as 2.0 mg/l of zinc is not toxic.

It has been observed that effects of zinc poisoning may not become apparent immediately so that fish removed from zinc-contaminated water to zinc-free water (after 4-6 hrs.) of exposure to zinc, may die within 48 hrs.

Fish could detect and avoid concentration of zinc sulphate as low as 24 mg/1 although the salt is not extremely toxic. In tests lasting for about 4 months, during which rainbow trout were exposed to concentrations between 0.6 and 2.0 mg Zn/1 in hard water, there was a small but significant mortality among fish (Edward and Brown, 1966).

Sprague (1964 a) observed that in soft water (Hardness, 18 mg/1 as $CaCO_3$), the lowest concentration causing average fish to show significant/avoidance were 0.053 mg zn/1 for zinc sulphate alone, and 0.006 mg/1 for zinc in presence of 0.0004 mg Cu/1 some more work on avoidance behaviour of fish to pollutants has been done by Sprague (1968), Sprague and Drury (1969).

Ray (1993)

Water flees *Daphnia-magna* are killed within 5-10 days at 0.01 mg/1 zinc.

Copper : (Water quality criteria, 1963)

Copper concentrations as low as 0.1 to 0.5 mg/1 have been reported to be toxic to bacteria and other organisms, and concentration as low as 0.01 to 0.5 mg/1 have been found to be interferred with BOD determinations and also self purification of streams.

Copper chelated by citrate or lignin is effective in controlling the algae as free copper ion, but chelated copper is not toxic towards fish.

Photosynthesis of *Microcystis Pyripera* found to be inhibiting at 50 per cent in 2-5 days and 10 per cent in 7-9 days with 0.1 mg/1 of copper (in any form either chloride or sulphate salts)

Toxicity of copper to fish varies greatly depending on the presence of magnesium salts and phosphates.

In case of fish food organisms, *Daphnia similes*, the toxicity value (LC_{50}) of copper was found to be 0.06 ppm is moderately hard water (Ray and Bandopadhaya, 1984).

Lead : Water quality criteria (1963)

Toxic concentrations of lead for aerobic bacteria is reported to be 1.0 mg/1; for flagellates and infusoria, 0.5 mg/1. The bacterial decomposition of organic matter is inhibited by 0.1 to 0.5 kg/1 of lead.

In water containing lead, a film of coagulated mucus forms, first over the gill and then over the whole body of fish, probably as a result of reaction between lead and organic constituent of mucus. The death of fish is caused by suffocation due to their obstructive layer. Similar results were also observed with lead by forming insoluble compounds with the mucus that covers the gill by damage to the epithelium or possibly acts as an internal poison. In soft water lead may be very toxic; in hard water equivalent concentration of lead are less toxic. Calcium in a concentration of 50 mg/1 has an ability to destroy the toxic effect of 1.0 mg/1 of lead.

Zinc and Copper

Doudoroff (1952) found that although fish could survive for 8 hrs at 8 mg/1 of zinc alone and 0.2 mg/1 of copper alone most fish died within 8 hrs when exposed to a mixed solution containing 1.0 mg/1 of zinc and 0.025 mg/1 of copper.

Zinc is somewhat less toxic than copper. It has a toxic effect towards protozoa and bacteria but not as per with copper.

Venugopalan (1979) from his experiments conducted to study the effects of copper and zinc on growth and photosynthesis of *S. costatum* and *N. Longissima* (marine species) remarked that copper concentration up to 10 mg Cu/1 (0.01 mg/1) increased growth and photosynthesis in both the species. Although cell multiplication was reduced at copper concentration above 25 µg/1 (0.025 mg /1), photosynthesis was not found affected even up to a concentration of 50 µg/1 (0.05 mg/1). He further observed that unlike copper, relatively a higher concentration of zinc, 75 µg/1 (0.75 mg/1) was tolerated by *N. congissima* without any detrimental effect on the growth and photosynthesis; On the other hand, *S. costatums* was able to tolerate only up to 10 µg/1 Zinc (0.01 mg/1). These results indicated the existence of ecologically significant intraspecific differences in the physiological and metabolic adaptions of the micro-algae, although they co-existed in the same niche and under the same environmental conditions. He also observed that diatoms and other algal forms would revive after exposure to zinc pollution when transferred to a normal medium, but chances of such revival were remote with copper pollution.

Mercury

Mercuric ions are considered to be highly toxic to aquatic life. For fresh-water fish, concentrations of 0.004 to 0.02 mg/1 of Hg have been reported to be harmful. For stickle back *(Gasterostem aculeatus)* the lethal concentration limit has been found to be 0.008 mg/1. Unstable compounds mercuric sulfate and nitrate have killed minnos at a concentration of 0.01 mg/1 mercury after 80-92 days. At concentrations of 0.05 and 0.1 mg/1 as mercury fish were killed in 6-12 days.

In contrast, Schweiger (1961) reported that 0.2 mg/1 of mercury was not harmful to one and two year old tench, carp, rainbow trout and char, nor to fish food organisms such as crustacea, worms and insect larvae. The toxic effects of mercuric salts are accentuated by the presence of trace amounts of copper. (Water quality criteria, 1963). Mercury is hardly affected by water hardness.

The peculiar symptoms of mercury as stated by Fujiya (1977) are :

(a) Atrophic change and disappearance of granular layer of carebellum.

(b) Changes of spinal ganglionic cells—shrunken nuclei and disappearance of granules

(c) Atrophy of ganglionic cells.

(d) Dilation and distortion of axon.

(e) Haemorrhagea around the falciform process in eye.

The above symptoms were observed both in marine and freshwater fish and it can be understood that the brain is affected seriously. The symptoms are found only in the cases of organic form of mercury compounds such as methyl mercuric chloride and never found in the case of inorganic form such as $HgCl_2$.

Chromium

Toxicity of chromium salts towards aquatic life varies widely with the species, temperature, pH, valency of chromium and synergistic and antagonistic effects especially that of hardness.

Fish are relatively tolerant to chromium salts but lower forms of aquatic life are extremely sensitive. (Water quality criteria, 1963)

Copper, Lead, Mercury, Nickel, Silver and Zinc

Stones (1976) stated that if these salts are present in 1.0 ppm concentration, can reduce BOD by 44, 23, 72, 25, 81 and 8 per cent or alternatively, if the water contains 0.2, 1.2, 0.07, 1.0. 0.03 and 7.0 ppm of the metals, BOD can be reduced by 25 per cent.

Importance of hardness, PH and disssolved organics

The degree of water hardness influences the toxicity and consequently the activity of heavy metals by forming insoluble carbonates or by adsorption on Calcium Carbonate. Thus in hard water, the concentration of heavy metal necessary to reach the level of lethal dosage must be greater or lower when the water hardness is less. But mercury is hardly affected by water hardness. By differentiating between calcium and magnesium hardness, Tabata (1969),[a] established that, for Daphnia, the influence of calcium hardness was dominant.

pH plays an important rôle in the interactions between heavy metals and parameters such as carbonate hardness and organic compounds. In experiments by Shaw and Brown (1974)[b] in which a copper solution in hard water (at pH-8.0) was not renewed, the concentration of dissolved copper dropped to 0.5 mg/1 from the original concentration of 1.1 mg/1 within two hours, thereby reducing the toxicity of the "nominal copper concentration". The exact opposite effect was recorded by Whitby (1963)[c] in his experiments with lead conducted on tubificid worms. Lead was found to be more toxic at pH value of 8.5 than at pH 6.5 In his experiments, lead was precipitated at a Ph value of 8.5 was most probably incorporated orally; This confirms the fact that lead has low distribution co-efficient in the aquatic environment (Koppe, 1973).[d] Hence it is of utmost importance to consider the metal species in solution when establishing toxicity limits.

In respect to organic substances changes in pH value in the water can strongly influence the adsorption or desorption of cations. For example amino acids which occur in substantial amount in both freshwater and in sediments in the eutrophied or polluted waters can adsorb or desorb cations due to the pH dependent amphoterie character of the acids.

The combined effect of temperature, oxygen content, hardness, organic compounds, pH value and salinity shows the many influences which vary the toxicity and the incorporation of heavy metals in aquatic organisms. The heavy metal content determined in an organism should by no means be regarded as a constant value but rather as a factor subject to the influence of varying biotic (endogenic and oxogenic) and abiotic environmental conditions.

Singer (1977) summarized the influence of dissolved organics on distribution of metals as follows :

Dissolved organic substances are capable of

(1) Complexing metals and increasing metal solubility.

(2) Altering the distribution between oxidised and reduced form of metals.

(3) Alleviating metal toxicity and altering metal availability to aquatic life.

(4) Influencing the extent to which metals are adsorbed on suspended matter.

(5) Affecting the solubility of metal-containing colloids.

Besides chelation, colloidal suspensoids absorb considerable amount of metallic pollutants and non-polluted ions settle mostly around the outfall region and thereby reduces their concentrations from the water. The settled suspensoids containing metallic pollutants by their gradual accumulation increase considerably around outfall, and in all likelihood form artificial zeolite type of material which may again take part in the exchange reactions, and by this process make certain metallic ions available to the overlying flowing water to be carried and re-adsorbed on colloidal suspensoids and settle to a new area where such metals have no reason to be present. Such phenonmenon were recorded by Ray (1988) in the Hooghly estuary. Further, such high metallic accumulation in all likelihood would have made the river bed (especially around the outfall area) partially biologically unproductive. But this does not happen possibly as the condition before it reaches almost to an alarming state, when bio-accumulation of heavy metals by benthoes takes place (such phenomenon

a—d referred from the book "Metal Pollution in the Aquatic environment" by Förstner, U and GTW. Wittmann, 1979.

is also observed by Jhingran and Joshi, 1987 in the river Jamuna; by Ghosh *et al.*, 1983 in the Bhagirathi-Hooghly) fresh silting of monsoon covers the river bed and biotic life is saved despite the fact that some bio-accumulation has already taken place in some biotic life. With the cessation of monsoon, the organic material (dissolved and suspended) generated from sewage and industries with the help of temperature and solar energy helps entrophication of the ecosystem, the basic requirement of benthic life. Thus, increase in algae and inverterbrate production could possibly influences growth, reproductive success and recruitment of fish in an ecosystem where self purification gradually improves. The cycle of ecological balance is thus attained though not completely so long the stress is within limit.

Process involved in the accumulation of toxic material

Fujiya (1977) has clearly explained the process by which accumulation of toxins take place in fish and other aquatic animals. According to him the main passages are :

(a) Accumulation through (I) digestive tract and the other is through (II) gill.

I. In the ecosystem, the accumulation and concentration occurs in low level organisms, such as, phyto-plankton at first and then the process rises up to higher trophic level.

II. Usually, the food conversation ratio of fish is 6-10 (average 8). Therefore, it may be possible to find 8 times concentration of accumulative substances in each step of food chain, when the substance is contained in the food and finally relatively high concentration would be occurred in fish body.

(b) On an average the requirement of the essential amount of oxygen in fish is about 300 mg/kg/hr. Assuming the dissolved oxygen content in water is about 10 mg/1 and approximately 30 litre/hr. of water should be filtered through the gill to take enough oxygen in the body which comes to 720 litre/day. If some accumulative substances are present in water, it should be possible to concentrate about 700 times in fish body. The phenomenon has been observed based on several experiments, and it is estimated that the accumulation through respiration is significantly higher than that of feeding for fish.

From the above experiments it can be understood easily that the problems of accumulation are more significant than that of direct and acute effects of such pollutants, though the direct effects are usually estimated as the source of damage of fishery products.

Extent of accumulation of pollutants

Fujiya (1977) has provided some examples of accumulation of metallic pollutants in the muscle of molluscan shell fish which are often harmful to people. The oyster cultured in seawater containing copper of 0.5 ppm have finally attained to 2.000 v/g-wet weight and in 0.25 ppm 1,900 v/g and in 0.13 ppm, 1,500 v/g in each muscle of oysters within three weeks. It shows that oysters is capable of accumulating copper approximately 4000-10,000 times from the environmental waters.

The belief that mollucs accumulate toxicants easily compared to fish may not be true as there is not much of difference.

Some insecticides like DDT and BHC get accumulated in the fish muscle from the ecosystem. The pollutants present in water initially get their pathway through the primary producers. Subsequently the concentration increases in herbivor and finally transmitted through secondary and tertiary food chain. Relatively in simple food chain DDT may concentrate to about 85,000 times than that of water.

Dieldrin is concentrated to 70,000 times in the liver of trout from water, when the fish are exposed to a sublethal concentrations.

Other effects

Appearance of deformity. Several types of deformity can be observed, and they are a pughead, a rough appearance of body surface with tumour, disappearance of external organs, such as fins and a deformity of spines *etc*. The deformity are of two types. One occurs during the embryonic development stages and the other is observed during the growing stages after hatching.

During the embryonic development a very small fluctuation of the environmental conditions like water temperature, pH, Dissolved oxygen content and salinity in the coastal area influence the results of the production, since the organisms in the developing stages are very sensitive and delicate. Sometimes, the toxicants accumulated in female breeders more in ovary in maturation process and consequently the eggs contain the toxicants. But the actual situation of deformation for embryonic development has not been investigated clearly.

Besides the developmental stages, a serious trouble can be found frequently in fish. The most serious symptom is the curvature of vertebra and the pesticides like organo-phosphates and carbamates cause the effects.

Experimentally it has been observed that fish exposed to concentrations of approximately 1/10 of the lethal doses are affected in most cases. At the first step, a convulsion occurs on the muscle of lateral side and then a contraction of muscle is observed. This is the main cause of curvature.

Appearance of anaemia

Fujiya (1977) has categorised anaemia as three types :

1. **Haemorrhagic anaemia :** Some toxicants hurt the surface of gills and skin. This results aeute loss of blood.
2. **Haemolytic anaemia :** In most cases some toxicants enter into the circulatory system and are mixed with blood. As a result haemolysis occurs to cause anaemia.
3. **Functional anaemia on Haematopoiesis** : Fish has no bone marrow and the kidney is compactly related to the function. On the other hand, kidney is also one of the most important organs for detoxication and it is easily affected by toxicants. Thus often the functional anaemia can be observed on fish as the effects of pollutants.

Among these three types of anaemia, the haemorrhagic one is affected by herbicide. This substance injures the gill tissue and a loss of blood through the injuries causes the serious effects within 4-6 weeks. Such injuries occur in carp culture, but the same is also reported in the case of rainbow trout. It is quite likely that similar results may be obtained for other species of fish besides the carp to have the same damage.

In the Ganga, heavy metals like zinc, copper, chromium and cadmium were estimated at three points, city side, mid river and opposite bank for the centres like Haridwar, Kanpur, Allahabad and Patna. Of all the centres, city side of Kanpur indicated highest values for zinc (285 ppb), copper (175 ppb) and chromium (2 ppb). This was followed by Allahabad and minimum at Haridwar. Cadmium, lead and mercury indicated low values.

In the Hooghly estuary around the outfall, the upper level of zinc ranged between 0.05 and 7.2 ppm, the average value of which was more than 6 times the upper level of Kanpur and chromium

0.2 and 20 ppm. copper and lead could not be detected. In the sediment however, the upper level for zinc was 381.6 ppm (Rayon industry outfall) followed by 160 ppm (outfall) of Paints and varnishes) and 80 ppm (outfall of Oil Refinery).

In the river Jamuna the ranges of zinc were 22-54.76 μgl^{-1} (0.022-0.05476 ppm) ; copper 2.34-18.03 μgl^{-1} (.00234-.01803 ppm); chromium, nd-1.32 μgl^{-1} (nd - 0.00132 ppm); cadmium, nd-0.38 μgl^{-1} (nd-0.00088 ppm); lead, 00.82-6.90 μgl^{-1} nd (0.00082-0.0069 ppm) and mercury, nd-0.17 μgl^{-1} (nd-0.00017 ppm) in the filtered water, between Delhi and Allahabad. Undoubtedly all such values are quite low.

In the sediment the values were much higher in the city side bank. The concentrations were zinc, 44.33-284 .8 μgg^{-1} (0.04433-0.284 mg/g); copper, 12.57-81.35 μgg^{-1} (0.01257-0.08135 mg/g); chromium 3.54-99.45 μgg^{-1} (0.00354-0.9945 mg/g); cadmium, nd-3.85 μgg^{-1} (nd-0.00385 mg/g); lead, 28.48-105.12 μgg^{-1} (0.02848-0.105 mg/g); and mercury, 0.06-1.052 μgg^{-1} (0.0006-0.001052 mg/g) between Delhi and Allahabad.

It is evident from the above that in the upper and middle stretches of the Ganga, metallic pollutants were comparatively higher in the city side than the middle or opposite bank. Main polluted centre was around Kanpur followed by Allahabad and was mainly from zinc and copper (Fig. 23), and compared to unpolluted centre (Haridwar) Zinc concentration was three times more, copper was more than seven times and lead was more than three times.

In the Hooghly estuary, zinc was six times more than the highest value recorded around Kanpur.

River Jamuna was affected mainly by zinc, copper, lead and mercury. But in all cases the concentrations were at a much lower level. It is interesting to note that the average concentrations of Zinc and lead were three times, while copper was five times more than the concentration recorded above Delhi (unpolluted centre). Concentration of mercury on the other hand was eight times higher than the concentration of unpolluted water. During extreme summer, the metal concentration in the water reaches its maximum due to its low dilution ratio but simultaneously the chemical constituents like calcium, magnesium, phosphorus *etc.* also get concentrated and due to their higher concentration the toxic effect of metals like zinc, copper, lead *etc* get neutralised to some extent by forming insoluble carbonates or by absorption on calcium carbonate. This is a common phenomenon in case of R. Ganga also. Again during monsoon the above salts get diluted. As a result there remains a probability of increasing the toxicity of metals under low hardness. But this does not happen, as simultaneously the metals released along with the wastes also get diluted and high colloidal suspension brought in the water body absorb the metals and settle. The magnitude of silt suspension brought during monsoon due to their precipitation also reduce the effect of metals to some extent absorbed earlier by benthes during pre-monsoon days. In case of late monsoon or draught the problem becomes more acute. Thus nature controls the ecosystem and protects it from any ecological disaster.

Thus although the toxic level increases, it remains at a comparatively lower level for the reasons stated earlier. Perhaps the entire river system (both in the Ganga and the Jamuna) remains contaminated by the presence of such metals and wherever fish migrates for feeding or breeding, accumulation of heavy metals occur in their system through their feed and respiration as according to Fujiya (1997) accumulation and concentration of toxins takes place through digestive tract and gills. Under such condition fish and other aquatic life may not be killed and can accumulate such metals in their system at a slow pace which gradually increases. These facts can be revealed by the accumulation of metals in fish and in its tissues, crustaceans like shrimp, carb and molluscs. These are common phenomenon both in the Hooghly estuary and in the Jamuna.

In the Hooghly estuary average Zinc accumulation in fish (50-185 mm) ranged between 11.1 and 70.6 ppm; Shrimp (40-85mm), 23-61.6 ppm; crab, 24.4 ppm and molluscs, 83 ppm; Copper 2-20 ppm in fish (54-115 mm); Shrimp (40-85mm) ranged 8.1-24 ppm; molluscs, 20 ppm and chromium in fish (50-149 mm) ranged nd-3.1 ppm in the Hooghly estuary. Besides, in longer specimen, *Silago panijus* (272-274 mm), Zinc accumulation was recorded in gill, muscle, kidneys, gut (intestines) and gonad except in liver, with highest concentration in the gill (40-60 ppm), and copper was present only in gut (nd-30ppm). In *Rita rita* (420-430 mm) zinc accumulated in all the tissues stated above, the highest concentration recorded being in the gonad (137.5-146.8 ppm); copper was recorded in muscle (2-4 ppm) and in Kidney (5-15 ppm) and Chromium in gill only (15-19.2 ppm). In *Hilsa ilisha* (374-493 mm) except kidney in all other bissnes and organs stated above accumulated zinc, the highest concentration recorded was in gill (22.4-26.0 ppm); copper was recorded in muscle, liver, and gut, the highest concentration recorded was in liver (11-17 ppm).

It is interesting to note that zinc content in the Rayon waste was highest which was around 8.5 ppm and in water around its outfall, the concentration came down to 7.2 ppm. Its accumulation recorded in small sized fishes, shrimp, crab and molluscs was several times more than what was estimated in water. Even copper which could not be detected either in water or sediment was found to be very high both in small sized fish, shrimp and molluscs and in different tissues of large size fish. Lead was not detectable in all the test organisms. It is interesting to note that Chromium accumulation indicated lowest value when it was abundant in the water around tannery waste.

Bottom dwelling animals like crab, shrimp and molluscs however accumulated a very high concentration of zinc, possibly some metal enriched sediments are ingested along with food material. This fact is revealed in two groups of fishes, plankton and bottom feeding crastaceans of average size 102 mm (6.9 gr.) in the case of former, the accumulation of zinc and copper were respectively 40.8 ppm and 14.0 ppm (average value); while in case of later, the bottom feeders of average size 67 mm (2.9 grs) only, the concentrations were almost similar i.e. 40.2 ppm and 14.4 ppm respectively. This shows the bottom feeding crustaceans even though were quite smaller in size, indicated accumulation of zinc and copper similar to larger sizes (plankton feeder). * This finding is based on the results obtained by Ghosh, *et al.* (1983). Possibly they might have ingested orally some precipitated from the river bed as opined by Whitley (1968).*

In the river Jamuna heavy metal accumulation in fish *Rita rita* and other miscellaneous fishes of the upstream location (above Delhi) was lower than the metal contents detected in the fish collected from the downstream locations (Delhi and below) (Fig. 41 & 42). The mercury content in the trash fishes collected from around Delhi showed violation of safe limits (FAO, 1983). The copper and Zinc levels in fish (muscle) and molluscs (whole body) without shell) exceeded the permissible limit of copper (0.2-3.6 μgg^{-1} and Zinc 2-17 μgg^{-1} wet wt. (ICES, 1974). The metal enrichment in molluscs is closely associated with the metal content of sediments. Aquatic weeds mainly *Eichhornia Crassipes* which was found around Delhi showed highest accumulation of Zinc (182-193 μgg^{-1} dry wt. Although much higher levels of DDT have been reported in water (0.60-3.42 ppb) and fish (0.06-36.9 ppb) in the Jamuna, near Delhi compared to tidal stretch of the Ganga (Jhingran, 1988), their adverse effects on the health of fish is yet to be studied.

Konar (1969) however observed damage of liver cell of *L. rohita* and *C. mrigala* exposed to DDT (0.005 ppm) for 30-60 days. Chronic exposure to low levels of DDT (5.2 ppb) and BHC (150 ppb) has

*In *Metal Pollution in aquatic environment* by Förstner, U. and G.T.W. Whittmann, 1979.

resulted in lower growth and fecundity rates in *O. mossambica.* Deleterious effects on reproduction, growth and haematological condition have been observed in fishes at sub-lethal condition (0.0017-2.950 ppm) of various organochlorine pesticides (Konar, 1969). Exposure to 0.1-1.0 ppb of DDT for 7 days inhibited the growth and induced sheel thinning in gastropod *Lymnea* Leuteola the results expressed by Konar (1969) are restricted to laboratory conditions only.

Calamari (1976) has established a nice correlation between environmental stress and living organisms. Aquatic ecosystem as calamari states are complex entities in dynamic equilibrium between the tendency to change and the capability of homeostasis. The result of these two conflicting properties of the living system is an environmental steady state with very slow, long term displacements, but the influence of human activities tends to alter the system by the addition of extrenous energy or matter which might also involve change to the abiotic part of he environment. The condition of life are largely determined by the properties of the abiotic part of the ecosystem, which in tern can be changed or influenced by the biotic components and thereby life can exist only within a narrow range of conditions.

In the presence of environmental stress any living organism has the tendency to maintain its internal environment within the limits permitting survival and reproduction but this property can be practiced only in a certain range of conditions.

Thus when organisms are subjected to increasing concentrations of a substance their homeostatic power is reduced. But when the impairment of the vital functions increases to a point of intoxication that does not permit recovery. The phenomenon is shown in Fig 45. Perhaps organisms of the Ganga—its estuary and the Jamuna are gradually heading towards such conditions.

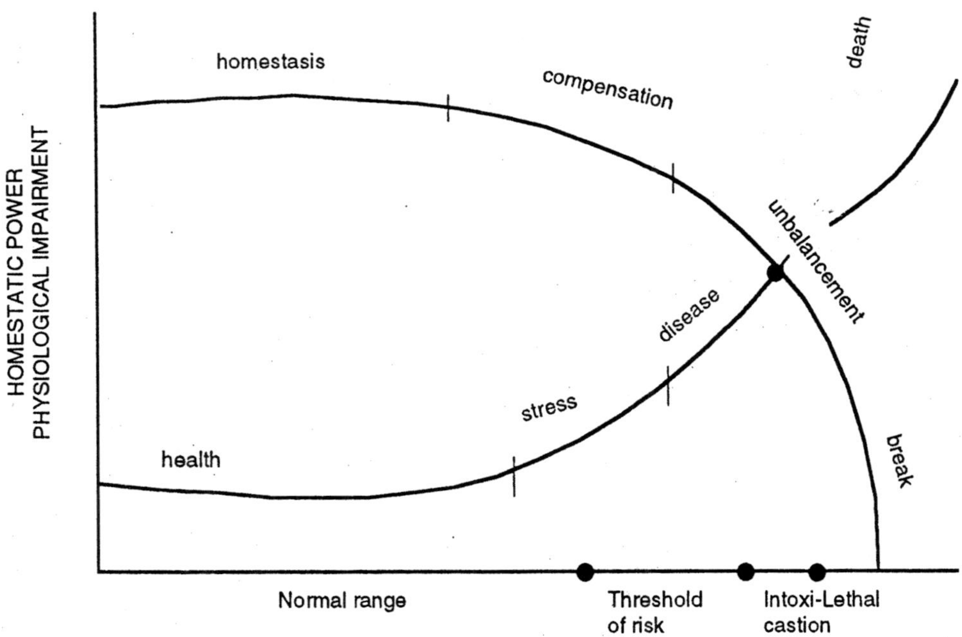

Fig. 45 : Environmental stress and aquatic life.
Source : Calamari, D. (1976).

Thus the abiotic and biotic factors affecting heavy metal concentration in aquatic organisms temperature and oxygen content, hardness, organic compounds, pH values, general physiological behaviour, life cycle and life history, seasonal variations, species specific and individual variability, contamination by food and intestine content *etc.* must be taken into consideration when evaluating the metal contamination of biologic system. This is particularly true for determining the degree of metal enrichment within the food chain. The omission of any single factor can lead to serious misinterpretation.

The accumulation of different metals/pesticides given in the above account must have occurred through digestive tract and gill as experienced by Fujiya (1977) & Whitley (1968). But how far such accumulation of pollutants has affected the commercial important fish (of different sizes) in their healthy growth and reproduction and other aquatic biomass including plankton have not been studied so far.

Bhargava (1981a) has however studied the water quality index of the Ganga and the Jamuna which may indicate simply to what an extent the water is recovering and fit for fish culture. The self purification of the Ganga as Bhargava (1981) opions takes place through exocellular polymers present in the Ganga which interact with colloidal matter present in sewage discharged into the river and this resulted in a considerable reduction of the BOD and hence turbidity also in a very short time. The fact that Ganga has the capacity of self purifying ability is shown by the fact that BOD rate constant values and reaeration rate constant are not only extremely high but perhaps highest so far accepted values for any other river, and that the colifom group of organisms are also found to reduce very significantly in a very short time. Thus it is clear that the Ganga water always remains in a highly fertile condition. The river being shallow during most of the time with low water velocity (up stream 0.35-0.9m/sec; mid river, 0.35-0.44 m/sec and 0.09-0.56m/sec in lower stretch) during May and July and pH, BOD, DO, COD are not found abnormal, produces moderate to high concentration of plankton in the entire fresh water zone.

Bhargava (1981a) has extra-polluted water quality indices (WQI) for the Ganga waters in the upper Ganga between Rishikesh and Varanasi for fish culture in different centres both during summer and winter on the bases of parameter such as temperature, transparency, DO, BOD, conductivity, chloride, hardness, alkalinity, pH, coliform MPN and NH_3-N and is indicated in Table 86.

Table 86 : Water quality index of the Ganga in the upper Ganga plain

WOIS	*Rishikesh*		*Narora*		*Kanpur*		*Allahabad*		*Varanasi*	
	S	W	S	W	S	W	S	W	S	W
Upstream	89	100	72	73	65	80	66	90	74	90
Down steam	85	99	66	72	23	91	66	86	53	71

Note *:* 90= excellent ; Between 65-89 = Good; Between 35-65 = satisfactory ; Between 11-34 = poor and below 10 = unacceptable.

S : Summer

W : Winter

It is evident from Table that water quality at Kanpur is at lowest ebb, particularly during summer months. Although complete recovery of water quality does not take place even up to Varanasi. All other centres are also affected to some extent.

It is also evident from Table 86 that in a stretch of about 800 km of the upper Ganga (Rishikesh to Allahabad) the average water quality index of unpolluted upstream of the five cities during winter is estimated to be 86.6 which during summer comes down to 73.2. This means that during summer water quality index at the unpolluted up-stream falls by 11.3 per cent. At the down stream (Recovery zone), the average values (WQI) of the above 5 centres estimated to be 83.0 during winter which during summer comes down to 58.6 *i.e.* a fall of 16.6 per cent against 11.3 per cent szone (down stream) during winter showed value lower by 4.5 per cent than upstream during winter which increases to 11.6 per cent during summer. This shows that during summer the stretch requires further extension downstream for more dilution to attain better WQI. Further impact of pollution depends on lateral, vertical and longitudinal mixing which is determined by turbulence. The turbulence is dependent on discharge, slope, depth, canal roughness, configuration and wind action (Ray, 1981).

Bhargava (1985) had worked out the water quality indices of the R. Jamuna for fish culture as 58, 50,35 and 88 at Delhi, Mathura, Agra and Allahabad respectively. He further added that in case metal contents of water were also taken into account the water quality indices would further go down.

Based on the same statement of Bhargava (1985) the water quality index of the Ganga stated above (which stands between satisfactory to good) would also go down further in case metal content of water which are higher than the Jamuna water are also taken into account. This shows that water quality of both Ganga at the usptream (Narora-Varanasi) and Jamuna for fish culture are far from satisfactory. The maximum admissible levels of zinc, copper, chromium and lead in fishery waters are 30,5,50 and 30 $\mu g/g^{-1}$ (Mc Neill *et al.* 1979). copper and zinc levels in fish and molluses (whole body without shell) exceeded the world base line level of copper, 0.2-3.6 μgg^{-1} and zinc, 2-17 Mgg^{-1} wet (ICES, 1974). The fishes and other crustaceans of the Hooghly estuary also accumulated much higher values than the world base line level. This indicates that fishes in both the rivers—the Ganga and the Jamuna are living in an ecosystem which is far from satisfactory. This is also evidenced by the fact that, major carps (Catla, Rohu and Mrigal) and Hilsa which once attained maximum lengths of 180 cm, 91.11 cm and 99.2 cm and 50.1 cm in the past has come down to 61.8cm, 50.0 cm and 55.8 cm, and 34.7cm respectively during late eithtees. Also during the severe drought of 1987 in the country as a whole and the Gangetic valley in particular a fall in length of major carp, 11.7 cm 7.0cm and 6.9 cm in the above order, and catfishes (0.9cm for *M.* acr; 5.7cm for *M. seenghala* and 7.4 cm for *W. attu* and 4.9 cm for *hilsa* recorded during 1988.

The overall picture of stress on the aquatic biomass in the Ganga rivers system are highlighted and are resulted from :

(a) Heavy siltload brought in the Ganga river system are resulted from intensive deforestation.

(b) Physiographical damage of the river system, namely shifting of river course, formation of remnants and loss of deep pools *etc.*

(c) High abstraction of river water for various uses resulting loss of water level and release of waste water.

(d) Destruction of spawning grounds, breeding of fish, loss of egg/spawn, bioaccumulation of metals and pesticides and capture of large scale juveniles and brood fishes.

(e) Ecological hazards due to poor dilution ratio, metallic and pesticide pollution.

(f) The resultant impact on bio-mass, stream recovering capacity unhealthy fish—their survival, growth, reduction in fecundity and reproduction, quality of fish and their composition and finally production potentiality of the river system as a whole.

16

Environmental Changes Vis-a-Vis Riverine Ecology

After meticulous examination of the foregoing account it is revealed that human activities for their own intereat are playing the major role in creating the ecological imbalance of the Ganga river system without considering the consequence and effort to solve them.

With the population influx, deforestation has become necessary for their settlement. Such deforestation without looking for adequate afforestation, leaves terrible impact on recurring annual soil erosion which has far reaching effect on silt precipitation in river bed. Water holding capacity of the river is gradually reduced. As a result overflows or breaching of river banks from flush floods become a common phenomenon these days. (Refer plates given in the document).

To meet the food problem wide scale irrigation in larger areas in different states have become essential. To help irrigation, dams, weirs, barrages are constructed. Monsoon plays its major role for 3/4 months. India being a tropical country, the rate of evaporation during non-monsoon, months is very high. Conditions are such that rain waters are retained as far as possible in reservoirs and also in deep pools of dams, barrages, weirs etc. Besides, some of the rivers including Ganga being snow fed receives additional flow during summer. But with all such efforts the demand for freshwater is gradually on the increase leaving low flow in the main river downstream which has no capacity to erode the precipitated silt. Such weirs, dams, and barrages on the other hand have adverse effect of silt precipitation over and above by arresting the flow of water many of the sluice gates become inoperative. Unfortunately desilting by dredging, either in reservoirs, dams or weirs from the main river bed has not been thought of or sometime it is partially done. Hence the problem is gradually heading from bad to worse. Thus on one side we find heavy silt precipitation during wet months while during the dry months there is appreciably low flows. Under such condition shape factor goes high during monsoon due to shallow depth. Velocity or water is very much arrested due to low slope. Large sand dunes or sand bars come up one after another in the river bed. Deep pools where Gangetic major carps normally harbour, get filled up due to silting and the commercially important major Carp fishery is gradually changed to non-commercial one suitable for their growth and reproduce in such shallow areas. The river course is changed leaving their remnants behind in the form disconnected pools Ox-bows lakes and other lentic water bodies etc. With the onset of monsoon large scale breeding occurs in such shallow areas. Spawn grows up to immature juvenile size within 2/4 months and are trapped with the receding of water. These fishes mostly cannot grow up to mature size. Such phenomenon area also not uncommon in the main river. All these not only lead to heavy economic loss but also dwindle the fishery of economic importance. These are the effect of chain action of deforestation and heavy abstraction of river water and are results from human interference.

Secondly, the population in cities and industries are growing fast along the river banks, dispose waste water in the river and complicates the ecosystem further, so much so, that the ratio of waste to river water is gradually reduced with the lowering of water level due to heavy abstraction. In Uttar Pradesh during summer the waste exceeds the river water. The industrialists (except a few cases) are not interested to treat the effluent. The ecosystem further deteriorates thereby affect the biomass production further.

Thirdly, under such critical physiographical stress and limnological conditions scientific organizations often go to the extent of providing erroneous results of waste and water analysis (Bhargava, 1992, and Mazumdar 1989). This has a far reaching effect on the standardisation of toxic substances present in the waste. Ecosystem of both river and tidal estuary are different and hence sampling also should be done in such a manner that it should be representative. When the situation did not show any significant improvement, Supreme Court, more recently with their own team of experts took up—the cases and requested Municipality and factory anthority (irrespective of private or Govt. owned) to treat their effluents before releasing into the river, as the highest judiciary of Govt.of India as they could not rely upon SWPCB report. Even they have imposed pollution fines against some Govt. Owned companies for not installing pollution Control devices (1 & 2). Such was the impasse prevailing upon in case of industrial sector and also in municipal organisations.

Besides, lock out, lay-off etc. due to labour unrest or other reasons, the factory gets completely suspended for certain period. Under such circumstances the ecological condition of river water changes from time to time along with the rise and fall of pollutants. Condition being such any conclusion drawn by the scientist could be erroneous unless a sort of stability is achieved in the ecosystem.

Thus from the foregoing account it is clear that human activities are the main source of bringing imbalance in the ecosystem through a chain action indicated above and which they cannot solve even after spending so much of fund, and on the contrary get entangled more and more with the problem.

Man lives under two environments. One is nature. The other is social, economic and political atmosphere which he himself creates. The reaction of the second on the first causes pollution. This pollution should be kept to the minimum by recycling the pollutants and maintaining the balance of nature. The environment must be understood as one and is to be shared by man, animal, plants and micro organisms. They are all dependent on one another and cannot exist themselves (Deoras, 1973).

1. Supreme Court : July 3, 1993 | Nov. 6, 1993 | Dec. 6, 1993 | Dec. 19, 1993 — Ananda Bazar Patrika, Calcutta
Oct. 18, 1993 —The Statesman, Calcutta
Feb. 4, 1995 —Bartaman, Calcutta
Nov. 25, 1995—The Statesman, Calcutta

2. Central Water Pollution : Sept. 19,1993, Bartaman, Calcutta Control Board.

17

The Ganga and the Bhagirathi-Hooghly—Their Past and Present

The Ganga, a perennial mighty river which was once known for its characteristic feature namely—depth, flow, volume of water, water quality, rich aquatic life (flora and fauna) including fish and for which it was known as life line for millions has lost its past glory for various reasons. With the population influx, deforestation which has become essential for habitation and for their comfort, urbanisation, industries and expansion of agriculture started at a rapid pace especially after independence. As a result deterioration of natural healthy ecosystem of the river started rapidly by way of heavy silt erosion and their precipitation in the river bed, resulting recurring floods, high abstraction of water, during non-monsoon months, release of toxic waste water from the industries and municipalities, and pesticide from agricultural fields etc. which affected the fishery wealth.

Simultaneously Bhagirathi-Hooghly has also met with similar fate. The main flow of the Ganga which once used to flow through this system is now shifted to the Padma river. The mighty river Damodar of West Bengal that used to contribute its flow in the Bhagirathi changed its course and joined further down. All these have brought a setback to the Bhagirathi-Hooghly. The bed is considerably elevated, so much so that Bhattacharya (1986) opined, "In the near future the bed of the Hooghly will raise to a level of 10/12 ft higher than the level of central Calcutta. As a result during rains Calcutta will remain inundated for days together and will be difficult to remove the water as Calcutta will be surrounded by high embankments". However, the entire ecosystem which was once known for its rich flora and fauna is now limping for its survival. Below are given some pictures of the past days and the alarming condition the entire Ganga including Bhagirathi-Hooghly is heading at present. The limited data of the past and early to that (fifties onward), whatever is available are presented here to show how the critical balance of the Ganga ecosystem which existed in the past is being gradually altered during the present days for which "Ganga Action plan" has become necessary to offer suggestions and implementation to save the river so that the past glory is revived.

It was during 1828, East India Company contemplated the idea of making a trip from Calcutta to Allahabad by steam ship. It was captain Henry Johnson who first made the trip in 23 days. He had no doubt about the flow and depth of the Ganga and hence could easily covered the distance. Subsequently in 1830 Captain Jonhson made similar trips to Allahabad and back. In this context in recent past some hydrological status of the Ganga as reported by Sarkar (1988) are indicated below. Since ecological study of the Ganga river system was initiated only around fifties when degradation of physiographical and limnological changes was already in the process any information prior to fifties on physico-chemical and biological status could not be provided.

Hydrological Status (1869)

Stretch of maximum depth in the Ganga : Between Allahabad and Rajmahal (843 km)

Maximum depth at Rajmahal :11' - 15'.

Maximum discharge : 15,00,000 Cusecs.

Velocity of water during summer : 3 miles/hr or 4.4 ft/sec

Velocity of water during other period : 7-8 miles/hr or 10.27-11.7 ft/sec.

(*Source* : Sarkar 1988)

Physico-Chemical Status

Physico-Chemical status of the Ganga during the above period is not available except around 1960. The stretch studied was between Kanpur and Rajmahal at seven Centres. The average values of lower and upper levels with their ranges are indicated below :

Parameters	*Range*	*Average values*
pH.	7.45 - 8.3*	8.0
Dissolved Oxygen (ppm)	5.3 - 9.7	7.4
Bicarbonate alkalinity (ppm)	82 - 195	135
Specific conductivity (μmhos)	257 - 328*	287.6
Total dissolved Solids (ppm)	130 - 170*	148
Chloride (ppm)	4.2 - 20.1	12.1
Phosphate-P (ppm)	0.07 - 0.14	0.1
Nitrate-N (ppm)	0.087 - 0.17	0.13
Silicate (ppm)	6.1 - 15.3	10.7

Source : Jhingran and Ghosh (1978)
*Jhingran (1989)

POLLUTION STATUS

Prior to independence (1947) the demand for water of the Ganga river system was not as high as it is today due to inadequate planning and development of industries, urbanisation and agriculture. As a result release of complex waste water besides municipal waste disposed into the Ganga was very limited and was practically negligible compared to the huge volume of water that used to flow in the Ganga. The dilution ratio naturally being high, the question of pollution was kept to the lowest level.

BIOLOGICAL STATUS

Plankton

Plankton composition and their abundance studied by different workers although was somewhat different in their methodology, the results obtained by them will atleast give some picture of the ecosystem. Systematic study of plankton was made from fifty onwards. Initially plankton studied by Ray *et al.* (1966) in the Ganga around Allahabad for Phyto and Zooplankton during 1958-59 showed concentrations 3588 and 230 unit/litre, respectively. The dominant species for phytoplankton were—Bacillariophyceae, Chlorophyceae, Myxophyceae and Dinophyceae while in case of Zooplankton they were Rotifer and Copepods. Subsequently Pahwa and Mehrotra (1966) studied between Kanpur and Rajmahal during 1960 at seven Centres.

Yearly average of Plankton (U/L) at seven Centres

Centre	*Phytoplankton*	*Zooplankton*
Kanpur	8521	363
Allahabad	606	372
Varanasi	2323	180
Ballia	944	51
Patna	143	18
Bhagalpur	557	53
Rajmahal	593	97

Source : Pahwa and Mehrotra (1966)

The dominant forms were :

Phytoplankton : Bacillariophyceae, Chlorophyceae and Mexophyceae

Zooplankton : Rotifer, Copepoda and Cladocera.

Fishery Status : Commercial important fish and their sizes in the Ganga and the Jamuna in the pest.

	Species	*Maximum length recorded*	*Source*
			(1871)
(a)	*Catla catla*	180 cm.	Day (1878)
			(1889)
			Raj (1939)
		123 cm.	Jhingran (1968)

Source : Synopsis of Biological data on *Catla Catla* (Hamilton 1922) FAO Fisheries Synopsis No. 32. Rev. 1, 1968

(b)	*Labeo rohita*	91.11 cm	Day (1878)
		73.7 cm.	Shaw and Shabbeare (1937)

Mean length 450-470 mm from February-July.

(Source : FAO Fisheries symposium No. 111, SAST. *L. rohita* (1975) and in case of(c) *Cirrhina mrigala :*

Maximum size 992.00 mm at the time of XII completed annuli (Jhingran, 1958). (In : FAO Fisheries symposium No. 120; 1979)

(d) *Hilsa ilisha*

Male : 22.1-46.5 cm	Jhingran (1957) recorded
Female : 32.5-48.5 cm	difference in sizes between
	sexes in the Ganga fishery.

Jhingran and Ghosh (1978) documented total average market arrivals of Carps, selected Cat fishes, hilsa and others (Miscellaneous) during 1958-59 at seven centres between Kanpur and Bhagalpur (963 km). The total landings (t) in the stretch stated above during 1958-59 and their subsequent landings after 10 yrs 1968-69 are indicated below to show the trend of rise or fall of fish production.

Total landings (in tonnes) between Kanpur and Bhagalpur

Fish species	*Total Landing*		*Per centage of increase (I) or Decrease (D)*
	1958-59	*1968-69*	
Major Carps	161.7	146.0	9.7 (D)
Cat fishes (selected)	119.2	155.6	30.5 (I)
Hilsa	63.2	252.9 (Maximum landing at Buxar)	300.1 (I)
Others (Miscellaneous)	192.5	284.5	47.7 (I)

Source : Modified from Jhingran & Ghosh (1978)

The above table shows that there is a change in the average landings. In the stretch between kanpur and Bhagalpur there is a fall of major carps by 9.7 per cent when hilsa showed a very high rise of 300 per cent. Catfish and others (miscellaneous group) showed increase by 30.5 and 47.7 per cent respectively. The rise of hilsa was shown at Buxar during 1968-69 (112.7 t).

Natarajan (1989) has further shown that the important centre, Buxar (upper stretch of middle Ganga) where mean annual landings of major carps, catfish, hilsa and others were 7.67, 9.33, 27.7 and 46.83 tonnes respectively, showed fall by 55.5, 49.4, 99 and 80.1 per cent in the above order of fishes during the course of 23 yrs, while at Bhagalpur (lower stretch of the middle Ganga) the species of economic importance—major carp and Hilsa showed a fall by 19.3 and 56.3 per cent respectively while catfish and others increased by 42 and 74.2 per cent during the course of 23 years, as evidenced from the following table.

Centres

Fish species	*BUXAR* Mean annual landings (t)			*BHAGALPUR* Mean annual landings (t)		
	1958-62	*1981-84*	*Rise/ fall (in per cent)*	*1958-62*	*1981-84*	*Rise/fall (in per cent)*
Major Carps	7.67	3.41	55.5 (D)	9.60	7.74	19.3 (D)
Cat fish	9.33	4.72	49.4 (D)	20.15	28.63	42.0 (I)
Hilsa	27.7	0.26	99 (D)	4.08	1.78	56.3 (D)
Others	46.83	9.31	80.1 (D)	43.18	75.22	74.2 (I)

Source : Natarajan (1989).

Thus the overall picture indicates that except hilsa which showed a rise during 1968-69, Ravishchandra *et al.* (1990) the general tendency is a decrease of major carps fish of economic importance and increase in Cat fish and other group of fishes (miscellaneous) within a period of 10 yrs in a stretch Kanpur to Bhagalpur. Similar trend was also indicated when data was analysed for the centres—Buxar and Bhagalpur separately. Within a period of 23 years, the entire fishery showed a decrease ranging from 49.4 to 99.0 per cent at Buxar while increase in percentage was maintained at Bhagalpur centre in the case of Cat fish and miscellaneous group, and decrease in case of major carp and Hilsa.

In this context a look into the physiographical condition may reveal some valuable information.

Considerable deterioration occurred in the Ganga as evidenced from physiographical condition indicated below :

Physiographical condition of the Ganga during 1976-77

Stretch (km)	*Slope*	*Mean annual rate of flow (m^3/sec)*
Source-Rishikesh (250)	1 in 67	850
Rishikesh-Allahabad (800)	1 in 4100	850-1700
Allahabad-Farakka (1005)	1 in 13,800	4000-10,200

Source : Das Gupta, (1984)

Flood Condition

Physiographical condition of the Ganga indicates how the river bed is affected due to heavy silt precipitation. This has encouraged increased flood and is evidenced from the data of the past (Fifties) and present days (Seventies) with an indication of recent flood condition of 1995 although data of the flood affected area could not be made available.

Flood affected area of the past and present (area in lakh ha) of the Ganga in Uttar Pradesh and Bihar

*Uttar Pradesh**		*Bihar**	
1950 s	*1970s*	*1950s*	*1970s*
0.7-41.3	3.1-73.4	0.4-25.0	2.2-42.6
(19.8)	(31.0)	(8.6)	(18.9)
Maximum 41.3 during 1955	73.4 during 1978	25.0 during 1954	42.6 during 1971

Source : Chaphekar and Mhatre (1986). Figures in the parenthesis indicate average values.

The above table clearly indicates that in Uttar Pradesh, the average value of flood hit area of 1970s are 56.6 per cent more compared to the values of 1950s, (*i.e.* in 20 yrs time) while in Bihar the flood hit areas increased by 119.7 per cent during the same period and is quite alarming. All these indicate clearly that the rate of silt percipitation in the river Ganga is abnormally high to cease rapid degradation to physiographical condition.

Hydrographical Status

Seasonwise ratio of waste to river water/month in different states

State	*Monsoon June-Sept.*	*Post monsoon Oct.-November*	*Winter Dec-Feb.*	*Summer Mar.-May*
Uttar Pradesh	1 : 1	1 : 1	×	×
Bihar	1 : 121	1 : 22	1 : 7	1 : 5
West Bengal	1 : 18	1 : 8	1 : 9	1 : 7

×Waste water exceeds the flow.

Source : Based on the data of Das Gupta (1984).

Pollution Status in Present Days

In recent past pollution studies reveal that the main polluted centre is around Kanpur followed by Allahabad and is mainly from zinc and copper and compared to unpolluted centre (Haridwar), Zinc concentration is three times more, copper, seven times and lead three times higher.

Fishery Status

Analysis of age groups in the major carp fishery between 1976 and 1983 revealed that the fishery consisted largely of age groups II for *C. mrigala* (45.8 c.m.); for catla II and III (51.4 and 71.6 c.m.); for *L. rohita* III (65.0) and III and IV for *L. Calbasu* (39.0 and 46.0 c.m) (Natarajan, 1988). These trends indicate clearly that the values are much lower than what was recorded in the past.

Impact of draught on fisheries

Besides high abstraction of water from the Ganga and its tributaries which leaves alarmingly low flow causing dilution ratio of waste to river water far from safe level additional stress caused by draught brings severe impact on the growth of fish and evidenced from the following table :

Impact of draught on the mean length of commercial important fishes in the Ganga and Jamuna at Allahabad

Species	*1987*	*1988*	*Decrease in length (c.m.)*
Major Carp			
C. mrigala	62.3	55.8	6.9
C. catla	73.5	61.8	11.7
L. rohita	65.0	58.0	7.0
L. calbasu	47.2	45.1	2.1
Major Cat fish			
M. aor	44.7	43.8	0.9
M. Seenghala	54.5	48.8	5.7
W. attu	54.5	47.1	7.4
Others			
Hilsa ilisha	39.6	34.7	4.9

Source : CICFRI, Annual Report 1988-89.

The severe draught experienced during 1987 in the country as a whole and the stretches of the Gangetic valley in particular and scouty erratic rains of 1988 appears to be greatly responsible for the fall in the growth during 1988. Even the hilsa fishery at Allahabad which was maintaining a steady trend of revival from 1984 to 1987 (Average 1.8 t) dropped to 1.2 t during 1988-89 and the declining trend continued during 1989-90 (0.7 t) CICFRI, Annual Report 1989-90.

Bhagirathi–Hooghly

The river which once used to carry the main flow of the Ganga had almost similar conditions as was in the main Ganga in the past. In the Ganga even around thirties the volume of water flow was 9,72,000 and in its important tributaries Damodar and Roopnarayan joining the Hooghly had 1,62,000 and 6,21,000 mf^3 of water (Anon, 1919; Oag, 1939) even though Damodar shifted its course of flow around 1770. The flow of the Hooghly thus maintained considerably high volume of water.

Further, from available discharge records for 5 years as stated by Majumdar (1953) the average discharges at Farakka on monthly basis had various between 2,99,000 and 79,000 cusecs during dry season (November-May) and between 16,15,000 and 1,75,000 cusecs during monsoon (June-October). Some hydrographical conditions of the Hooghly during 1969 as obtained from the record of Sarkar (1988) are as follows. Physico-Chemical and biological data are available only from fifties and fishery data from sixty onwards.

Hydrographical status of Bhagirathi–Hooghly during 1869

Discharge of water at Jalangi	:	1,40,160 cusecs
Discharge of water at Baharampur	:	1,40,762 cusecs
Prior to fifties the number of bore days as recorded by Bhattacharyee (1986)	:	20-22/ year

Quoting Oag (1934) and Anon (1919), Bose (1956) indicated some hydrological conditions of the Hooghly estuary of the past indicated below :

Approximate depth and breadth of the Hooghly estuary at different points (Anon, 1919; Oag, 1939)

Centre	*Breadth (miles)*	*Maximum depth at HT during monsoon (in feet)*	*Maximum depth at LT during Winter (in feet)*
Barrackpore	0.34	60	40
Calcutta (Khidderpur)	0.47	51	30
Uluberia	0.47	71	50
Falta	0.76	68	46
Diamond Harbour	2.91	91	69
Kulpi	2.87	56	34

Source : Bose B.B. (1956)

Physico–Chemical Status

The range of Physico-Chemical status are available only from fifties and were recorded by Bose (1956) between Dumurdaha and Birlapur during 1953-55 and are indicated below :

Water temperature	:	18 — 33°C
Maximum difference between surface and bottom temperature	:	1.5°C
Dissolved Oxygen (ppm)	:	4.43 — 8.4
pH	:	7.9 — 8.4
Total alkalinity (ppm)	:	100 — 350
Salinity	:	Freshwater — 30 % (Calcutta & above) (Uluberia-Kakdwip)

Pollution Status

In view of high water flow in the Bhagirathi-Hooghly in the past in addition to high tidal influence the delution ratio of wastes (municipal and industrial) was very high as was in the main river. The pollution problems was obviously negligible.

Biological Status

Data pertaining to biological status are also available from fifties. Phyto and Zooplankton concentration (μ/1) with their dominant groups recorded by Dutta *et al.*, (1954) are indicated below :

Year	*Phytoplankton*	*Zooplankton*
1949–50	3,781–77,194	10,494–17,485
	(31,956)	(14,007)
1950–51	4,378–1,58,613	7,120–12,304
	(56,603)	(10,002)

Note : Values in the parenthesis are the averages.

Owing to the mingling of waters in the Hooghly system from a larger number of tributaries and to the banking of water during the flood tides from Bay of Bengal the phytoplankton in the areas surveyed was varied and composite in character.

Relative abundance of phytoplankton were :

Diatoms, Green algae, Blue green algae and

Residual (mainly Englenineal and Dinoflagellates)

The Zoo-plankton in the Hooghly estuary consisted mostly of fresh and brackish water species. The freshwater species were washed down from the upper reaches. A few marine forms were also observed between Falta and Diamond-Harbour. The dominant forms were : Protozoa, Rotifer, Cladocera, and Copepoda Nauplic.

Fishery Status

Fish landing data of old days i.e. prior to twenties or around the period when Bhagirathi-Hooghly used to carry enormous volume of water, are not available except some stray reference. It is from sixty onwards regular data became available when considerable damage to the physiographical condition was effected in the river. The progressive changes in the fishery composition have been summarised as follow :

Hamilton (1822)	:	Purely freshwater fish except Hilsa : Calcutta and below.
Hora (1943)	:	Considerable changes in fishery composition noticed. Estuarine groups were available around Barrackpore.
Hora (1953)	:	Mainly brackiswater fishes around Barrackpore
David (1954)	:	——— do ——— (Salinity 0.204-1.156 ppt)

Source : Nandy *et al.* (1983).

Mc Dowell (1955) in a review stated that the situation existed in 1964 was that the distributary rivers which once fed the western delta from the Ganga were decaying. As a result those feeding the Hooghly had been cut off from the Ganges during much of each dry season since 1915. The extensive construction of embankments intended for land reclamation and control of flooding had confined tidal flow to the main estuary channels causing tidal range to increase in the upper reaches of all the estuaries in the western delta. Besides, a series of dams constructed in the Damodar Valley affected greatly by reducing peak flows and mass of sediments brought down to the Damodar and Rupnarayan rivers and into the Hooghly at Hooghly point. The piers for the bridge (constructed across the Rupnarayan) would have appreciable effect on the local flow resistance and tidal propagation. Finally, the construction of Farakka barrage should have ensured free flow of freshwater down the Bhagirathi during dry season. But the dry season abstraction of water from the Ganges upstream of Farakka has reduced the quantity available at the barrage for this purpose.

Indication of deterioration of the Bhagirathi-Hooghly started by the end of 19th or earlier twentieth (Bose, 1972) after the change of main flow into the Padma and shifting the flow of Damodar further down. The bed of the Hooghly river was raised to 7.3 m higher (due to formation of a sand bar) than the main bed of the Ganga (Bera, 1984). The seriousness of Bhagirathi-Hooghly has never been felt as acutely as the present time, more so since the partition of the state in 1947. Hooghly has lost its maximum discharge capacity by 40 pc in 1959 compared to the discharge figure of 1954. (Bhattacharjee 1986). Duration and intensity of the upland supply from the main stream Ganga into Bhagirathi Canal was limited to the months of mid-June to mid-November. The dry weather flow is negligible. Volume of supply diminished from 75-80 thousand cusecs to nil (Bose, 1972). Further, at the time of construction of Farakka Barrage, the flow of the Ganga Padma did not even reach 40,000 cusecs during summer. Average flow of Bhagirathi at Jangipur being 61 cusecs only for seven months against 1,40,160 cusecs (during 1869). Bhagirathi-Hooghly received strong tidal inflow from the Bay of Bengal throughout the year and during 1986 the number of bore days were more than 180/year against 20-22 days/year prior to fifties (Bhattacharjee, 1986). In the stretch Farakka to Nabadwip (230 km) the slope is 1 in 23,000 when rate of flow was variable due to flood. (Das Gupta, 1984). The slope is 3" inches per mile. (Bera, 1984).

Pollution Status

Alongwith the pesticide (DDT and BHC -V) heavy metals were also estimated not only in the water and soil but also in fish which includes fish muscle and different organs and crustaceans like crab, shrimp and molluses.

Pesticides :	In sediment	DDT	(17 – 89 ppb);	BHC	(21 – 73 ppb)	Joshi, H. C. (1985)
	– Fish	–	(91 – 460 ppb);	–	(46 – 210 ppb)	
	– Molluscs	–	(65 – 963 ppb);	–	(40 – 850 ppb)	
	– Plankton	–	(15 – 130 ppb);	–	N. D.	
Heavy metals	In sediment,	Zinc 381 ppb	upper level			
	–	Chromium 9.6				
	In Fish –	Zinc	(2.5 – 120.4 ppm;	Copper	(2 – 20 ppm)	Ghosh B. B. *et. al.* (1983)
	Shrump	–	(14.8 – 112.2 ppm;	–	(2 – 38 ppm)	
	Crab	–	24.4 ppm	–	14.0 ppm	
	Molluscs	–	83.0 ppm	–	20.0 ppm	

Their concentrations are appreciably high. In the Hooghly estuary, the estimated zinc was found to be six times higher than the highest value obtained around Kanpur.

Biological Status

Plankton estimated (μ/1) by Nandy *et al.* (1983) in Zone I (Kalna and Barrackpore) and Zone-II (Uluberia and Nurpur) during pre, 1960-61 and post (1975-77) Farakka discharges is indicated below.

	Pre Farakka discharge (1960–61)		*Post Farakka discharge (1975–77)*	
	Zone I	*Zone II*	*Zone I*	*Zone II*
Phyto-plankton	68.04	32.18	219.38	162.16
	(Average 50.11)		(Average 190.77)	
Zoo-plankton	26.88	7.97	84.99	60.62
	(Average 17.42)		(Average 72.80)	

The above table indicates that there is considerable increase in the average value of phytoplankton (280.77 p.c.) and Zoo plankton (317.9 p.c.) in both the zones (I and II) during the post Farakka discharge and is dominated mainly by fresh water forms. The excess water released from Farakka barrage has pushed the brackishwater zone further down.

The dominant forms were diatoms (Bacillarocophyceae and Chlorophyceae in phytoplankton and copepod (Calanoid and cyclopid) among Zooplankton.

However, these values are definitely much lower than the values estimated during 1949–50.

Fishery Status

The estimated total yield of fish in a period of 16 years at the two zones, I (unpolluted : Nabadwip-Bandel, 86 km (4509.84 ha) and zone II (polluted : Bandel to Batanagar, 72 km (4782 ha) between Babadwip and Baj Baj of the Hooghly estuarine system covering the periods 1962-63 to 1977-78 has been estimated which states that average annual yield in the respective zones were 719.25 and 124.94 tonnes and the catch/ha were 159 and 26 kg, which shows that ecological condition in zone I is more favourable for biotic life than zone II. Further, there is a year to year increasing tendency of fish landing in zone I and on the contrary in zone II (polluted zone) it does not follow the same pattern and shows a decreasing trend (Ray and Laha, 1981).

Mainly brackish water forms around Barrackpore was recorded (Ray, 1981). According to Nandy *et al.* (1983) the immediate impact on fisheries resources of the estuary after the commissioning of the Farakka barrage has not been studied in details. So far, although some trends have been recorded in the upper estuary (Kalna to Nurpur). Some of the typically euryhaline species had been showing a declining trend.

Fish landings before and after Farakha Barrage discharge in a stretch Kalna to Nurpur

Annual average landing (tonnes)		*Landings after the commissioning of the Farakka Barrage*					
1972–73 to 1974–75		*1975–1976*		*1976–1977*		*1977–1978*	
Zone-I	*Zone-II*	*Zone-I*	*Zone–II*	*Zone-I*	*Zone-II*	*Zone-I*	*Zone-II*
107.71	328.71	851.18	421.84	711.78	175.00	518.06	110.35

Source : Nandy, *et al.*, (1983).

Zone I : Kalna and Barrackpore. Zone II : Uluberia and Nurpur.

Additional information on the Zonewise annual fish landing (in Kg) during 1966-67 to 1970-71 of the Hooghly-Matla estuarine system of Six Zones (Zone I-IV, Dutta *et al.*, 1973) stated below and subsequent landing during 1990-91 although shows a steady rise, a fall (13.7%) in the total fishery was noticed in the following year (1991-92), while hilsa fishery registered a sharp decline (42.2%) in 1991-92 compared to the previous year (1990-91).

Annual flish landing (in tonnes) in the Hooghly-Matla estuarine system during 1966-67 to 1970-71 and subsequently from 1990-91 and 1991-1992.

Year	*Fish landing (in tonnes)*	
1966–67	5900.8	
1967–68	8376.7	
1968–69	7592.2	
1969–70	7978.1	
1970–71	10754.9	
1990–91	40981.0	During this 2 year's period Hilsa fishery *(T. ilisha)* dwindled
1991–92	35353.9	by 42.9 per cent, and the total fishery by 13.7 per cent.

*Zone I—Nabadwip to Khusiogoli; Zone II—Khusiogoli to Diamond Harbour; Zone III—Lower Zone including entire estuarine area of Sundarban and tract below Diamond Harbour; Zone IV-Rupnarayan estuary and Zone V—Small Zones in the Matla estuary.

Summing up the above results both in the main Ganga and Bhagirathi-Hooghly, the parennial mighty Ganga which in the past had high flow, water velocity, depth with adequate aquatic biomass round the year and free from any appreciable pollution problem, created an environment where fish could harbour safely, move freely and reproduce. Under such healthy ecosystem fish could grow well as is reflected from their sizes. With the changes of time population influx, high silt precipitation resulted from deforestation recurring floods abiotic changes resulted from industrial growth and urbanisation the physiographical conditions together with water quality is now changed. The reduced volume of water, flow velocity, depth, slope which encourages the pollution problems by reducing the dilation level cannot retain the same water quality of the past. As a result a gradual change in the fishery composition, especially fishery of economic importance and their sizes are distinctly noticed, and fishes of economic importance is rapidly changing into non–commercial and miscellaneous group of Fishery.

Bhagirathi-Hooghly is also facing similar problems. In the past when the main flow of water was through this system, purely fresh water fish except Hilsa was recorded around Calcutta (Hamilton, 1922), has now been converted into its tributary with minimum flow during non-monsoon months. A gradual change in the fishery composition *i.e.* estuarine group around Barrackpore was distinctly predominant with the increase in salinity (Hora, 1943), which later changed mainly into brackish-water fishes (Hora, 1953; David, 1954).

Added to it the release of more toxic metallic wastes which caused metals to accumulate in different organs of fish possibly has affected the growth and reproduction and ultimately fishery as a whole.

After commissing of Farakka barrage from 1970 entire biotic life indiated gradual changes due to freshwater influx, therby the fauna and flora have been shifted to zone II down stream. Nandy *et al.* (1983).

Besides natural calamities like draught or late monsoon *etc.*, which effects the growth and Farakka Barrage constructed across the Ganga has affected free migration of commercial important fish like Hilsa in the upstream. This has resulted dwindling of Hilsa fishery appreciably in the upstream. The estuarine fishery is also considerably affected on account of their restricted migration.

ADDENDUM

The recent 30 yr.Water pact on Ganga water at Farakka between India and Bangladesh has invited the following comments from different experts. (Statesmen, 13 December, Calcutta, 1996)

1. During the two cycles of 10 days period in the crucial month of April, Calcutta will get from Farakka barrage between 25,000 and 28,000 cusecs, which according to C.P.T. experts is "too low a head water flow to flush silt into the sea".

   The agreement in no way reverses the process of deterioration of the Hooghly. Heavy siltation will increase the intensity and frequency of tidal bores which will seriously impede navigation.

2. C.P.T. experts and Mr. Debesh Mukherjee, Former General Manager of Farakka Barrage seriously questioned the sharing formula worked out on the basis of average total flow at Farakka between *1949 and 1988.*

   "The average data on water flow does not reflect the ground reality. Flow of water varies from day to day. Under the agreement the average flow in April has been shown to be between 60,992 cusecs and 63,180 cusecs where as the *actual flow* during the month for the *past decade* has been about *54,000 cusecs.* It *would have been somewhat proper if the average flow had been arrived at on the basis of data of the past one decade rather than the past 40 years.* Naturally, the basis for the sharing formula is flowed as also the quantum" they said.

3. A senior port official said "*Calcutta will get much less than what has been stated in the agreement* as the *offtake in the upstream, especially in Uttar Pradesh and Bihar,* has *been growing fast in recent years.*

   The above statements of hydraulic experts indicate the fate of the Bhagirathi-Hooghly is heading infuture and its resultant impact on biological life.

18

Conservation and Management of the River System

It is an uphill task to suggest means for conservation and management of the river system whose problems are enormous. However some probable suggestions are put forward keeping in view of the present economic situation of the country.

1. Considerable fund as shown below (cf. Table 87) are spent every year towards the flood relief measures resulted from breaching of elevated river bed during monsoon. Enormous volume of silt load are brought down every year during monsoon due to high deforestation. Such phenomenon will continue increasing every year until some positive steps are taken to control them.

Table 87 : Statewise details of the funds allocated under State Plan during 1993-94 and 1994-95 (in Rs. Crore)

States	*1993-94*	*1994-95*
Punjab	11.24	15.7
Delhi	10.0	12.0
Haryana	8.9	9.1
Uttar Pradesh	9.0	8.0
Bihar	27.0	44.7
West Bengal	15.0	36.0

Source : Ministry of Water Resources (in : The Statesman, April 14, 1995).

Deforestation has to be controlled and proper scheme for afforestation is of vital importance for further control of erosion. The engineers should suggest the remedial measures in case dredging is not possible or partly possible for increasing the velocity of water. Since this is a tough task and very expensive to put the entire system in order, the only alternative is to make afforestation scheme very successful.

This will help to protect the reproductive feeding guilds, established breeding grounds, destruction of eggs/larvae from being carried away along the spilled water of river during monsoon, filling up of deep pools due to rise of river bed, save large scale juveniles and brood fishes from their easy catch, improvement of dilution ratio etc. The control of recurring heavy silt precipitation is thus essential as far as possible.

2. Reduction in flow of the river system including the estuary due to silting and increased abstraction of water for agriculture, industrial and urban uses, cause serious problem as sufficient

water for dilution does not become available during the dry months. Added to it, the uncertainty about the onset of south-west monsoon in some years—Hence serious attention should be given to effluent treatment for conservation of the river water.

3. Special attention may be given in case of tributaries (particularly originating from the hilly terrain) joining the main river at right angle (*e.g.* the Kosi, the Son, of Bihar; the Ajoy and the Damodar of West Bengal etc.) have torrential flow, during monsoon such rivers bring enormous silt along with the water and thwarts the main river water already in spate or vice versa (depending upon the volume of discharge and velocity) where silt gets settled and in the long run the course of the river is shifted. The river thus forms a number of zigzag turns which encourages further silt precipitation and recurring floods. Perhaps it may be beneficial if the tributary is made to flow further down diverting its course through a Canal. The same operation may be done by digging separate canals in the areas where the river has developed a number of zigzag turns. In such case during monsoon the river will flow down with greater velocity when not only the silt load will be carried down along with the flow but may eraze the deposited silt since the velocity of water exceeds 5'ft/Sec.

4. The river system receives various complex waste water discharged from various industries. Pesticides and heavy metals in the waste make the ecosystem complicated. Such pollutants retard the self purifying capacity of the river besides its accumulation in the fish (liver, gill, kidney, gut, muscle, gonad etc.) and in other biota. Hence, either adequate dilution or the treatment of waste water is adequately necessary to avoid such impasse.

There are two basic types of criteria of standards that have been used for control of water pollution. One is the control of water quality of stream while the other is the control of waste water to be discharged from a given plant and is called effluent standard. In case of the former, to maintain the water quality, it is difficult to have a control over the flow particularly in an agricultural country like India where demand of water for agriculture, industry and cities are always on the increase and the sources are limited. Hence the effluent standard is the only solution. Equitable standards for each individual effluent are to be established after careful consideration, both the effluents and the receiving stream, as the later changes (due to contribution from tributaries) so would the standards have to change. It is reasonable to have separate standards for each region (upper, middle and lower Ganga plain) due to wide climatic and geographical differences. Estuarine zone should be separately treated for its standardisation due to its different tidal influence. Thus effective and economic abatement of pollution should be the logical goal. While monitoring the stream from pollution, adequate reasonable sampling points at proper intervals in the entire stretch are to be established from where at proper intervals biotic and abiotic studies are to be made. Magnitude of pollution in terms of chemical, biological—which include benthos, fish population their healthy survival, growth and reproduction, accumulation of metals and pesticides in fish flesh, of different sizes, their disease and in different tissues and other bio-mass, requires a systematic study along with water and soil. The entire data when pooled together will reveal valuable information on the fishery status in the river system.

5. The effluents after their treatment may be disposed to the river tributary through a long pipe drawn almost to the mid river. Because littoral zones are rich in reproductive feeding guild where biotic life can grow and hence should be protected at any cost. Mere treatment of effluents may not produce satisfactory results, unless simultaneously some effort is made for the river to hold more water with adequate flow.

6. In view of the presence of high silt load in the water, the brick industry may be encouraged for their operation along the banks of the river away from the city without damaging the river banks. The

water can be drawn to a little interior, allow the sediment to precipitate (even by adding some chemicals to enhance the precipitation) and send back the clear harmless water to the river by pump. Such operation can be done in a more scientific way occupying minimum space. According to an estimate Calcutta Metropolitan District (CMD) has a production capacity of 90,000 bricks/day in its only brick field near Calcutta. (Bhattacharya and Chatterjee, 1972).

Recently, Walter Craven Group, a German-British-Dutch consortium has taken up a project to produce brick from silt of the river Hooghly, mud inside drains and ashes of Thermal plants. For this about 100 silt banks of the river from Bandel to Budge Budge (around Calcutta) will supply clay to the plant. City garbage will also be mixed with the raw material. It would also make hollow bricks and engineering bricks and would produce 80 million bricks annually. The production cost per thousand will be only Rs. 800 and will be cheaper than traditional ones now available in the market. This may solve the problem to some extent arising from silt precipitation in the river bed and also city garbage.

Besides, there are a number of private Brick fields on both sides of the estuary (stated earlier) and are being operated regularly. Simultaneously the river bed may be leased our stretch by stretch to a number of private enterprising organisations to remove the earth free of cost. These organisations will remove the earth simultaneously in different states. This will not only improve the condition of the river but also save crores of rupees that are being spent every year towards flood affected victims as flood relief. Such deposited alluvial soil are supposed to be rich in nutrients (rich organic matter and also metallic ions) and is used by agriculturists at a low cost in their fields may fetch a good crop. The Government agency by dredging may remove the earth but will place them on the banks which will again find their way in the river during rains. Whether such soil have any adverse effect on the crop may be tested and then used. Webber (1972) studied effects of toxic metals in sewage on crops opined, that so long as the soil retained a high organic matter content and the lime status was kept high no problem might arise. But if the soil was allowed to become acid, and if the organic matter content fell following a change in the cropping system, trouble could arise. Further, the contaminated soil can be diluted with uncontaminated soil from below by deeper cultivations. It is worthwhile to mention that solid garbage of Calcutta Metropolitan city were largely used as fertilizer for raising good crop (Ray, 1993) which were being supplied regularly to the city and its suberbs. All these efforts may solve the problem to a great extent arising from silt precipitation in the river bed and also city garbage, especially when crores of rupees are spent every year towards flood relief measures.

Besides, some ancillary reservoirs constructed at a number of suitable points of the rivers or tributaries as suggested by the eminent scientist M. N. Saha (Chattopadhaya, 1993) and more recently by Sanyal and Chakrabari (1995) for holding maximum water during monsoon and releasing them during off seasons according to the requirement to maintain a constant flow would have helped the situation.

7. Ray (1981) has already indicated the process of recovery of biproducts such as sodium hydroxide from sulphite waste, calcium oxide from sulphite waste, and oil from hydrogenated vegetable oil and soap and celex from magnesium bisulphate of sulphite waste are also practised. It is difficult to treat large volume of waste water and hence as far as possible more attention should be paid for economic recovery of biproducts. Recovery of potash (Charkraborty and Bhaskaran, 1966) and methane gas (Paul, 1969) from distillery waste; alkali from kier waste of textile industry and resin from lac waste have been found to be economical on pilot project basis, may be attempted on a large scale. The waste (distillery) can be used for reducing turbidity in water bodies. Distillery waste has been found quite useful as fertilizer. With little modification in the N : P : K ratio it can be

used to prepare a standard manure. Dried cake contains high manurial constituents than liquid waste and can be used not only as cattle feed but also as quality fish food (Banerjee and Pakrasi, 1977, Ray unpublished). Pulp which cannot be easily recovered, is being trapped at the outfall area of sulphite waste and used for manufacture of card board. Finally, fly ash generated from the thermal power station which was once an ecological hazard and was difficult to dispose until recently is being used in manufacturing bricks. It is reported that more than 3500 metric tonnes of fly ash that the four 210-MW units of the Kolaghat thermal power station of West Bengal produces daily, are being used to make high quality bricks and are sold while they are still in the Kiln (The Statesman, 1992). Monre recently, the Chief Minister of West Bengal while inaugurating the Pulver fly-ash Project adjacent to Bandel Thermal Power Station said that high quality bricks from Fly ash, a waste product would manufacture about 80,000 bricks daily and plans were being framed to manufacture low cost cement, tiles and stone-chips from fly ash. (The Statesman April, 17, 1994). Besides its use as filler, it can be widely used for extraction of aluminium and germanium used in photo cell. Where treatment of waste material is not economical its recycling for other use should be examined.

8. Technical fault in treatment of industrial and municipal wastes and use of such faulty effluent-standard are serious offences. This was experienced by Bhargava (The Statesman Report, 4 Nov. 1992) professor of environmental energy, Roorkee University and Mazumdar (The Statesman Report, Aug. 5, 1989).

As stated by them various state pollution control Board adopted effluent standards that had been arbitrarily set without any scientific rationale. Such practice hamper effort at cleaning the receiving water and intern the health and reproduction of its biomass. Adequate precaution is necessary against such practice.

It has been observed from the recent past that though steps taken by the Government to put pressure on the industrialist including the industries run by themselves for treatment of waste water, they could not bring any fruitful result. Industrialists are not taking the matter seriously. Realising the fate of suffers the supreme court has taken stern action against some private as well as Government run factories without depending upon the state Government action. They send experts to collect waste water samples from different industries and get them analysed. This together with the report supplied by the state water pollution control board are examined by them and accordingly issue directives to the concerned authorities of the state. In this process the Supreme Court has already taken action against some industries irrespective of Govt. run or private.

Undoubtedly this may be a correct step taken for future to come. Their effort is not to close the factory but to see that polluted waste water is properly treated before it is released, otherwise it will jeopardise the whole system.

9. Mere determination of pesticides and heavy metals on biota (fish and their different organs, fry of different sizes, eggs etc.) as is normally done to highlight their bioaccumulation may not serve any purpose unless their specific effect on the physiological activities affecting the growth and reproduction is revealed. For this regular periodic growth trend of fishes (maximum size attained with dominant size group) compared to the past particularly in case of fishes of commercial importance are to be recorded to know whether growth is maintained or a fall is noticed along with the time space. A similar record may be made before and after the draught hit period. It is from this point of view only the importance of treatment is necessary.

In this context, based on studies made by Fujiya (1977) referred earlier on the requirement of essential amount of oxygen in different sizes of commercial important fishes *i.e.* mg of oxygen required per kg of fish in a particular time may reveal valuable information.

In warm water, quality of fresh water that will support a good mixed fauna should not contain less than 5 ppm of DO along with other factors such as pH, specific conductance, carbon dioxide, ammonia and suspended solids (Ellis, 1937, 1944 and 1946). Doudoraff and Shummy (1967) also gave similar opinion *i.e.* 65-75 per cent saturation of water with oxygen is required. This means when the oxygen level remains 5 ppm or above fish can take oxygen from water without any strain. But when DO falls below 5 ppm, the fish has to struggle to filter more of water through the gill to attain the required level of oxygen from the water for their system.

Accumulation of toxicants in fish is directly proportional to the volume of water filtered through the gill and fall of oxygen level in water is inversely proportional to the accumulation of toxicants present in the water, as more water is required to be strained through the gill.

In the Ganga, the usual concentration of DO is about 5 mg/ℓ and above, with some exceptions around the outfall of the industrial belt which even comes down to nil. Water with low concentration of oxygen extends in a wider areas across the river and moves down stream depending upon water flow, velocity and physiographical conditions of the river. When fishes more in such areas in search of food, may face problems of accumulation of toxins injurious for this system. Hence, some studies on the line might be useful in revealing valuable information on the probable zones where fish might accumulate more of toxins which is harmful for their system.

10. The breeding of major carps in the upper stretches last from July till the end of August, while in Bihar and West Bengal breeding starts in June. It is therefore recommended that the closed season for fishing be totally stopped during June to end of August in the states like Uttar Pradesh, Madhya Pradesh, Rajasthan, Haryana, Bihar and West Bengal.

11. Based on biological studies of major carps on their habit and habitat the minimum size limit for capture should be fixed at 50 cm for Rohu and Mrigal, 55 cm for catla and 30 cm for calbasu by making lagislation in all the states or from Agricultural Ministry, Govt. of India. (Jhingran, 1959; Natarajan and Jhingran, 1963; Kamal, 1969; Gupta and Jhingran, 1974 and Khan and Jhingran, 1975).

12. Actual charting and the conservation of deep pools in the main Ganga and most of its important tributaries are essential by proper survey. The paractice of leasing out such waters to commercial fishermen should be abandoned and pools declared as Sanctuaries.

13. The fish-pass constructed in dam or barrage should be done scientifically (even if necessary suggestions may be obtained from some foreign experts) so that fish migration is not affected. This will maintain proper fish migration and natural recruitment in the entire stretch of the Ganga and its tributaries will not be a problem any longer.

14. Intensive fish culture in floating cages may be taken up in perennial running-water-canals used for irrigation purposes. Hatchlings, fry and early fingerlings can be reared in cages of the frame 2.20 m × 1.60 m (width) and 1.45 depth. Natarajan and Saxena (1978) got encouraging results in lentic water. Their study reveals that Catla, Mrigal, Rohu and Bata grew to a size 192mm from 63mm, 120.5mm from 40mm, 116mm from 44.5mm and 102mm from 53mm within a period of 89 days. Thus cage-reared fingerlings can be stocked in running waters, rivers and reservoirs or in a bigger cage (2m × 1.5m × 1.5m) or still bigger size for a longer period for further growth.

In a personal communication P. R. Sen of Central Institute of Freshwater Aquaculture (ICAR) informs that young fry/fingerlings get injury in their snout when they try to pick up the algal matter or periphitic organisms attached to the nylon net fixed in the cage/pens. Such injury causes fungal attack. According to him such cage/pen culture studies may not face such problems in saline water. Natarajan and Saxena (1978) did not mention any such problems during the course of their studies.

However this fact has to be kept in view during cage/pen culture studies.

15. In ox-bow lakes, mender loops, dead arms, chars, tals and ruins of settlements etc. under natural condition, the composition of fish remains uncertain. In one year major carp dominates while in the next year the composition may be dominated by uneconomical species. Besides, their production per hectare is not encouraging. Mass killing of immature juveniles from such areas reduced the productive potential of the river in a particular area. These water areas are seasonal (4-6 months). Such areas can be exploited to a great extent profitably by culturing spawn, fry or fingerlings in floating cages (stated elsewhere) or by pen culture. Such type of culture may be recommended to get a crop from derelict impoundment to get some profit 1988-89, 1989-90 (CICFRI Report Barrackpore, 1988-89; 1989-90).

16. Seeds may be raised by hypophysation. Young juveniles reared in such cages and pens may be transferred in rivers, reservoirs, ponds and in other culturable waters. From the investigation it is revealed that quality of water and soil of such lakes are enriched with nutrients where benthos can grow well (Table 73A & B).

19

Suggestions for Future Investigation

Detailed investigations on the seasonal dynamies of both riverine and estuarine biota are urgently needed. Data on various aspects of the river system (river and estuary) would be extremely useful in formulating effluent standards (river and estuary separately) with due consideration to seasonal variations.

Systematic monitoring of both freshwater and estuarine zone of the Ganga river system must be carried out to collect the base line data on physical, biological and chemical characteristics of the river system. In view of the fact that plankton, fish and other benthos are affected by metals like, Zinc, Copper, Chromium, Lead and also by pesticides, through their bioaccumulation, studies on heamatological, serological, histopathological and cytochemical tests are essential to find out any changes in their physiological aspects which affect healthy growth, survival, maturation and reproduction. The treatment of such pollutants may be given top priority. A systematic monitoring programme is necessary for a proper assessment of man's impact on the estuary.

The effect of pesticides and metals on the biota (fish and spawn of different sizes, eggs etc.) needs through investigation. Synergistic and antagonistic effects, and chelation of metal ions with organic compound in relation to their effects on aquatic life remain to be studied. Bioaccumulation tests should be carried out in fish samples collected from different outfall regions to determine index of heavy metals accumulated in living system.

Diverse opinion are on record about the hatching success of eggs of different fishes. For example :

1. Rainbow trout eggs failed to hatch in soft water (4.8 mg/1 $CaCO_3$) containing 0.04 mg/1 zinc.
2. With Rainbow trout eggs, there was no significant mortality at 0.06, 0.2 and 0.6 mg Zn/1 in water of hardness 12.5, 50 and 200 $CaCO_3$.
3. Hatching of Salmon was only 30 per cent at concentration of 0.04 mg Zn/1 at 14mg/1 $CaCO_3$ and pH 6.4 while Branin and Paulsson (1971) found that in water of similar quality, 13mg/1 $CaCO_3$, pH, 6.4 the hatching developed successfully even at 0.1 mg Zn/1 (EIFAC/T21, 1973).

In view of the above it is worthwhile to undertake a separate study on this aspect specially with commercial important species at various stretches of the Ganga river system including the estuary as this will reveal valuable information on breeding success, since the matter is related to production potentiality of the ecosystem.

Studies on avoidance behavior of fish to toxicants on the lines suggested by Stott (1975), Sprague (1963) and Höglund (1975) or modified, under natural ecosystem may reveal valuable information.

Critical studies are needed to determine the retention period of effluents in the estuary. This would help to determine the pollution levels and self purification capacity of the estuary. Environmental impact on assessment of heat-waste disposal by six thermal power plants on the estuarine ecosystem deserves urgent attention.

Technology and industrialisation must come but they should not affect the critical balance of nature endangering the very basic needs of life.

20

A Word on "GAP"–A Scientific Study

The Gang Action Plan launched in 1986 was planned to improve the overall water quality of the river including its tributaries and to restore their productivity and biodiversity.

Although considerable work to this effect has been done and data presented in the document "The Ganga—A Scientific Study" reveal some important information, but it is noticed that the major work is restricted to the upper and middle stretches of the Ganga and adequate attention for the study of the Bhagirathi–Hooghly estuarine complex ecosystem is not given. Further, physiographical conditions (Depth, slope, shape factor *etc.*) which is responsible for water holding capacity, flow characteristics, discharge figure, velocity of water *etc.* some valuable characteristics of the ecosystem of the Ganga, has failed to draw any conclusion.

Nature of industries, their characteristics, pollution load and probable dilution ratio *etc*, (state wise) some important aspects have not find place in this document. Metals and pesticides studied are for a short stretch only (Unao to Phaphamao). Data presented on this seems to be inadequate to draw any conclusion appropriately for the entire river. Bioaccumulation of metals and pesticides by aquatic life including fish are not adequately studied to show their impact on survival, growth and reproduction, except that some references are cited. Various aspects on fishery breeding, growth and reproduction some vital aspects are not given any attention.

The document "GAP—A Scientific Study" depicts results of a number of investigations carried out in different aspects (in some cases not data are provided) mainly on plankton, microbiological and biochemical analysis, fungi, coliform bacteria, species diversity, metals, pesticides, haematological studies etc. In different stretches of the Ganga during different periods. Although each study has its own merit and individuality, mere presentation of data does not lead to any conclusion unless the WQI of the entire Ganga is determined for different uses. The overall impact of pollution on the biomass like plankton, benthos, fish population of the entire Ganga requires to be indicated, as, future plan of action will depend on this. Further, no clear cut indication between the conditions of the Ganga of the past and the improvement that has taken place now are not indicated.

More stress is given to reduce the BOD values as indicated in the article "Rejuvenating Ganga" (Sharma, 1994) for which adequate steps are taken in the upper stretches.

Nevertheless it may be remembered that BOD itself is not a pollutant and exercises not direct harm. Only by depressing DO content to levels that are inimical to fish life and other beneficial uses, does BOD exert an indirect effect. Where reaeration, dilution, and or photosynthetic action offset or minimise their depletion, BOD does not interfere to a reasonable extent for uses of its water.

BOD is important only in so-far as it produces septicity or decreased DO or subsequent growth of saprobic bacteria which increases the turbidity or other undesirable characteristics of the streams. In a slow sluggish stream a 5-day BOD of 5 mg/1 might be significant to produce deoxygenation resulting in anaerobic conditions, where as a swift mountain stream can easily handle 50mg/1 of 5-day BOD without appreciable depletion of DO. For this the study of physiographical and limnological conditions are important, as these will provide information on slope, discharge figure, velocity of water, dilution, shape factor *etc.* These are the most important parameters which contribute in the self purification of a stream.

The stream characteristics along different sections of the Ganga as indicated by Das Gupta (1984) in his document, "The Ganga Basin II" is a clear example of such situation indicated below which are quite alarming.

Stretch	*Section*	*Length (km)*	*Av. Slope of land*	*Mean annual rate of flow m^3/s*	*Stream velocity m/sec*
Source Rishikesh	Mountaineous	250	1 in 67	850	May : 0.09–0.6
Rishikesh-Allahabad	Upper plain	800	1 in 4100	850–1700	July : 0.22–00.9
Allahabad-Farakka	Middle plain	1005	1 in 13800	4000–10200	
Farakka-Nabadwip	Deltaic non–tidal plain	230	1 in 23000	1000–1300	
Nabadwip-Outfall	Deltaic tidal plain	240	1 in 24000	Variable due to tides	

Natural calamities like heavy rainfall, soil erosion from deforestation (Recently 650 sq. km. area of forest in Uttar Pradesh was affected by fire, (The Statesman, August 5, 1995) and from other causes like evaporation (mean monthly evaporation being much higher than mean monthly rainfall at selected stations of upper Ganga plain) associated with high abstraction of water may have far reaching effect on the riverine ecosystem by bringing ecological imbalance.

This will not only encourage recurring flood in the lower plains but also in the places of higher altitudes like Punjab, Uttar Pradesh, Haryana, Delhi, Assam etc. due to breaching of banks on account of shallow depths of river arising from high silt precipitation. Besides, draught or late monsoon also contributes appreciably to the problem.

Hence, instead of giving adequate attention on these aspects if only treatment of municipal wastes is given importance a time may come when the river, during non-monsoon months may be converted into several disconnected large water spread areas as happened in the Kosi and other rivers due to continuous silt precipitation and large scale use of river water and we sill have to remain satisfied with the sewage treatment only when simultaneously on the other side, a grim picture will gradually appear in the river for which we are appending so much of fund for its improvement. We are to see that the river also holds sufficient water with adequate flow. May be that, three fourth of the pollution on the Ganga is from untreated municipal wastes (as opined by CPWC) but the river has its own self purification capacity and by that not only the wastes will be purified, but simultaneously such wastes may cause eutrofication to the river water that may result good growth of flora which is the ultimate goal of our study.

Bhargava (1981 b) in his article "Ganga, the most self purifying river" reported that the rate of exocellular polymers exereated during endogenous growth phase of various species of bacteria is considerably reducing BOD and turbidity. The interaction of exocellular polymers present in the river with colloidal matters present in seepage results in a very rapid and significant reduction of the BOD. The author further states that the data analysis has shown that Ganga has the highest BOD rate constant value and the reaeration rate constant value, both of the which are higher by an order of magnitude than the so far accepted values for any river. The coliform group of organisms are also found to reduce significantly in very short time.

Results obtained by various workers (Bilgrani and Dutta Munshi, 1991 Bhargava, 1981, Das Gupta, 1984 and also scientists of CIFRI, now CIFRI, Barrackpore) indicate that BOD except around Kanpur were below 20 ppm when DO in most cases was well above 5 ppm. When conditions are like this (except on special cases-around the outfall) it is worthwhile to give more attention for the treatment of metals and pesticides, and simultaneously see that the river gets adequate depth, flow and discharges even during the lean months.

Based on this Bhargava (1981) using summer data on river Ganga in a stretch, Rishikesh to Varanasi prepared the water quality indices (WOI) and classified the Ganga for various individual uses. According to him around Kanpur the stream showed poor water quality both for drinking and fish culture purposes. The rest of the stretch appears to be satisfactory to good, provided the water is not further contaminated with metals and other pollutants. Thus we find that it is metal toxins which are more harmful and requires to be treated earlier than sewage. The three-fourth of municipal waste may not cause that severe problem, as the rest one-fourth of the industrial waste containing metallic toxicants and pesticides which are not easily biodegradable. These are the main problems.

From the little study so far made in the Ganga it is noticed that heavy metals are present in the water and sediment in the upper stretch of the Ganga. In the Bhagirathi-Hooghly (lower stretch of the Ganga) their concentrations are more and fishes and crustaceans have accumulated much more in their tissues.

Fujaya (1977) has observed that if the normal requirement of DO in water for fish falls, the fish has to gulp more oxygenated water to replenish their DO deficiencies and by doing so more of water containing metal toxins or pesticides are likely to enter the system to cause damage. The problem is more in the estuary (where the effluents spread in a wider area due to tidal influence) than the other stretch of the Ganga where the water is flowing in the littoral Zone. If adequate flow is there the pollution will be restricted to the marginal area only for a short stretch. The optimum requirement of DO in warm water fish is believed to be 5 ppm. (Doudoruff and Shumway, 1967) which would require at least 65-75% saturation of water with oxygen. Around outfall especially where the flow is minimum and in the Hooghly estuary (Lower stretch of Ganga) where DO falls below 5 ppm / similar situation may arise as observed by Fujiya (1977). Hence, bioaccumulation of metals / pesticides in fish in their different organs should be studied to see the impact of pollutants on growth and reproduction of fish and entire biotic population.

Although BOD (out of 42 parameters used to assess the water quality) has come down to 3.0 mg/1 and slight improvement in DO has become possible by treating the sewage (as claimed by Virendra Vats, 1994) it is not possible to find out the difference between the past and present condition of the Ganga to assess the overall extent of recovery of the river water. Further, the sludge which settles down in the treatment plant is dried and compacted into a compost like manure. Such manure prepared from dried sludge, in all likelihood contains adequate metallic pollutants which may be harmful for any agriculture use directly.

Hence, under the present degrading physiographical and limnological condition of the Ganga which is heading from bad to worse, how-far only such treatment (sewage) will help the river condition to improve where biomass has to survive, grow and reproduce, as they did in the past, are the aspects, which are yet to be seen in future.

Appendices

APPENDIX I A

Classification of nets used in the middle stretch of the Ganga River System
(Figs. 1–14)
Source : Saxena, R. K., 1966

Drag Net	*Gill net*	*Purse net*	*Cast net*	*Scoop net*	*Hook & line*	*Traps*
With Pocket (Local name) Chanta Without pocket (Local Name) 1. Mahajal 2. Darwari 3. Karia 4. Chundhi 5. Do-dandi	With foot rope (Local name) Gochail Without foot rope (Local name) Tiar	*(Local name)* 1. Bara Kamel 2. Chhota Kamel	*(Local name)* Bhawnar jal	*(Local name)* Jali	*(Local name)* Jor	*(Local name)* 1. Kuriar 2. Gopal Jal 3. Bandal 4. Sirki 5. Ujhki 6. Kohni

Drag net

Dragnets without pockets

Under this head 5 local nets have been described in order of this frequency of use.

1. Mahajal : This net is rectangular in shape and is composed of 50-55 pieces made up of hemp. Each piece is 7-14 meters long and 3-8 metres wide with mesh size 5.8 to 6.5 cm (stretched). The number of strips in each piece varies from 2-3 and the number of meshes in strips in each pieces varies from 200 to 400 cm across and 22 to 107 meshes in depth. A selvage of 12 to 22 meshes in depth of nearly double the size of mesh of main strips run all along the length of the net. 8-9 floats consisting of bundles of reeds measuring 30 cm in length and 6-7 cm wide at the middle covering 8 meshes across the net are tied at regular intervals along the head rope. To distinguish the central piece, 8-9 dried gourd are used as floats instead of reeds. In each piece of net on an average 10 oval shaped baked earthen sinkers measuring 8 cm long and 5 cm in diameter at the middle are woven through the foot rope. Generally 12-16 men are employed for the operation of the net. One end of the net is left on the shore in charge of half of the party and the other half gradually pay out the rest of the net from the boat in a semi-circle enclosing a body of water depending on the total number of pieces of the net. The entire party helps to haul ashore the net by dragging from both the ends. The catch is collected in the central pieces from where it is removed to a small carrier boat. A party of fishermen with complete paraphernalia is called 'Katra'.

Mahajal is operated from November to July. The frequency of use is maximum during December to February. *Hilsa ilisha*, big sized carps, *S. silondia*, *Mystus* spp. and *N. chitala* are the dominant species caught by this net. The detailed measurements of 1st and central pieces are given in Fig. 1.

Fig. 1. Mahajal :

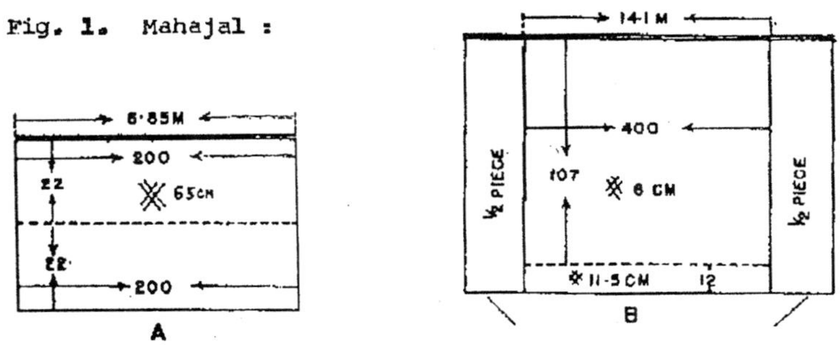

A – 1st piece

B – Central piece.

2. Darwari : (Also called Poorai in Ganga) : This is also rectangular in shape and is similar to Mahajal in all respects excepting the mesh size which varies from 2.4-3.8 cm. The net made of cotton thread, is composed of 40-60 pieces, each piece 5-9 metres long and 5-11 metres wide. The number of strips in each piece varies from 2-5 and number of meshes in strips in each piece varies from 40-80 in depth whereas, across the number is 400 meshes. The mesh of the lowest strip is generally bigger (4.2-4.6 cm) this merges with the selvage of $2-\frac{1}{2}$ meshes. While the same type of floats as in Mahajal are used for this net too, the sinkers, though of baked clay, are smaller in size. Sometimes drum shaped iron sinkers are also used. The operation of this net is similar to that of Mahajal. This net is used almost throughout the year and from July to September it is the main net of the commercial fishery. Mostly juveniles of *Mystus* spp. small and medium sized fishes much as *C. garua*, *Chela* spp. *A. morar*, and *S. phasa* are mainly caught besides major carps and catfishes. The detailed measurements of 1st and central piece are given in Fig. 2.

3. Karia Jal : This net largly used in Varanasi-Mirzapur area is a new entrant to the fishing waters of Allahabad. It is identical to Mahajal in shape and mode of operation except in the size of mesh which is more than double the mesh size of Mahajal. The net has 50 pieces on an average each 9-15 meters long and 6-11 metres in depth with a mesh of 16 cm. The number of strips in each piece varies from 2-4 and the number of meshes in strips in each piece varies from 100-110 meshes across and 10-25 meshes in depth. Since the meshes are bigger in size the selvage is absent. Floats and sinkers are same as used in Mahajal. The operation of this net is identical to that of Mahajal and Darwari. One

Fig. 2. Marwari :

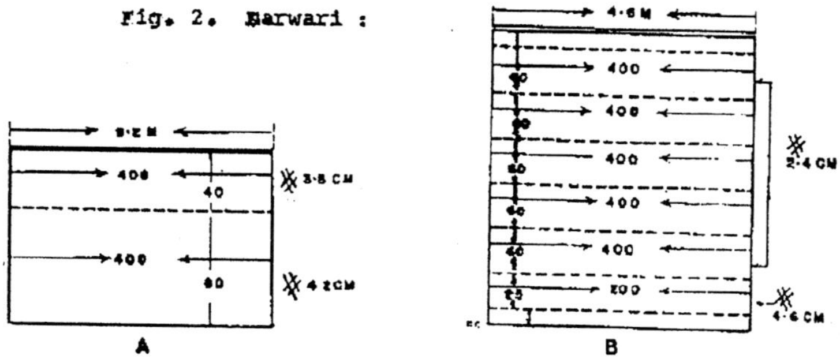

A – Corner piece B – Central piece

big boat and one or two carrier boats with 20-22 men are required for the operation of Karla Jal. It is operated in sufficiently deep waters from November to July in Jamuna River only. Big sized carps and catfishes are caught with this net. The measurement of 1st and central piece is shown in Fig. 3.

Fig. 3 Karia Jal :

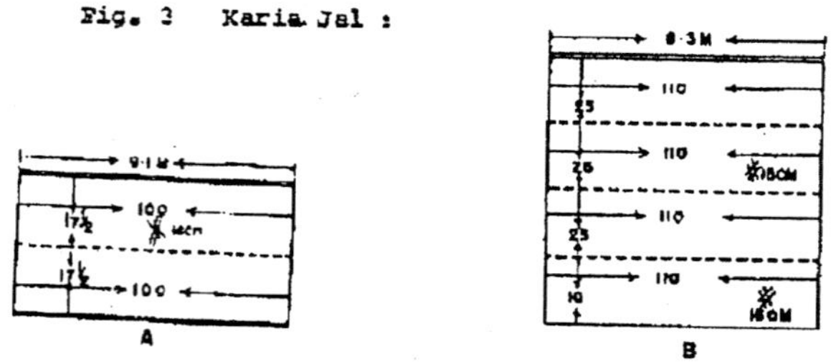

A – 1st piece B – Central piece

4. Chaundhi : This is a small drag net of cotton consisting of 14-15 pieces, each 24 metres long and 1.12 metres wide with a uniform mesh of 10.2 cm (stretched). The pieces are uniform having 600 meshes across and 11 meshes in depth. About 64 floats of 1.5 cm in diameter and 21 cm long are attached to the head rope at an interval of 18 cm. Small round balls of baked clay weighing about 200 g each are woven in a thin rope and is attached to the lower part of the net with the help of an additional webbing placing one sinker in each loop. The net is laid out in the water from a small boat in the usual way like that of any other drag net with the help of 3-5 men in shallow waters. The net is dragged closer in a circle and when the fish congregation is located, one of the men jumps over the catch with the kuriar (see description further) which is an essential accessory of the net. The net is operated all through the year except in the months of August and September when the river is generally flooded. *Mystus aor* and *M. seenghala* are mostly caught but other bottom feeders like *C. mrigala* and *R. rita* are also caught. (Fig. 4).

5. Do–Dandi : This is a simple drag net made up of cotton usually composed of one or two pieces, each 9.6 metres long and 4.35 metres in depth. There are two strips of main webbing each having 800 meshes across and 100 meshes in depth of size 2 cm (stretched). A selvage of 3 – ½ meshes in depth joins the main webbing from upper and lower sides. About 6-8 gourd floats are attached to the head

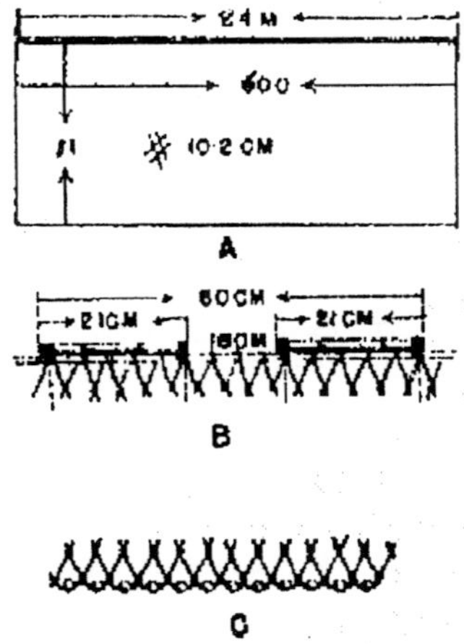

Fig. 4 : Chaundhi : A. A single piece Chaundhi
B. Arrangement of floats
C. Arrangement of sinkers

rope and drum shaped iron sinkers about 20 in number each 1.8-2.0 cm. long, 1 cm. in diameter are attached to foot the rope of each piece. The net is dragged by two men along the shallow waters of the river with the help of two bamboos attached vertically at either end of the net. This net is mostly used from May to September. The catch is mainly of Prawns and small fishes. (Fig. 5).

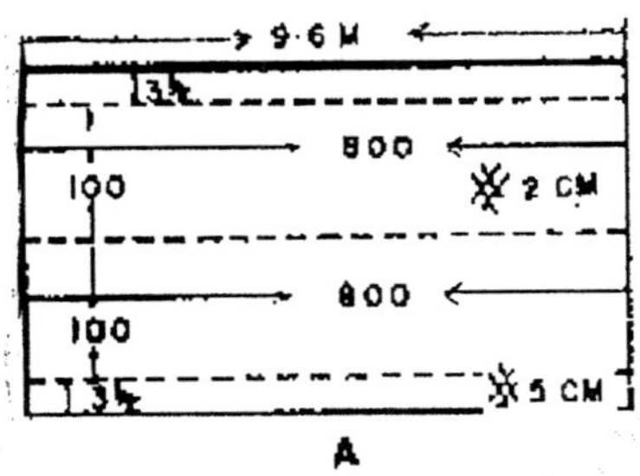

Fig. 5. Do-Dandi : A. A single piece of Do-Dandi

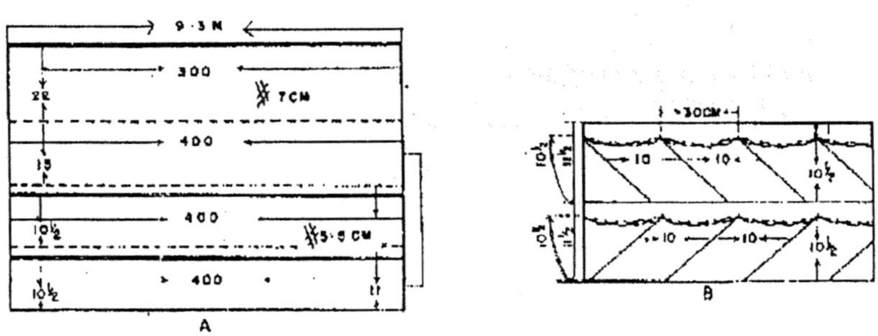

Fig. 6. Chhanta : A : A single piece of chhanta
B. Showing arrangements of pockets

Drag Net with Pockets

Chhanta: It is similar to other drag nets with the difference that it has rows of pockets at the lower end. Generally, Chhanta with two rows of pockets is prevalent. Hornell (1923) has described several types with different names. Faruqui and Sahai (1943) have recorded this net as Chanti. The net, made of cotton, is usually composed of 4-5 pieces each 9.3 metres long and 3.57 metres wide with a mesh size of 7.0 cm (stretched), in the first strip and 5.5 cm in the lower three strips. The arrangement of different strips and pockets is shown in Fig. 6. The number of meshes across in 1st and lower three strips is 300 to 400 respectively. In depth, the number of meshes in 1st, 3rd and 4th strips is 22 meshes and in second trip the number is 15 meshes. The 3rd strip is folded from the middle so as to leave 11 – ½ meshes in the back and 10 – ½ meshes in front side leaving one mesh space free. The fourth strip is stitched to the lower end of the fold of third strip and it is also folded like 3rd strip. The two folds form a big pocket in each of the lower two strips. This big pocket is divided into a number of small quadrangular pockets by sewing the upper fold with the back fold obliquely at regular intervals so as to leave 10 meshes across in each pocket. Four iron cylindrical sinkers are attached to the loose end of each pocket with the help of a thin rope. Dried gourd floats about 5 in number in each piece are attached to the head rope. The operation is similar to other drag nets. When the net is dragged the mouth of pockets remain open due to weight of sinkers. Since the pockets are narrow and oblique there is no space for the fish for turning and it gets inextricably lodged in the pockets. This net is operated from July to March. *H. ilisha, W. attu, M. aor* and *M. seenghala* and major carps are mainly caught.

Gill Net Without Foot Rope

Tiar: Made of hemp, a single piece of this net has 400 meshes across and 10 – ½ meshes in depth with a mesh size of 18 cm. The head rope of one piece measures 39.20 meters. About 50 floats of thick reeds, each 40 cm long and 1 cm in diameter are attached vertically to the head rope of the net. The upper margin of the net is also tied with the head rope at these places leaving 7 meshes between the floats. This is for ensuring a constant hanging coefficient all through the net. To one end of the head rope is attached a big gourd float while sinkers are totally absent. Two men with a small boat operate this net. The net is laid out in the water at sufficient depth. After some time when the net is settled in the water, from a sufficient distance away from the net, is beaten by bamboo stick and the scared fish run toward the net. The season for Tiar is December to June when there is little current in the river and the water is more of less undisturbed. Medium sized *M. aor* and *M. seenghala, P. pangasius, S. Silondia* and major carps are largely captured by this net (Fig. 7).

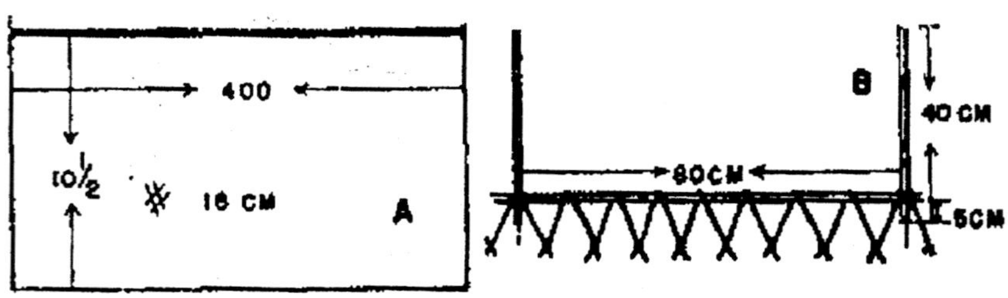

Fig. 7. Tiar : A. A single piece of Tiar
B. Showing arrangements of floats

Gill Net with Foot Rope

Gochail : Made of hemp, a single piece measures 41.4 meters with 340 meshes across and 10 – ½ meshes in depth with a mesh size of 26 cm. Tin floats of shape shown in Fig. 9 are tied with the head rope with the help of a separate net piece. In one piece 6 such floats are tied at a distance of 6.9 metres or 56 meshes across. Every seventh mesh is tied with the head rope with the help of a twine to maintain an uniform hanging coefficient all through the net. There are 8 such knots between two floats. Sinkers are absent but unlike Tiar, foot rope is present. The net is laid out in the water in a straight line across the flow of current with the help of a boat and two men. Since it is operated where the current is very fast, the use of this net is generally restricted to Ganga river only. This net is operated from February to July. Very big sized *B. bagarius, S. silondia, P. pangasius, L. rohita, C. catla, M. aor* and *M. seenghala* are caught with the help of this net (Fig. 8).

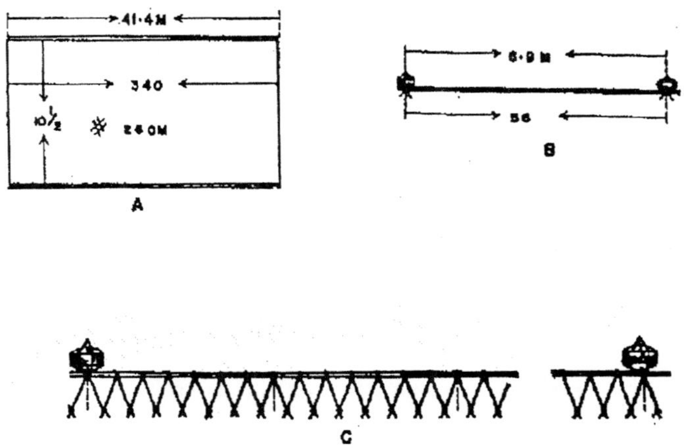

Fig. 8. Gochail : A. A single piece of Gochail
B. Showing arrangements of floats
C. Showing arrangement meshes between two floats.

Purse Net : This is the same net used in the lower reaches of the system where it is known as Shangala (Hornell 1923, Government of India 1949, 1961). The shape of this net is like a purse. The mouth is of two long flexible bamboos hinged at the two ends which can be shut instantly when desired.

Kamel : Two types of Kamel, one with a mesh of 14 cm of stronger hemp for big sized carps and the other with 8.5 cm mesh of finer cotton or nylon exclusively used for *H. ilisha,* locally known as 'BARA' and 'CHHOTA' kamel respectively are in use at Allahabad. Two flexible claps 5.2 metres long are hinged together firmly at the two angles. In the middle of upper clap is attached a brass ring of 2 cm diameter and to the lower clap is attached a weight which is either a brick or stone. A stout rope 10-15 metres long is tied to the lower clap and passes through the ring of the upper bamboo and another thin rope is tied with one of the meshes of the upper fold of the net. This string works as an indicator and is called, 'telephone' by fishermen. The net, both in Bara and Chhota kamel, is rectangular in shape having 45 meshes across and 32 in depth in Bara and 70 across and 54 in depth in Chhota Kamel. The two flexible claps which can be used with both type of nets, form the mouth of the purse. Two men each with a set of kamel sit in a small boat and both row with one hand and manipulate the two strings of the net with the other in such a way that the mouth of the purse remains wide open and the lower clap of the net touches the bottom of the river. The moment any fish enters the net, a jerk or quiver is felt by the fisherman through the 'telephone' and the net is instantly closed by releasing the indicator chord and pulling the main rope. The net is hauled up with the help of the thick rope and emptied. It is operated throughout the year. *Hilsa ilisha* is largely captured by this net during October to January and large sized carps. *P. pangasius* and *S. silondia* are mainly caught in Bara Kamel from May to July (Fig. 9).

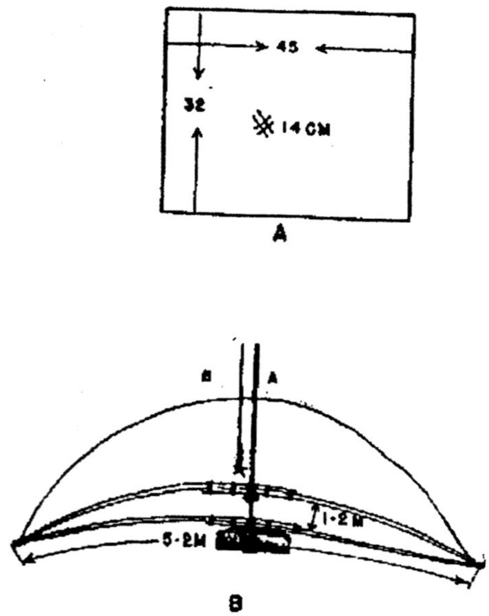

Fig. 9 : Kamal : A. Net of Kamal
B. Arrangement of bambo claps and the two string.

Cast Net

Bhawnar Jal : This is a circular net made up of cotton with a wide variety of names like Khepla (Hornell, 1923), Ghagria (Faruqui and Sahai, 1943). The number of meshes at the apex is 50 and at the periphery 1000. The mesh size is 2.5 cm. At the periphery, circular pockets are made by folding the net inwards to about 6 – ½ meshes in depth. There are 91 pockets in all each of 11 meshes having two iron cylindrical sinkers. At the apex a thin rope of 3 mm diameter is woven through the 50 meshes and are tied to close the end. The free end of the central rope is tied to the fore finger of the left hand. The net is roughly divided into three equal parts in depth, the portion near the rope is held by the left hand, the middle portion is kept loose and the third portion is held with the other hand. With a swing the net is lifted up over the head and dropped to a distance reachable by the attached rope trapping the fish in the net pockets at its circumference. The net is operated in shallow water all through the year. Fishes mostly small in size *B. stigma, Chela* spp. *Ambassis* spp., *Roti cotio, M. cavasius, R. rita* and fingerlings of major carps and catfishes are captured.

Scoop Net

Jali : This is a very common type of net described under Dip net (Hornell, 1923) Hand Net

(Faruqui and Sahai, 1943), Scoo ner (Govt. of India, 1961). A piece of net is attached to two bamboo poles each about 2.75 metres long, forming a 'V' shape structure. A third bamboo, 50 cm long joins the two bamboo poles 40 cm away from the angular portion. The net consists of 10 simple strips 9-16 mm mesh size. The meshes across and in depth are 240 and 40 respectively in each strip. The strips are joined in the pattern shown in Fig. 10. The net is operated by a man from a small boat or without boat in shallow water. The tail portion is held by one hand and the middle bamboo is firmly held by the other. The water is gradually scooped and the catch of individual operation is collected in the tail end till it is finally taken out. During rainy season this net is operated and small fishes and shrimps including juveniles of *Mystus* spp. other catfishes and major carps are scooped by this net.

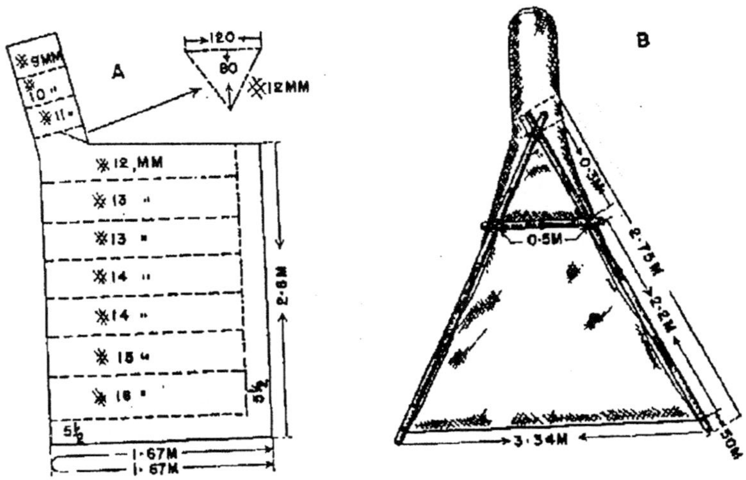

Fig. 10 : Jali : A. Net of Jali.
B. Arrangement of net and bamboos.

Hook and Line

Jor : This is an ordinary fishing and is popular with fishermen who can not afford to have any other net. A thin but stout rope generally 155 metres long is used to which about 150 hooks are attached with a small string. One end of the line is fixed to some bamboo or anchored boat and the other end is left in the river with a float of gourd or tin attached to the rope and a sinker of some stone or brick is fixed with a subsidiary rope for keeping the line stationary. Snails, earthworm, cockroach or small fishes are attached to the hooks as a bait. The line is inspected by the fishermen at regular intervals and hooked fishes are collected on the boat. This is operated throughout the year and during the rainy season it forms a major source of fishing. *Rita rita, Clupisoma Lgarua, Mystus* spp., Eutropiichthys *vacha* and *Wallago attu* are generally caught.

Traps

1. Kuriar : This trap is similar to Chakjal described by Hornell (1923). Six bamboo sticks each 87 cm long are tied at one end to an iron ring of 22 cm diameter, as shown in Fig. 11. A conical net

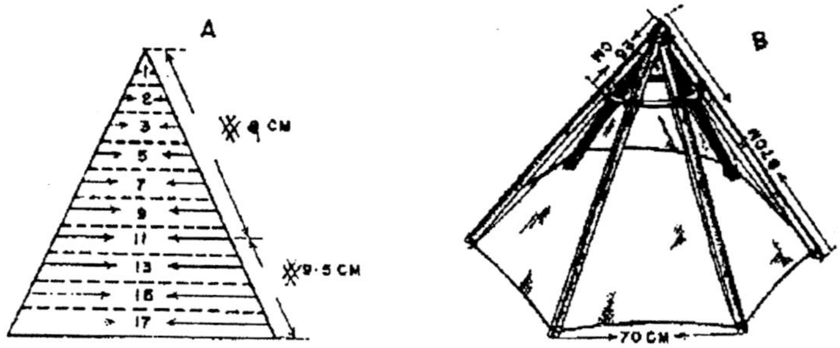

Fig. 11. Kuriar : A. Net of Kuriar
B. Arrangement of Bamboos.

having six sides, is attached to the free ends of bamboo skeleton form the inner side. The size of mesh varies from 9-9.5 cm from top to bottom. The net is operated in shallow water either with or without a boat. When a fish is sighted, the fisherman jumps with the Kuriar over the fish, the free end of the net which is held by the fore finger is dropped and the fish gets entangled in the meshes. Bigger Kuriar are used in Ganga river. With Chaundhi and Gopal jal, Kuriar form an important accessory and the fish is ultimately trapped in it. The net is operated from November to July. Big sized *C. mrigala, C. catla, L. rohita, P. pangasius* and *B. bagarius* besides *M. aor* and *M. seenghala* are mostly captured.

2. Gopal Jal : This trap is exclusively used to take *M. aor* and *M. seenghala*. The trap consists of (i) enclosure of net and a (ii) rope for dragging fish. The net is made up of hemp of 360 meshes across and 30 meshes in depth with a mesh of 8 cm. To this net are attached 17 bamboo stocks, each 1.5 metres long, at regular intervals. The rope is made of worn out nets of Mahajal or Darwari and is about 300 meters long and 12 cm in diameter. A rectangular enclosure of the net is made by fixing the poles in shallow part of the river. One end of the rope is attached to one stick of the gate enclosure and the other end of the rope is brought to the other stick of the gate by laying out in a circular area. The circle of the rope is made smaller and smaller without disturbing the water much. Big sized *M. aor* and *M. seenghala*, which are inside the circle of the rope are ultimately scared in the enclosure. The fish is taken out with the help of a Kuriar. This is in use from February to June when the water level goes down considerably (Fig. 12.).

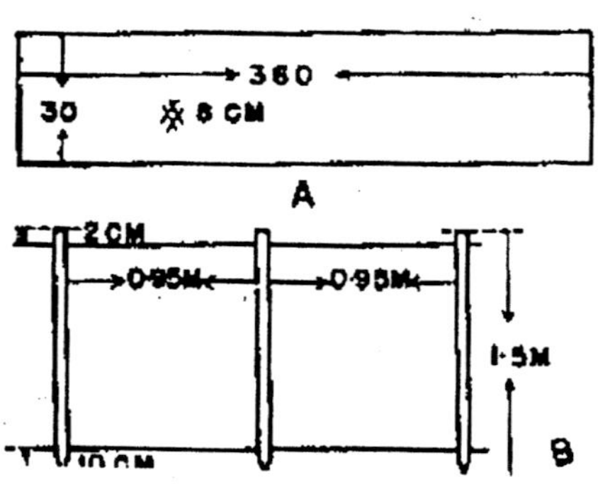

Fig. 12. Gopal Jal : A. A single net.
B. Arrangement of net and bamboos

3. Bandal : This trap is exclusively used for *Hilsa ilisha*. Shallow water generally, 2-3 feet deep over a considerable area with fast current, is preferred on account of which this trap is mostly in operation below the confluence of the two rivers. It is however, also operated above the confluence in Ganga river where the required site is available. The trap consists of two parts : (1) Barrier of bamboos and (2) Ghanch for lifting the fish. About 600-800 bamboo poles are fixed firmly at the selected sits at an interval of 25-35 cm apart as shown in Fig. 13. Ghanch is fixed at the end of the 'V' shaped barrier with the help of two stout bamboos. A simple net of cotton of mesh 6.2 cm (stretched) is fixed to the bamboos of the Ghanch. Ascending *Hilsa ilisha* which moves against the current on finding the passage barricaded are directed along the barrier leading to Ghanch. When the fish is within the scope of the net, the back part of the Ghanch is pressed by the hand and along with it the net with fish is lifted up from the water. The fish is taken out with the help of a scoop net. For descending fish, which moves with the current, the two wings of the 'V' shaped barrier is removed and the fish is directly led to the Ghanch. One boat and two men to work in shift are required for the operation of the trap. Immediately after floods when the water recedes down considerably the Bandal is fixed and as soon as the water level starts rising the trap is removed. *Hilsa ilisha* is mainly harvested but sometimes *M. aor, M. seenghala, P. pangasius, W. attu, L. rohita* and *C. catla* are also caught.

4. Sirki : Taking advantage of the peculiar habit of *Mugil corsula* shoaling at the surface and leaping from water to evade danger, this trap is used specially for catching this species. Chali or Chanchi trap described by Hornell (1923) is very much similar to Sirki described here. Several pieces of rectangular reed matting 8.54 metres long and 2.44 metres wide are stretched and supported by split bamboos tied along its length. To one side of it is attached a narrow net 10 cm in height and supported by upright pieces of bamboos. The mat is floated flat on water in moon–lit night with the help of two ropes attached to each corner of the opposite side of the mat. One of the ropes is held by a man on the bank and the other by another man on the boat. The passing shoal of mullets, frightened by shining reed mat, jumps over the mat. The marginal net prevents the fish from leaping off. The mat is drawn towards the bank and the catch is collected. The trap is in operation from February to May. Along with *Mugil corsula, Chela* spp., *Aspidoparia* spp. are also caught.

5. Ujhki or Cheer : This is another trap, seen operated in the downstream of the confluence of rivers Ganga and Jamuna, used to trap *Mugil* spp. A long barrier of dried grasses with the help of bamboo pieces, is fixed in the river upto a short distance across the stream. Near the bank a square dip net 120 cm long on each side, is manipulated by a lever system. The fisher-man hides on the bank behind bushes or cut branches of trees. The shoal of fish is directed towards the barrier by throwing small pebbles with the help of a sling called 'GULEL' On finding the way blocked, the fishes are directed to the dip net and when they are within the net the string of the net is pulled by the man lifting the fishes with the net. This trap is operated from February to May.

6. Kohni : Hornell (1923), Job and Pantulu (1953) have described this trap under cage traps. It is a conical structure made of fine reeds of size 1.2 metre long and 24 cm in diameter at the wider end. The conical end can be closed and opened when desired. At the two sides of the wider end a barrier of dried weeds supported by six pairs of bamboo sticks are fixed in the river with cones or 'Kohni'. Five to six such Kohni traps are fixed in shallow water near the bank generally one after the other. The mouth faces the current. The traps are fixed in the evening and the catch is taken out in the morning by opening the conical end. It is operated only in rainy season mainly for shrimps but small fishes are also trapped along with it.

Fishing Boats : For fishing , two types of boats are used; big boat known as 'Katra' and small boat or Dongi.

Big Boat : It is mostly used with big drag nets like Mahajal, Karia and Darwari. The over-all,

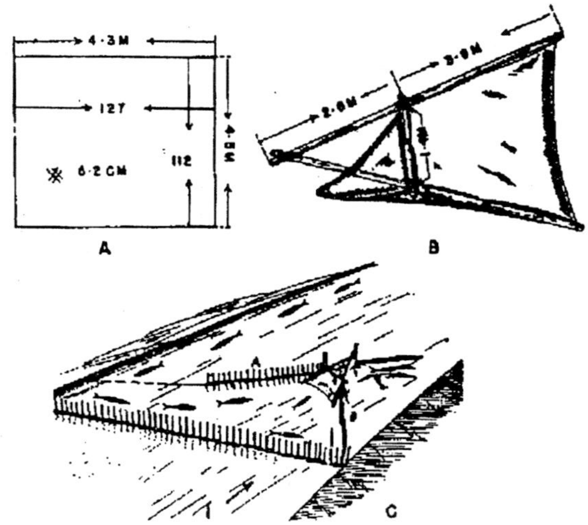

Fig. 13 : Bandal : A. Net used in Ghanch.
B. Complete Ghanch.
C. Arrangement of Ghanch and bamboos forming barrier.

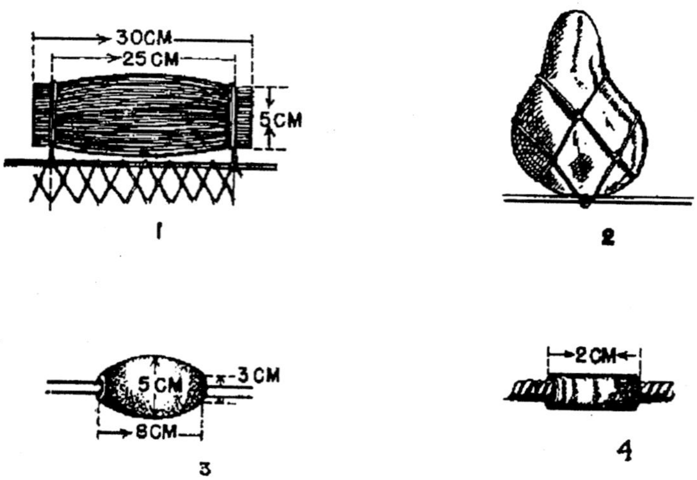

Fig. 14 : Floats and Sinkers : 1. Float used in dragnet.
2. Gourd float.
3. Baked clay sinkers used in dragnet.
4. Iron sinker used in Do-Dandi.

length of the boat is generally 9.95 metres with one end tapering and the other flat. The depth is 90 cm. The flat end is 3 metres wide with a space of 2 metres in length mounted with tin sheets for keeping the net. Three or four raised platforms about 15 cm wide are made on which the net is kept. The middle part, about 4 metres in length is of wooden planks and is used by the fishing party of 10-25 men. Near the tapering end 2-4 men sit and row the boat.

Small Boat: The small boat or Dongi is used with almost all other types of net. It is about 6 metres long, 1.5 metres wide in the middle and 45 cm in depth. The boat is very light and handy in use. The different types of floats and sinkers used are shown in Fig. 14.

APPENDIX IB
COMPARATIVE ACCOUNT OF MAJOR NETS

Type of Fishing Nets	No. of Pieces	Dimensions: Wing: Length	Breadth or depth.	Size of mesh	No. of pieces	Central: Length	Breadth or Length	Size of depth	No. of pieces	No. of floats in a piece
1	2	3	4	5	6	7	8	9	10	11
I. Drag Net (Without Pockets)		Metres	Metres	Cms.		Metres	Metres	Cms.		
(i) Mahajal	50-55	7	3	6.5	20,26	14	8	6	4	8-9
(ii) Derwari	40-60	9	5	3.8	15,22	5	11	2.4	5	8-9
(iii) Karia	50	9	6	16	18,25	8	11	15	7	8-9
(iv) Chaundhi	14-15	24	1	10.2	—	—	—	—	—	64
(v) Do-dandi (with pockets)	1-2	10	4	2	—	—	—	—	—	6-8
(i) Chhanta	4-5	9	4	5.5	—	—	—	—	—	5
II. Gill Net										
(i) Tiar	1	39	2	18	—	—	—	—	—	50
(ii) Gochail	1	41	3	26	—	—	—	—	—	6
III. Purse Net										
(i) Bara Kamel	—		45 Mesh 32 mesh	14	—	—	—	—	—	—
(ii) Chhota Kamee	—		70"	54"	8.5	—	—	—	—	—

Type of Fishing Nets	No. of sinkers in a piece	No. of men required	No. of boats	Period of Operation	Species caught	REMARKS
	12	13	14	15	16	17
I. Drag Net (Without Pockets)						
(i) Mahajal	10	12-16	2	Nov. – July	Major Carps, major catfishes, *H. Ilisha*, *N. Chitala*, other Clupeids & minor, carps & Catfishes.	
(ii) Derwari	10	12-16	2	Throughout year	Juveniles of *Mystus* spp. *C. garua*, *E. vacha*, *Chela* Spp. *A. morar*, *S. phasa*, *G. Chaora*, Major Carps & catfishes.	
(iii) Karia	10	20-22	2	Nov. - July	Big sized carps & catfishes	Kuriar is used for trapping the fish.
(iv) Chaundhi	600	4-5	1	Throughout year except flood months	*M. Aor* *M. Seenghala* *C. mrigala* *R. rita* Prawas & small fishes	
(v) Do-dandi (with pockets)	20	2	—	May-September		
(i) Chhanta	4 in each pocket	3-4	1	July-March	*H. ilisha*, *W. attu*, *Mystus* spp. & Major carps.	2 Rows of Pockests.
II. Gill Net						
(i) Tiar	—	2	1	Dec.-June	Big Sized *Mystus* spp. *P. Pangasius*, *S. silondia* & Major Carps.	
(ii) Gochail	—	2	1	Feb-July	Big sized *B. bagarius*, *S. silondia*, *P. pangasius*, *L. rohita* & *Mystus* spp.	
III. Purse Net						
(i) Bara Kamel	1	2	1	May-July	Major carps, *P. pangasius*, *S. silondia*.	Two split bamboos 5 metres long form the flexible clap.
(ii) Chhota Kamee	1	2	1	Oct.-Jan.	*H. illisha*	

APPENDIX IC

Gear used in the Ganga River (Fresh Water Zone)
Source: CIFRI (MS) 1970
(MAJOR DRAG NET)

Gear with mesh Size	Area of Operation	Season of Operation	Species Caught with sizes
1. Mahajal (5-12 cm)	Upper Ganga	Intensively used in winter Oct/Nov. to March/April	Large size Catla, Rohu, Mrigal, Calbasu, Seen-ghale, Aor, Pangas *etc.*
2. Paundhi Jal (Similar to Mahajal) (4-8 cm)	Varanasi to Patna at middle stretch only	October-July	Mostly major Carps, Cat fishes, Hilsa, Pangas, Selon-dia etc.
3. Kadhiya Jal (Similar to Mahajal) (12-15 cm)	Operated mostly in deep waters during summer around Allaha-bad, Mirzapur & Varanasi	Summer	Mainly major Carps and larger catfishes.
4. Par Jal (5-8 cm)	Jamuna	—	Mostly major carps and cat fishes.
5. Darwari (1.5-3.5 cm)	Allahabad Zone	Winter, and summer November-June	Small fishes like Garua, Vacha, Banspati, Juveniles of Carps, minnows.
6. Chaundhi Jal (8-10 cm)	Most prevalent in upper stretches of Ganga around Allahabad	Round the year except during monsoon. Maximum during summer.	Besides Mystus sp Mrigal is also Captared.
7. Chhanta Jal (4-8 cm)	Because of its entangling properties it is suitable ever in deep waters.	Round the year	All species, but mostly juveniles of Carps and Cat fishes.
8. Joha Jal (4-8 cm)	Deep pools of Ganga where currents are mild	November to April	Mostly Mrigal, Catla, Aor and Seenghala.
9. Ghanaili Jal (1-4 cm)	Restricted to monsoon sea-son only in Bihar and in UP-round the year.	—	Almost all fishes that come in the sweep.
10. Berjal (largest of dragnets) (5-10 cm)	Below Rajmohal in Bihar	After monsoon when water level starts receding	Mostly Hilsa, Silondia, Pangus and major carps.
11. Tana Ber Jal (Smaller dragnet) (4-8 cm)	Lower reaches of Ganga around Bhagalpur.	—	—
12. Chandi Jal (2-3 cm)	Upper Stretches of Ganga system, shallow clear and calm water towards mid river.	—	Mullets, Garua, Vacha, Chela spp banspati etc.
13. Patia Jal	Upper stretches of Ganga narrow and shallow nalas		Only small sized fishes caught.

MINOR DRAG NET

Gear with mesh Size		Area of Operation	Season of Operation	Species Caught with sizes
1.	(a) Dondi Jal (2 cm)	—	—	Carp juveniles, smaller, Cat-fish and others like chela etc.
	(b) Hala Jal (2-3 cm)	West Bengal	—	—
2.	Chote Jal	Lower stretches of Jamuna around Allahabad Narrow sheet of water or shallow island	During first monsoon when breeders ascend the fields to breed	Mostly matured Mrigal, Catla, Rohu, Kalbasu, Mystus sp., Wallago *etc.*
3.	Bayali Jal	Stretch of Jamuna around Allahabad.	—	Kalbasu and rita taking shelter in creeks of rocky bed.
4.	Jhingur Jal	Middle Stretch of Ganga around Patna	October to May	—
		GILL NETS		
1.	Gochhail Jal (10-30 cm)	—	January to July also during monsoon	Big Catfishes and Carps, Bagarius,Wallago, Silondia, Mystus sp. *etc.*
2.	Dharmo Jal (20-30 cm or) even more	Ganga around Allahabad	Monsoon	Major Carps move for spawning.
3.	Tiar Jal (15-30 cm)	Shallow waters	December to July	Mystus sp. commonly captured.
4.	Phasla Jal (16-25 cm)	Entire Ganga system	—	Rohu, Catla, Mrigal Aor, Seen-ghala, Wallago and other bigger fishes are captured.
5.	Chandi Jal (4-5 cm)	Mostly in West Bengal for hilsa fishing	—	Hilsa fishing
6.	Panjia Jal (15-20 cm)	Patna, Monghyr and Bhagalpur of Bihar		Major Carp and Catfishes.
7.	Saraila Jal (15-30 cm)	Between Varanasi and Patna.	—	Cat fishes. Sometimes Catla, Rohu and Mrigal.
8.	Phansa Jal (5 cm)	Buxar-Hilsa (Bihar)	November-March	Hilsa
9.	Phans Jal (15-25 cm)	Lower Stretch in Murshidabad Dist.	—	Large Size Rohu, Catla and other fishes.
10.	Bhasa Kona Jal (5 cm)	—	July-October	Primarily Hilsa.

Gear with mesh Size	Area of Operation	Season of Operation	Species Caught with sizes
PURSE NET			
1. Sangla Jal (5-10 cm)	Allahabad region increases in concentration along Ganga down stream	Mostly monsoon and early winter months	Exclusively for Hilsa. Besides, Phase, Garua, Garua and Vacha are also rarely caught.
2. Sangila Jal (8-18 cm)	Middle and lower stretches of Ganga	October-December	Mostly bigger specimens of major carps and cat fishes.
3. Kharki Jal Similar to Sangla Jal	—	During monsoon	Hilsa
SCOOP NET			
1. (a) Ghanch Jal (4-8 cm)	Below Confluence (Allahabad)	—	Main catch-hilsa
(b) Kharra Jal (similar to Ghanch Jal in measurement)	—	October-March	
2. Paunti Jal (5-8 cm)	Lower stretch of Ganga below Patna	Fast current during monsoon and post monsoon months	Hilsa, Goonch, Silondia, Major Carps.
3. Farhara Jal like Ghanch Jal (1-1.5 cm)	Allahabad and Varanasi	Round the year	Carp minmous and other small fishes.
4. Mang Bisara Jal (Biggest scoop net) (1-3 cm)	Buxar-Patna	—	All fishes big and small sizes.
5. Thehra Jal (Big scoap net)	Santal Parganas and Murshidabad districts of Bihar & W. B.	During monsoon when hilsa ascend for spawing	Mostly hilsa
6. Bisari Jal (common scoop net) (0.5-2 cm)	Entire stretch of Ganga.	Off seasons in shallow waters near bank.	Chela, Morar, Chapra, and other small fishes like carp, smaller cat fishes shrimps etc.
7. Lokani Jal	Around Allahabad	Summer and Winter months.	Mullets only.
8. Bhesal Jal	Lower stretch of Ganga in West Bengal	—	Major carps, Hilsa cat fishes occasionally.

	Gear with mesh Size	Area of Operation	Season of Operation	Species Caught with sizes
		CAST NETS		
1.	Jhinguri Jal (0.5-2 cm)	Throughout the Ganga, generally from bank also from low, Varanasi & Ballia.		Small fishes like Carp minnows, Catfishes, *Chela* Spp. shrimps etc. Fingerlings of major carps and bigger prawns.
2.	Adhaukhi Jal (5-8 cm)	R. Yamuna also between Patna-Baichal (Santal Parganas) and West Bengal.		Rita, Silone, Kalbasu
		SET BARRIERS		
1.	R & K	Upper stretches of Ganga and Jamuna 80% of summer landings of the Jamuna (around Etawah and Agra)	Summer	Large sized major carps and cat fishes.
2.	Baudal	Below the confluence of the Jamuna and Ganga	Oct-April	Exclusively for hilsa fishing.
3.	Gopal	Operated from bank		Specific for *M. aor* and *M. seenghala.*
4.	Ujhuki			Mullets.
		TRAPS		
1.	Kuriyar (5-10 cm)			Mainly major carp, carp, cat fishes, shrimps, Chela etc.
2.	Sirki	Common in the middle stretch of Ganga from Allahabad-Patna.		Exclusively used for capture of *Mrigal Corsuls Chela* spp., *Aspidoparia* morar and other carp minnows are also occasionally caught.
		LONG LINES		
1.	(i) Jor	Used when most of the gears become in effective		Garua, Vacha, Rita, Notopterus, Seenghala and some times Silong and Pangus.
2.	(ii) Dori			All Major carps.
3.	(iii) Dauni	Ganga near Bhagalpur		Exclusively for garua catch.

APPENDIX II A

Gear used in the estuarine Zone
Source : Dutta, *et. al.* 1973.

Gear with local name	*Area of Operation*	*Season of Operation*	*Species Caught*
1. *Trawl nets* (Moi, Kelta, Buro, Kantni and Kachhi)	Upper and Middle Stretch of Zone I	Nov./June	Prawns, also miscellaneous.
2. *Sine nets* (a) Large : (Jungla, Kachal and Jagatber)	Coastal region of Zone III	Nov.-Feb. (Occasionally in Monsoon)	*H. ilisha*, *T. jella*, *P. pama* *Setipinna* spp. & *S. biauritus*
(b) *Medium & small* (1) Ber (2) Charghers, (3) Kachal and (4) Chatber	(1) Zone III (2) III also upper area of Zone I (3) Upper area of Zone I (4) Upper area of Zone I	(1) All seasons, (2) All seasons, (3) March-July. (4) Nov. - August	(1) Mullets, Perches, Polynemids and sciaemids (*P. pama*) (2) Mullets, perches (*S. panijus*), Polynemids and sciaenids (*P. pama*) (3) *H. ilisha*, (young) and Miscellaneous fishes. (4) *H. ilisha* (young). *S. Phasa*, Prawns and Miscellaneous fishes.
3. Purse nets : (sanglo, Dar and Khorke)	Zone - I	July - Oct. and Jan. - March.	*H. ilisha* only
4. Drift nets* (1) Chhandi, (2) Dholi (3) Kona (Bhasa-Kona, Ghai-Kona, Ghalo-Kona (4) Shele and Bhasa-larang,	(1) Zone III, I, II (2) Zone I (3) Zone I (4) Zone III	(1) Monsoon and Winter (June to February) (intensity greater in Monsoon) (2) Nov. - May (3) Feb. - Sept. (4) Nov. - May	(1) *H. ilisha*, other (*I. elongata*, *H. toli*, *Setipinna* spp.) and *S. Cinereus* (2) *H. ilisha*. & *S. phasa* (3) *H. ilisha* (4) *P. indicus* and *S. biauritus*
5. *Lift nets* : Bhesal, Gara-bhesal, Nauka-bhesal, Pata, Thela, Seitki, Thapa, Dhain, Tana, Char and Koli	Zone I & III	All seasons	Mullets, *I. elongata*, *Setipinna* spp. and Prawns.
6. *Cast nets* : Bachari & kepla	Zone I & III	All seasons	*P. paradiseus*, *P. pama*, *P. pangasius* and *Prawns*

Contd. ...

Gear with local name	*Area of Operation*	*Season of Operation*	*Species Caught*
7. Bag nets : (Stationary) (1) Been (2) Behundi (3) Thor (4) Atone and (5) Gopa	(1) & (2) All Zones (3) Zone II, also main channel of Zone III (4) Zone III	All seasons	(1) (2) and (3) Almost all species especially *H. nehereus*, prawns, *P. pama*, *Trichiurus* Spp. *Setipinna* spp. *I. elongata*, mullets & *S. Paningus* (4) and (5) *L. Calearifer* and *S. biauritus*.
8. Set gill nets : (1) Barang (Med- barang, Bhetki-barang) (2) Nangar, (3) Khota, (4) Khuti (5) Patang, (6) Geba- Chandi	(1) Zone III (2) (3), (4) Zone III (4) & (5) Zone I	(1) Nearly all seasons (2) — (5) Monsoon & winter.	(1) *L. calcarifer* and *S. biaurіturs* (2) — (5) *H. ilisha* & other clupeids.
9. *Set- Barrier nets* (1) Pata, (2) Char-para (3) Khat pata, (4) Khal ghera (5) Kumor.	Zone III	All seasons	All species, particularly Prawns, *M. parsia*, *Setipinna* spp., *P. pama*, *E. tetradactylum*, *L. calcarifer*, and *I. elongata*.
10. *Traps* : Bitti, Dwar-bitti, Chaibitto and Bhasa-pata	Zone I	Monsoon May-October	Prawns
11. *Hooks & Lines* (Barsi)	Zone I, II, III	All seasons	Catfishes, Perches, Polynemids, Scienids and *Setipinna* spp.

* Except (3) all are gill nets. (3) is a labyrinth form of net which drifts.
Zones I, II and III are indicated in Fig. 24.

APPENDIX IIB

Gill nets used in the Hooghly Matla estuarine System
(Figs. 1–6)
Source : Dutta, S. N. 1973.

- Gill Nets
  - Without foot rope
    - Without sinker
      1. Shele jal
      2. Chandana jal
      3. Ilish-dholi jal
    - With sinker
      1. Phase-dholi jal
      2. Khoira-dholi jal
  - With foot rope
    - Without anchor but with sinker
      - Chhandi jal
    - With anchor
      - Without sinker
        1. Barang jal
        2. Khota jal
      - With sinker
        1. Nongar jal

A. Drift gill net

(a) Gill nets without foot rope

1. Shele Jal (Fig. 1) : Made of either cotton or nylon, a fleet is formed by 150 to 200 shots. A single shot has 70 meshes across and 13 – ½ meshes in depth with a mesh size of 20 cm. The head rope of each piece measures 6.65 m. The main webbing is directly hung on the head rope by the reeving method. At the joining of 2 shots, a bamboo float of 1 m long is attached vertically to the head rope of the net. Sinkers are totally absent. The hanging coefficient of the net is 0.48 (or 48%). The net is operated as bottom drift by 7-8 men with a larger sized boat, near coastal area of the middle zone of the Hooghly. The operation is restricted between October to middle of March, when the water remains comparatively calm. Large sized *Polydactylus indicus* and *Otolithodes biauritus, Epinephelus* spp., *Tachysurus* spp. *Pampus* spp. and *Pristis* spp. are largely landed with this net. Other local names are bhethi jal, bhasapatan jal, karal jal and bhasa-barang jal.

2. Chandana Jal : Made of cotton and very rarely nylon, is formed by 50 to 100 shots. A single shot has 320 meshes across and 120 – ½ meshes in depth, with a mesh size of 8.70 cm. The length of head rope in each shot measures 14.3 m. Attachment of webbing with the head rope and arrangement of bamboo floats are same as in the case of shele jal. The hanging coefficient of the net is 0.51. The net is operated as surface drift. Number of men, type of boat, area of operation and season of operation are more or less same as in the case of shele jal. *Hilsa sinensis, Tachysurus* spp., *Osteogeniosus militaris, Chirocentrus dorab* and *Pampus* spp., constitute the major species in the landings. Another local name is kajal jal.

3. Ilish-dholi Jal : Made of either cotton or nylon, a fleet of net is formed by 4 to 6 shots. A single shot has 720 meshes across and 51 meshes in depth, with a mesh size of 8 m. Either reeving or stapling

Gill nets used in the Hooghly - Matla estuarine system (Dutta, S.N. 1973)

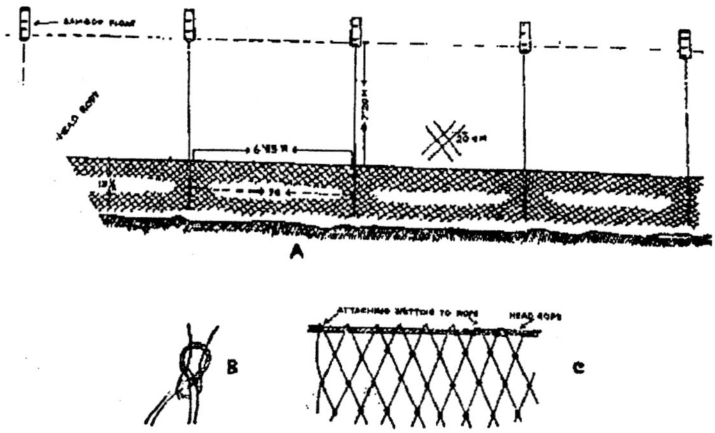

Fig. 1 : Shela Jal ; A. Way of setting net
B. Type of Knot used
C. Attachment of webbing with the head rope

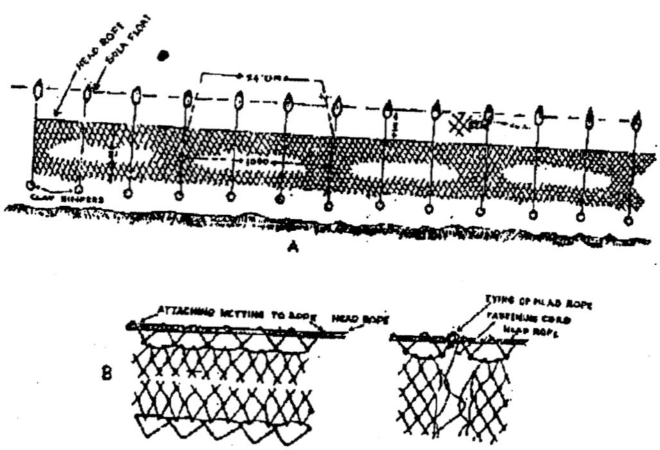

Fig . 2 : Phase-Dholi Jal : A. Way of setting the net.
B. Attachment of webbing with the head rope.
C. Joining the two pieces.

method is followed in the attachment of webbing to the head rope. Triangular shola floats (Eschynomene aspera), 7 in number, are attached to the head rope of each piece at equal distances. Sinkers are absent. Hanging coefficient is 0.35. The net is operated as surface drift. One to two men with smaller boat operate this net in the upper stretches of the Hooghly estuary during December to June. Medium sized *H. ilisha* forms the major catch. Another local name is dholi jal.

4. Phasa-Dholi Jal (Fig. 2) : Made of cotton, a fleet of this net is formed by 4 to 6 shots. A single shot has 1080 meshes across and 81 meshes in depth, with uniform mesh size of 6 cm. The head rope of each shot measures 24 m followed by stapling type of attachment with the webbing. Oval shola floats, 4 in numbers are attached to the head rope of each shot, similarly 4 burnt clay sinkers are attached to the (bottom) side. The hanging coefficient is 0.37. The net is operated as column drift. Number of men, type of boat and area of operation are same as in the case of ilish-dholi jal. Season of operation is from October to February. *Setipinna phase, Eutropiichthys vacha, Clupisoma garua, Pama pama* and small sized *H. ilisha* are caught with this net. Another local name is phase jal.

5. Khoira-dholi jal : This net, constituting 4 to 6 shots, is almost similar to phase-dholi jal, except for the measurements. A single shot comprises 1080 meshes across and 101 in depth, with uniform mesh size of 4.1 cm. The head rope of each shot measures 17.85 m. The hanging coefficient is 0.40. The net is operated as surface drift. Season of operation is from January to July. Small sized *H. ilisha, Gudusia chapra, E. vacha. C. garua* and *S. phasa* are caught.

(b) Gill Nets with Foot Rope

6. Chhandi Jal (Fig. 3) : These days the net is mostly made of nylon and is one of the very important gear of this estuary. A fleet is formed by 25 to 150 shots and a single shot has 260 meshes across and 35 meshes in depth, with a regular mesh size of 11 cm. The length of the head rope in each shot is 8.85 m. followed by reeving type of attachment with the webbing. At the joining of 2 shots a bamboo float and in between 2 bamboo floats a fin float are attached to the head rope. Similarly sinkers are attached to the foot rope. Hanging coefficient is 0.31. The net is operated as surface drift by 4 to 8 men, with a chhandi naoka, almost throughout the entire. Hooghly and Rupnarayan estuaries. The net is operated from June to February to cover both monsoon and winter runs of *H. ilisha. H. ilisha, H. sinensis, Pampus* spp., *C. dorab* and *Tachysurus* spp. are caught in this net. Other local names are bhasa jal, bhasa chandi jal, ilish jal and dora jal*

(C) Set Gill Nets With Foot Rope

These nets are kept stationary with the help of anchors.

1. Barang Jal (Fig. 4) : Made of synthetic material or cotton, a fleet of net is constructed by joining 4 to 8 shots. A single shot has 280 meshes across and 29 meshes in depth with a mesh size of 21.5 cm. The head rope of each shot measures 29.4 m. Reeving method is followed in the attachment of webbing with the head rope. 6 bamboo floats, are attached to the head rope of each shot at equal distances. Sinkers are absent, but iron anchors are tied to head rope and foot rope with the help of anchor lines in order to keep the net fixed at one place. Empty tins or drum floats are used to join the

* (Dora jal is similar to chhandi jal, but prepared from cotton twine and for preservation of the net decoction is not applied. Soda and soap are used to clean the net. The net is used to catch *H. ilisha* during winter months (November-February) and is sometimes quite effective because during this period water remains clean and as the net is also white, it is not easily noticeable in the water. Sometimes, even nylon made gill nets are called dora jal.)

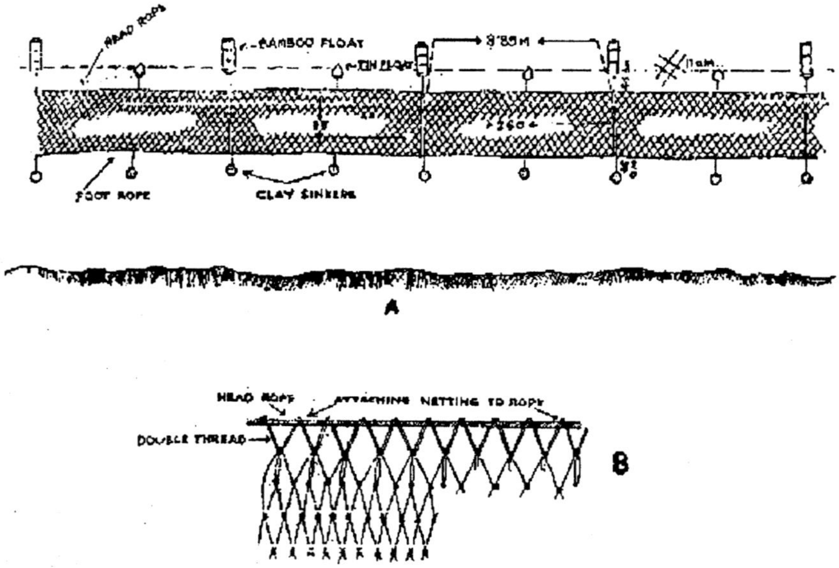

Fig. 3 : Chhandi Jal : A. Way of setting the net
B. Attachment of webbing with the head rope.

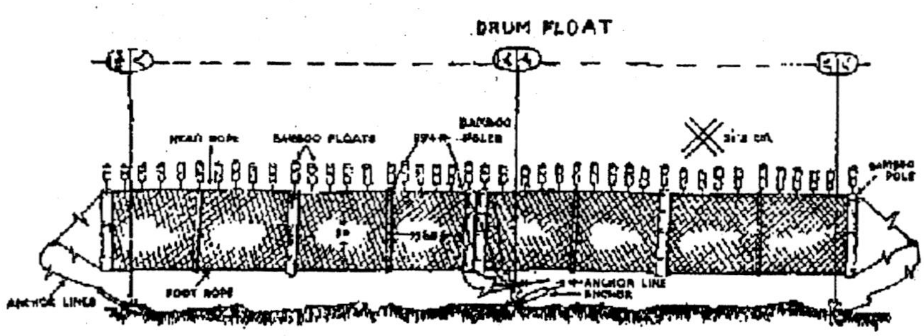

Fig. 4 : Barang Jal

anchors as shown in Fig. 4. One or two bamboo poles, at an interval of 2 shots, join the head rope and foot rope to support the net while in operation. A hanging ratio of 0.49 is always maintained. Eight to twelve men with a larger size boat operate this net in the lower zone of the Hooghly estuary between October to March as bottom set net and *lates calcarifer P. indicus, O. biauritus, Eleutheronema tetradactylum, Tachysurus* spp. and *C. dorab* are generally captured. Other local names are shele barang jal, bhetki barang jal and daura jal.

2. Khota Jal (Fig. 5) : Mostly made of nylon, 4 to 7 shots are joined to make a fleet of gill net. A single shot has 500 meshes across and 76 meshes in depth with a mesh size of 10 cm. The length of float line in each shot measures 17.2 m and is made of wire. The attachment of webbing with the float line is of reeving type. Empty drum floats are attached to the float line at an interval of 2 shots. Sinkers are absent. Two special type of anchors are used at both the ends of the net, joining the float line and lead line through anchor line. The lead line and anchor lines are also made of Wire. These anchors are bigger and heavier than ordinary iron anchors and are made of hollow bamboo poles filled with sand and with openings closed with wooden pieces. Bamboo poles, at an interval of 2 pieces, join the float line and lead line in order to support the net. The hanging coefficient is 0.34. The net is operated as column set. Two to three men with a medium size boat operate this net in upper and lower zones of the Hooghly estuary. Seasons of operation extend from January to March and June to October. *H. ilisha, Pangasius pangasius* and *S. phasa* form the bulk of the catches. Another local name is khuti jal.

3. Nongar Jal (Fig. 6) : The chhandi jal when operated as a set type of net with the help of anchors is termed as nongar jal. Iron anchors, at an interval of 4 pieces, join the head rope and foot rope through anchor lines. The net is operated as surface set and is operated in the upper zone of the Hooghly estuary for catching *H. ilisha.* Other local names are longar jal, nongar-chhandi jal and patan jal.

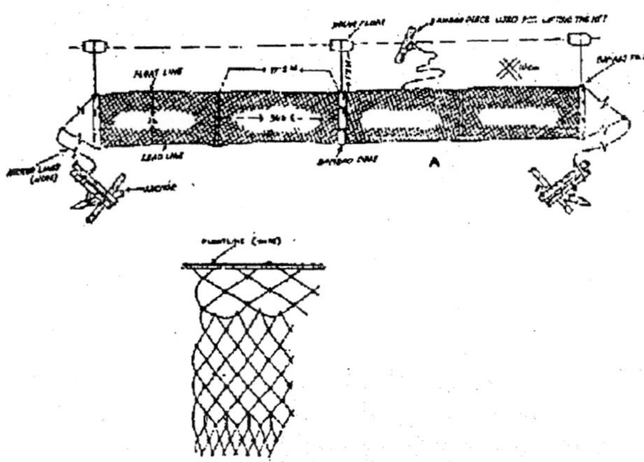

Fig. 5. Khota Jal : A. Way of setting Khota jal B. Attachment of webbing with the head rope.

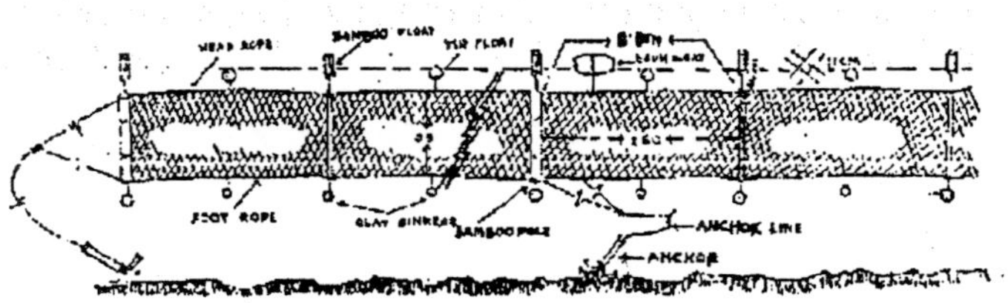

Fig. 6. Nongar Jal

APPENDIX III

List of gears used in the Hooghly estuary

Source : De, D. K. 1980

De (1980) further enlighted the fishing gear and their use in the estuarine system. Based on the mode of operation De (1980) has classified the fishing gear into nine groups. The classification followed here is that of von Brandt (1958 and 1964). A list of gears with their local name together with the method of operation are presented here except Gill net which is already discussed by Dutta (1973).

	Type of gear		*Local name*
1.	Long lines	–	Barsi
2.	Traps & Barrier	–	Bitti, Charpata, Khal pata
3.	Bag net	–	Been, Thor
4.	Dragged net	–	Buro Jal
5.	Seine net	–	Kochal jal, Gaiber jal or Tanaber jal, Berjal, Jaget berjal, Chatber jal, Kona jal.
6.	Dip or lift net	–	Gara besal, Nanka besal
7.	Falling net	–	Khepla jal
8.	Gill net	–	Discussed earlier by Dutta (1973) in detail and hence excluded here.
9.	Clap or Purse net	–	Kharki jal, Sangla jal.

Description of Fishing Gear and Method of Operation

1. Long-lines

The principle used in line fishing is to offer the fish a bait to entice it to bite.

Borsi

Long lines, known as barsi, are employed extensively throughout the estuary. It is a stout line made up of cotton. The length of the main line ranges upto 1000 m, armed with a large number of 450 to 500 short snoods, set at frequent regular intervals. To each snood a hook is tied at the free end. Small fishes, dried fishes and flour balls are attached to the hooks as a bait. Lines are fished in different ways, they can be anchored or left adrift. They may be fixed in any position from the surface to the bottom. The size of the line, snoods and hooks depend on the fishing depth and species of fish caught. Longlines are operated throughout the year. Catfishes, perches, polynemids, scienids and anchovies are generally caught. This gear contributes 3.4% of the total catch of the estuary.

2. Trap and Barrier

The catching principle is based on allowing the fish to enter the trap.

Basket

They attain a diversity of shapes and sizes. Generally they are rectangular, cylindrical or sub-cylindrical in shape having only a silt like mouth through which fish can enter but can not come out. The traps are generally made of fine bamboo splinters. Sometimes a bait is also kept inside the trap. The type of fish caught in basket trap are chiefly prawn in the upper zone and mullets, *Eleutheronoma tetradactylum, Lates calcarifer* and different prawns spp., are in lower zones of the estuarine system.

Charpata

In tidal regions of lower estuary, large areas may be surrounded by a long length of netting forming a wall at the time of high tide. At the time of ebb tide, all the fish present are stranded at the bottom, from where they are caught. Charpata has a number of pieces, each about 7 m long and 3 to 4 m wide. The mesh size at upper area is 5 cm and at lower zone it is 1.3 cm. The total length of the net varies from area to area.

Khalpata Jal

The mouths of shallow tidal creeds that completely drain out at low tide may similarly be cut off during flood tide by a wall of bamboo screens called Khalpata.

Both the traps are operated throughout the year and fishes caught are prawns, *Liza parsia, Setipinna* sp., *Pama pama, E. tetradactylum., L. calcarifer* and *Ilisha elongata.*

3. Bag net

A bag net has a totally or partially framed mouth. The fixed form of bagnet is kept open by the force of the water flowing through it . The fish are actually filtered from the water.

Beenjal or Behundjal

Bag net with a wide mouth is narrowing towards the cod end. Bag nets have wings extending from the mouth. There is a flap inside near the cod-end to prevent the fish from swimming out. The size of the net varies in different parts of the estuary. Generally, the size of the bag portion is 20 m long, mouth 6 m wide and wings 9 m long. Mesh size near the mouth is 4.0 cm and narrowing to 0.5 cm at cod-end. Number of meshes around the periphery of the net varies between 700 and 1000. The nets are operated with the help of heavy anchors and floats. The mouth of the net, facing the current of water, remains open due to force of water. In Hooghly-Matlah Estuarine System about 4000 bag nets are in operation and constiture more than 60% of the total catch of the estuary. The bag nets are used throughout the year. More than half the catches of the bagnets consists of *Harpodon nehereus* and different prawn spp. Remaining catches are dominated by *P. pama, Trichiurus* spp., *I elongata, mullets* and *Silago panigius.*

Thor jal

It is similar to been jal but smaller in size. The number of mesh around the periphery ranges between 300 and 700. These nets are operated in pairs from boats with the help of wooden framework.

4. Dragged net

This group contains all nets and gear which are towed through the water.

Buro jal

The net has a bag and a pair of short wings. Its shape is identical to that of a bag net and operation is similar to that of a typical boat seine. This net is mainly operated in the upper stretches of estuary during November to June for catching prawns.

5. Seine net

In seining, one end of the net is shot from a fixed point, a certain area is surrounded and the other end of gear returned to the starting point; after which the gear is then hauled.

Kochal jal

It consists of 22-24 pieces and each piece has a length of about 12 m and depth of 30 m. Overall length of a net is about 280 m. Mesh size varies from 8-11 cm. The head rope is provided with floats. On sighting a shoal of hilsa the net is laid out from two boats encircling the shoal. It is mostly operated between the November and February in the lower estuary.

Gaibor jal or Tanaber jal

This is a very long rectangular seine net measuring 300 metres or more with puckerings in the lower portion. The depth of the net ranges from 3 to 9 m. Floats are attached to the head rope. The net is operated as encircling net, either in the middle of the river, or from the shore enclosing a large area of water. It is used primarilly for hilsa.

Berjal

The simple type of seine net without puckerings is known as berjal. It is operated during October and May for catching mullets, perches, polynemids and scienids.

Jagat Berjal

Long berjals are known as jagat berjal measuring over 1600 m in length. The operation lasts for several days and a large variety of fish are caught.

Chatber

Fine meshed seine nets that are operated in the upper stretches of the estuary for catching fry and fingerlings of hilsa, anchovies, prawns and miscellaneous fishes during November to August.

Kona jal

This is a boat seine with a series of bags, each with a funnel shaped pouch placed at intervals of 9 to 12 m., provided with a valve to prevent the escape of trapped fish. It measures about 30 to 300 m in length and 9 m in breadth. The number of bags varies according to the length of net. Floats are attached to the head line but sinkers are absent. The mesh size of the net is about 5 cm.

The mouth of the bag is rectangular in shape, 6 to 9 m high and 4 to 5 m wide and the mesh is about 2.5 cm. The mouths of each bag is kept distended by two diagonally placed poles. The net could be operated as a drag net from shore or fixed to stakes in the river where there is a strong current. The catches are mainly hilsa and the season of operation is from May to October.

6. Dip or lift net

This method is to submerge a hanging net, then pull it rapidly out of the water so as capture any fish which happen to be over it.

Gara besal

This is a fixed lever net. It is roughly triangular and is tied to a "V" shaped frame of bamboo. The size of the net varies from about 11 to 14 m in length and 8 to 11 m across the cord. The mesh is about 4 cm towards the middle. The nets are fixed in the river bed close to the bank.

Nauka besal

This is similar to gara besal but smaller is size and is operated from a boat.

Both the nets are operated throughout the year and fishes caught are mullots, *I. elongata, Setipinna* spp. and prawns.

7. Falling nets

This method is to cover the fish with a cone-shaped net with a stiff opening.

Khepla jal

It is a conical shaped net, widely used in almost all parts of the estuary. These nets occur in different sizes. A standard kepla jal measures about 4 m in length and peripheral circumference is about 20 m. The height of the pocket measures 0.5 m. The fishes get caught by the pockets provided at the periphery of the nets. The peripheral pockets are provided with sinkers. Nets are operated throughout the year and catches consist of *Polynemus paradiseus, P. pama, Pangasius pangasius* and prawns.

8. Clap or Purse net

The principle is to purse fishes in bag shaped nets. The shape of this net is like a purse and the upper and lower lips of the mouth are framed with two long flexible split bamboo. Since there is every chance of the fish escaping, the closing mechanism of these nets is very well doveloped with the help of two lips which can be shut instantly as and when it is desired.

Kharki and shangla jal are commonly used in the upper stretches of the estuary during the period from July to December and January to March for catching migratory hilsa.

Kharki jal

It is rectangular in shape and is operated from a boat which is drifted with the current. When fish enters the mouth of net is closed by pulling up the bamboo pole which is vertically attached to lower lip and passing through the upper.

Sangla jal

It is an improved modification of kharki jal. The bag portion is rounded in shape and slightly larger than Kharkijal. In outline the mouth is semicircular and lips are often over 25 feet in length.

The size of net is about 100 feet from front to back. Net can be set at any desired depth from a boat. Mouth is kept open by weight tied to lower lip and can be closed by pulling rope attached to middle of lower lip which passing through the upper.

Materials used for fishing gear

Nets are made of either cotton or hamp twines. Presently synthetic twine is in extensive use for all kind of gears. Pieces of Bamboo (*Bamboosa* spp), Sola (*Aechynomena aspera)*, various kind of light wood, earthen pots, scaled tin cans of various sizes are used as floats. Pieces of stones and brick, burnt clay, lead and iron of various shapes and size are also common.

Preservatives applied for net made of either cotton or hemp are extracts of 'Gab' fruits (*Diospyros embryopteris)*, Coal tar, bark extract of Babul (*Accacia arabica)* are also used as preservatives.

References

Allen, K.R, 1969, Distinctive aspects of the ecology of stream fishes, A review *J. Fish Res. Board Can* 26 1429-1438.

Ananda Bazar Patrika, 1998 FARAKKA BIRODHITAYA MATSAJIBIRAO (Fisherman community of Bihar in protest against Farrakka Barrage) November 30.

Anon, 1919, The report on the Hooghly river and its headwaters, 1 and 2.The Bengal Secretariat Book Department, Calcutta. In : Observation on the Hydrology of the Hooghly estuary—Benoy B. Bose (1956) vol. 3 No. 1.

Anon, 1949, *Preliminary guide to Inland fish, Fisheries, methods of Fishing and Curing*. Manager of Publications, Govt. of India. Delhi (Marketing series. 66, 137 pp.)

Anon, 1965, Report on fish spawn prospecting investigations, 1964. I Uttar Pradesh and Gujrat. *Bull. Cent. Inland fish Res. Inst.* Barrackpore 4; 191 p.

Anon, 1984, Water Management : Coals, Policies, Objectives and implementation procedures of the Ministry of Environment, Ontario, Canada. In : Heavy metals in Water, Sediments and Fish in the river Jamuna—Jhingran Arun G. and H. C. Joshi. *J. Inland Fish. Soc. India* 19 (1) 1987 : 13-23.

Annual Report, 1988-89, *Central Inland Capture Fisheries Research Institute*, Barrackpore.

_____, 1989-90 *Central Inland Capture Fisheries Research Institute*, Barrackpore.

Babulal, 1990, Impact of Farakka Barrage on the Hydrological Changes and Productivity Potential of Hooghly Estuary. *J. Inland Fish Soc.* India 22 (1 & 2) : 38 – 42

Balon, E. K., 1975, Reproductive guild of fishes : a proposal and definition. *J. Fish Res Board Can.* 32 : 821-864.

Banerjee, B. N., 1972, Navigation in a tidal river with particular reference to river Hooghly. *The interdisciplinary symposium Organised by the Department of Geography*. University of Calcutta in Collaboration with other organisations. (ed) Kanan Gopal Bagchi, December.

Banerjee, R. K. and B.B. Pakrasi, 1977, Recycling of distillery wastes. December 8-9. National Environmental Engineering Research Institute and Calcutta Metropolitan Development Authority Calcutta.

Banerjee, R. K., 1988, The Damodar River System— A worst victim of man and environment. *Lecture delivered at the Summer Institute, Central Inland Capture Fisheries Research Institute.* Barrackpore, West Bengal.

Banerjee, S. M., M. P. Motwani and S. J. Karamchandani, 1956, A case of heavy fish mortality in the river Son at Dehri-on- Son. Bihar, Caused by the wastes of the Rohtas Industries Ltd. Dalmianager. *Indian J. Fish* 3 (I) : 186-96

Basu, A. K., 1966, Studies on effluents from Puls and Paper mills and its role in bringing the physico-Chemical Changes around Several discharge points of the Hooghly River Estuary. India. *J. Instt.* of *Engrs.* India, Vol. XVI no. 10 pt. PH. 3.

Basu, A. K. (MS), 1966, A note on preliminary studies of Industrial pollution in and around the Damodar river basin between Bokharo and Panchet Hill Dam.

Basu, A.K. and B. B. Ghosh, 1970, Observations on diurnal variations in some selected stretch of the Hooghly estuary (India) *J. Hydrol.* 32 (1). 271-283.

Basu, A. K., B. B. Ghosh and R. N. Pal, 1970, Comparison of the polluted Hooghly estuary with Unpolluted Matla estuary. India. *J .Wat. Poll. Contr. Fed.* 42 (10) 1771-1781

Basu, A.K., C.S.G. Rao, R. S. Dhaneswar, G. Mukherjee, M. K. A. Kutty, D. Ray and R. C. Trivedi, 1973, Studies on Hooghly estuarine Pollution in the vicinity of Tissue Mill. *Proc. Symp. on Environmental Pollution*, pp. 26-40. National Environmental Engineering Research Institute. Nagpur.

Bera. S., 1984, NADIBIGYANER KATHA (The science of rivers) 10, Raja Ram Krishna Street. Calcutta- 700006

Berkman, Hillary E. and Charles F. Rabeni, 1987, Effects of siltation on stream fish communities *Environ Biol. of Fishes* **18**. (4). 285-291 pp.

Bhanot, K. K., 1973, Observations on the spawning of *Hilsa ilisha* (Ham) in the Hooghly estuary. *J. Inland Fish Soc. India* Vol. V. December.

Bhargava, D. S., 1981 a, Pollution control strategy for Ganga. Water Resources Conservation Pollution and Abatement (ed) Uday Chitranshi, Roorkee. December. 11-13, pp. 31-33.

Bhargava. D. S., 1981 b, Ganga the most self purifying river : *International Symposium Water Resources Conservation, Pollution and Abatement*. December 11-13, Roorkee. India.

Bhargava, D.S., 1983, Use of water quality index for river classification and zoning of Ganga river. *Environmental Pollution Ser. B.* (England) 6 (1) pp. 51-67.

Bhargava, D. S.,1985, Water quality variations and control technology of Jamuna river*Environmental pollution (Series A)* 37 (4): 355-376. *In* : Heavy metals in water, sediments and Fish in the river Jamuna. Jhingran, Arun G. and H. C. Joshi. *J. Inland Fish. Soc. India* **19** (1) 1987 : 13-23.

Bhargava. D. S., 1992, Ganga work hit by scientific flaws. *The Statesman*, November, 4

Bhaskaran, T. R., R. N. Chakraborty, and R. C. Trivedi, 1965, Studies on River Pollution I. Pollution and self purification of Gomati River near Lucknow. *J. Instt. Engrs. India.* **45** (6) : 39-50.

Bhattacharya, S. K. and S. M. Chatterjee, 1972, Brick fields and pollution All India conference Abatement of environmental pollution proceedings Calcutta, December 14-16 Organised by Calcutta. Metropolitan water and sanitation Authority, Jadavpur University, in collaboration with CPHERI (Calcutta); ISI; AIIH & PH; Geological Survey of India, IPHE (India) and Indian Standard Institute.

Bhattacharyya, R. and De Sankar, 1972, Some aspects of Agriculture in Damodar-Hooghly inter fluve. "The Bhagirathi-Hooghly Basin" *Proceedings of the Interdisciplinary Symposiun Organised by the Department of Geography.* University of Calcutta. 1967 Rep. FAO/UNDP (TA) (2449) Reported by E. W. Surber.

Bhattacharya, Kapil, 1986, *SADHIN BHARATE NADNADI PARIKALPANA* (Planning of river system in independent India) Kalam, 52 B, Braa Nath Lane, Calcutta-700012.

Bhowmik. M. L., 1988, Pen culture—A means of more production of fishes from OX- bow lake (abstract) *National Seminar on Management of Fisheries in Inland open water system of India.* Inland Fisheries of India. December, 14-16 Central Inland Capture Fisheries Research Institute, Barrackpore.

Bilgrami, K. S. and J. S. Dutta Munshi, 1979, Limnological survey and impact of Human activities on the river Ganga (Barauni to Farakka range) *Technical Report prepared under Man and Biosphere Programme.* Project 5. UNESCO. at P. G. Departments of Botany, Zoology, Bhagalpur University. Bhagalpur during June 1976 to May 1979.

Bose, B. B., 1956, Observations on the Hydrology of the Hooghly Estuary *Indian J. Fish.* 3 (1). 101–118.

Bose, N. K., 1972, The Bhagirathi–Hooghly—A few remarks. *In* : The Bhagriathi-Hooghly Basin *Proceeding of the interdisciplinary Symposium.* Organised by the Department of Geography, University of Calcutta, in collaboration with the Geographical Society of India. Indian Association of Geohydrologists, Indian Statistical Institute, Regional Science Association (India), The Commissioner for the port of Calcutta. The Institute of Engineers (India) and River Research Institute,West Bengal. Edited by Kanan Gopal Bagchi in collaboration with Sunil Kumer Mandi. December.

Brungs, W. A., Chronic toxicity of zinc to the Fathead minnow. (Pimephales promelas Rotinesque) *Trans. Am. Fish. Soc.* 98 : 272-9.

Brusven, M. A. and K. V. Prathar, 1974, Influence of stream sediments on distribution of macrobenthos. *J. Entomol. Soc. BC.* 71: 25-32.

Calamari, D., 1976, Value and limitation of Chemical and toxicological data in the assessment of the water quality for fisheries. Instituto di Ricerca Sulle Acque (CNR), Via delle cascine Occhiate Brugherio, Milano, Italy.

Chakrabarty, R. D., P. Ray and S. B. Singh, 1959, A quantitative study of the plankton and the physico- chemical conditions of the river Jamuna at Allahabad in 1954-55. *India J. Fish.* **6** (1) : 186-203.

Chakarabarty, R. D. and M. K. Banerjee, 1966, On the depth conditions in the Hooghly estuary as recorded by echo sounding. *Proc. Nat. Acad. Sci*, India vol *XXXVI.* Sce. B : 36. Pt. IV 401-416.

Chakraborty, R. N. and T. R. Bhaskaran, 1966, Recovery of Potassium salts from spent waste of molasses distillary. *J. Inst. Engrs. India.* **46** (10) 37-46.

Chandra, K., S. B. Singh, G. N. Srivastava and S. N. Mehrotra. 1981, Pollution from wastes of Industries manufacturing nitrogenous fertilizer—a case study from river Ganga near Allahabad (India)*International symposium water Resource conservation, pollution and Abatement.* December 11-13, Roorkee, India.

Chaphekar, S. B. and G. N. Mhatre, 1986, Human impact in Ganga river system—An assessment *Concept Publishing Company* New Delhi.

Chatterjee, S. P., 1972, The Bhagirathi-Hooghly Basin Proceedings of the interdisciplinary symposium. Organised by Dept. of Geography, University of Calcutta in collaboration with—The Geographical Society of India, Indian Association of Geohydrologists, ISI, Regional Science Association, the Commissioner for the port of Calcutta, The Institute of Engineers (India) and River Research Institute, West Bengal. (ed) Kanan Gopal Bagchi.

Chattopadhaya, Enakshi, 1993, MEGHNAD SAHAR NADI-CHINTA (The thought on rivers of Meghnad Saha.) *In*:"DESH" Magazine, October 9.

Chutter, F. M., 1969, The effects of silt and sand on the invertebrate fauna of streams and rivers. *Hydrobiologia* **24** : 57-76.

CICFRI, 1992-93, Research Project Programmes CICFRI (ICAR), Barrackpore. (West Bengal).

CIFRI, 1970, Fisheries of the Ganga River System. A Report (MS), Barrackpore.

CIFRI, 1988, A Report on the Survey of North Bihar in relation to effects of Gandak and Koshi river valley Projects on the fisheries of the area. *Survey Report No. 7.* January. CIFRI (ICAR), Barrackpore, West Bengal (India).

Cordone, A. J. and D. W. Kelley, The influences of inorganic Sediment on the aquatic life in streams. *Calif, Fish Game.* **47** : 189-228.

Dakshini K. M. M. and J. K. Soni, 1979, Water quality of sewage drains entering Yamuna at Delhi. *India J. Environ.* **21** (4) : 321-331. *In* : Heavy metals in water, sediments and Fish in the River Yamuna—Jhingran, Arun G. and H. C. Joshi *J. Inland Fish Soc.* India *19* (1), 1987—13-23.

Das Gupta, S. P., 1984, *The Ganga Basin Part II* Centre for study of man and environment for the Central Board for the prevention and Control of Water Pollution. *In* : Assessment and Development study of River Basin Series : Adsorbs/7/1982-83.

David, A., 1954, A preliminary survey of the fish and fisheries of a five mile stretch of the Hooghly river near Barrackpore. *India J. Fish. 1* (1 & 2) : 231-255.

____, 1959, Observations on some spawning grounds of the Gangetic major carps with a note on carp seed resources in India. *Indian J. Fish 6* (2) 327-341.

Day, F. 1889, *The Fauna of British India including Cylon and Burmah. Fishes* I. Taylor and Francis, London, 187p.

David, A. and P. Ray, 1966, Studies on the pollution of the river Daha North Bihar by sugar and Distillary wastes. *Envir. Health. 8* (1) : 6-35.

De, D. K. 1980, Maturity, fecundity and spawning of post-monsoon run of Hilsa, *Hilsa ilisha* (Ham) in the upper stretches of the Hooghly estuarine system. *Jour Inland Fish Soc.* India **12** (1).

De, D. K., 1980, Fishing Gear and Fishing method of Hooghly-Matla Estuarine System. *J. Inland Fish. Soc.* India *12* (1) : 54-63.

De, D.K. and R.R. Sen, 1986, Observations on the embryonic and early larval. development of India shad, *Tenualosa ilisha* (Hamilton) *J. Inland Fish. Soc. India.* **18** (2), 1-12.

De, D. K., and B. N. Sehgal, 1989, Spawning of Hilsa, *Tenualosa ilisha* (Ham) in the Hooghly estuary, *J. Inland Fish Soc. India* **21** (2) : 46-48.

De, K. C., 1910, *Report on The Fisheries of Eastern Bengal and Assam* Shillong : 77p.

Deoras, P. J., Pollution, *Hindusthan Times.* August 20.

Deoras, P. J.,1973, What is pollution? *Hindusthan Standard,* March, 25

Department of Scientific (DSIR), 1961, Some effects of pollution on fish. Notes on water pollution, and Industrial Research No. 13 June. 1-4.

Devassy, V. P. (MS), Seasonal fluctuations in the abundance of plankton in the Ganga river system in the vicinity of Allahabad with special reference to copepods. *Ph. D. thesis submitted to the University of Agra.*

Dhaneswar, R. S., C.S.G. Rao, M. K. Kutty, D. Ray and R. C. Trivedi, 1972, Approach for pollution abatement in lower Damodar Valley. *All India Conference Abatement of Environmental Pollution Proceedings,* Calcutta, December 14-16 Sponsored by CMDA, the Institute of Engineers (India) and CMPO.

Dhirendra Kumar, V.R. Chitranshi, and M. P. Singh 1987, Fisheries of River Ganga around Patna and its conservation Abstract. *Symposium on the impact of current land use pattern and Water Resources Development on Riverine Fisheries.* April, 25-27, Central Inland Capture Fisheries Research Institute, Barrackpore, W. B.

Doudoruff, P. and D. K. Shumway, 1967, Dissolved Oxygen Criteria for protection of Fish. *Am. Fish. Soc.* Spec. Publ. 4 : 13-19.

Dutta, N., J. C. Malhotra, and B.B. Bose, 1954, Hydrology and seasonal fluctuations of the plankton in the Hooghly estuary. *Symp. on marine and freshwater plankton in the Indo-Pacific.* 1-13.

Dutta, P., G. C. Laha, P. M. Mitra, D. K. De, A. Chaudhury, P. K.Pandit, A. R. Chaudhury, R.N. De, B.K. Saha, H. S . Mazumder, N. D. Sarkar, N. C. Mondal, G. P. Bhattacharya, A. K. Namasudra, S. P. Ghosh and A. R. Paul, 1973. Fishery Resources of the Hooghly Matla Estuarine system. *Bull No. 19* December. Central Inland Fisheries Research Institute (ICAR). Barrackpore. West Bengal.

Dutta, S. N., 1973, On the Gill nets of the Hooghly-Matla estuarine system (West Bengal) with special reference to their specifications and operation. *J. Inland Fish Soc.* India vol *V.* December

Edwards, R. W. and V. M. Brown, 1966, Pollution and Fisheries. A Progress report *J. Inst. Water Pollut. Control Land.* 66 : 63-78.

Ellis , M.M., 1986, Erosion silt as a factor in aquatic environment. *Ecology* : 17 : 29-42.

EIFAC, 1978, Report on Zinc and Freshwater Fish European Inland Fisheries Advisory Commission Working. Party on Water Quality Criteria for European Freshwater Fish Technical paper No : 21 FAO, Rome, 1973.

Faruqui, A. J. and R. Sahay, 1979, On the methods of Catching fish in the united provinces and scope of establishing Inland Fisheries in these provinces. *Proc. Nat. Acad Sci. India* (B) *13* : 198-214.

Forstner, U. and G. T. W. Wittmann, 1979, Metal pollution in the Aquatic environment. Springer- Verlag. Berlin. Heidelberg. New York.

Fujiya M., 1977, Pollution indused fish diseases and thier effects. *Freshwater Fisheries Research Laboratory* Hino-Shi, Tokyo, Japan.

Gammon, J. R., 1970, The effect of inorganic sediment on stream biota. U. S. Environ Protection Agency. *Water Pollution Cont. Res. Ser.* 18050. DWC 12/70 1-141.

George, M. G., N. K. Kaushik and S. K. Srivastava, 1965, Bio-assay of a DDT factory waste by fishes. *Proc. Symp. Wat. Pollut. Control* 2 : 57-78p

George, M. G., S. Z., Qasim, and A. Q. Siddiqui 1965, A limnological survey of the river Kali with reference to fish mortality *Envir. Hlth. 8* (4) : 262-269.

Ghosh, A. N., 1976, Observations on the Hilsa fishery of the river Jamuna during the years 1955-62 at Allahabad *J. Zool. Soc. India* *17* (1 &2) : 135-149.

Ghosh, B. B. and A. K. Basu, 1968, Observations on estuarine pollution of the Hooghly by the effluents from a chemical factory complex at Rishra, West Bengal. *Envir. Hlth. 10* (3) : 204-218.

Ghosh, A. N. and T. D. Nagpal, 1968, On the winter breeding of *Hilsa ilisha* (Ham.) in the Ganga river system. *Proc. Indo-Pacific. Fish Coun. 13* (11) : 132-139.

Ghosh, A. N., R. K. Bhattacharjee, and K. V. Rao, 1968, On the identification of sub-populations of *Hilsa ilisha* (Ham.) in the Gangetic system with a note on their distribution. *Proc. Nat. Inst. Sci. India* Part B. *34* (1) : 44-59.

Ghosh, B.B., P. Ray and V. Gopalakrishnan, 1973, Survey and characterization of waste waters discharged into the Hooghly estuary. *J. Inland Fish. Soc. India* 4 Dec. 82-101.

Ghosh, B. B., P. Ray, D. D. Halder and M. M. Bagchi, 1976, Some observations on the pollution of the Hooghly estuary in the vicinity of a Rayon industry with a note onthe toxicity of effluent. *In* : ACMRR/IASO. Expert consultation on Bio-assay with aquatic organisms in relation to pollution problem, Dubrovnik Yugoslavia 16-19 Nov. FAO publication. FI : ACMRR/BE/7, 18 October.

Ghosh, B. B., P. Ray, M.M. Bagchi and D. D. Halder, 1977, Effects on the Biotic condition of the Hooghly estuary due to disposal of effluent from sulphite pulp and paper Industry. *Proc. of the seminar on Industrial wastes.* December 8-9, pp. 167-201, Metropolitan Development Authority and National Environmental Engineering Research Institute, Calcutta.

Ghosh, B. B. and M. M. Bagchi, 1979, Some observations on oil pollution in the Hooghly Estuary with special reference to oil refinery Complex at Haldia. *National seminar on Protection of Marine Environment.* November 13-15. National Institute of Oceanography, Dona Paula, Goa (sponsored by the Department of Science and Technology. Government of India).

Ghosh, B. B., P. Ray, M. M. Bagchi, D. D. Halder and R. K. Banerjee,1979 a, Biotic and abiotic condition of the Hooghly estuary in the vicinity of a Rubber Industrial Complex. *Proc.Symp. Environ Biol* pp. 139-156 Academy of Environmental Biology. India.

Ghosh, B.B., M. M.Bagchi and P. Ray,1979 b, Studies on organic production in the Hooghly estuarypolluted by industrial and domestic wastes. *In* : *National Seminar on Protection of Marine* Environment. November, 13-15. National Institute of Oceanography, Dona Paula. Goa. Sponsored by the Department of Science and Technology. Government of India.

Ghosh, B. B., M. M. Bagchi and S. B. Saha, 1980 a, Characterization of Tannery Wastes and their effect on the Primary Productivity of the Hooghly estuary near Calcutta (India) *3rd. International Congress on Industrial waste water and wastes.* Stockholm, February, 6-8, 1-9.

______, 1980 b, Effect of Industrial pollution by Soda process pulp and paper Mill waste on the Primary Productivity of the Hooghly estuary near Tribeni. *Symposium on Coostal Aquaculture,* Cochin, India, January, 12-18.

Ghosh, B. B., and M. M. Bagchi, 1981, Observations on Environmental pollution caused by zinc-borne wastes disposed from Rayon Industry into the Hooghly estuary. *Proc. Internal. Symp Wat. Resour Conser. Poll & Abatement. Dept. of Civil Engineering.* University of Roorkee. 11-13 December, 131-140.

Ghosh, B.B., M. M. Bagchi and D. K. Die, 1983, Some observations the status of pollution in the Hooghly estuary (West Bengal) with reference to heavy metal disposed through Industrial wastes. *J. Inland Fish Soc. India.* *15* (1-2) : 44-53.

Ghosh, B. B., H. C.Joshi, M.K. Mukhopadhyaya and M. M. Bagchi, 1988, Biomonitoring of hazardons substances in the Hooghly estuary. *In* : Workshop on Industrial and sewage effluents : Environmental Hazrdons and Remedies. The Society for the Advancement of Environmental Science and Department of Science and Technology. Vigyan Bhavan. New Delhi (Feb. 22-24, 1988) Abstract.

Ghosh, B. B., M. K. Mukhopadhyay, and M. M. Bagchi, 1990, States of Organic Pollution in the Mathabhanga Churni River System. West Bengal. *J. Inland Fish, Soc. India* 22 (1 & 2) pp. 16-21.

Ghosh, J. J., 1993, *In* : "GANGA DUSANER BHAYABAHA KAHINI" (The alarming news of the Ganga pollution) by Roop Kumer Basu. Bartaman, 9th November.

Gopalakrishnan, V., R. N. Pal and P. K. Chakraborty, 1966, Observations on breeding of major carps in the Tilaya and Panchet reservoirs. *Bull. Central Inland Fish Res. Inst.* Barrackpore 9 : 17p (Mimeo)

Gopal Krishnan V., P. Ray and B. B. Ghosh, 1970, Problem of estuarine pollution with special reference to the Hooghly estuary. *Proc. of the seminar on Pollution and Human environment.* 313-324 pp. Bhabha Atomic Research Centre. Trombay.

Gopalakrishnan, V., P. Ray and B. B. Ghosh, 1978,Present status of pollution in the Hooghly estuary with special reference to the adverse effects observed on the fishery resources. *Proc. of the symposium on environmental pollution.* pp 1-8. Central Public Health Engineering Research Institute, Nagpur.

Gorman, O. T. and J. R. Karr, 1978, Habital structure and stream communities. *Ecology* **59** : 507- 515.

Government of India, 1949, *Preliminary guide to Indian fish, fisheries, methods of fishing and curing.* Manager of Publications. Delhi, 137 p.

Government of India, 1961, *Marketing of fish in India 3rd ed.* Directorate of Marketing and Inspection, Nagpur, 212 p. (Agricultural Marketing Series, 126).

Government of India, 1966, *Report of the Fish Seed Committee, Ministry of Food and Agriculture.* Department of Agriculture. 211 p. (Mimeo) New Delhi.

Gupta, S. D. and A. G. Jhingran 1974, Aging Labeo Calbasu (Ham) through its scales. *J. Inland Fish. Soc.* India. 5 : 126-128.

Gupta, R. A. and R. K. Tyagi, 1992, Analytical approach to analysis of fish stocks of the Ganga River System. *J. Inland Fish. Soc.* India **24** (2) : 20-27 pp.

Hamilton, Francis, 1922, *An account of the fishes found in the river Ganges and its branches.* Edbinburg : Archibald Constable and Company. 405 pp.

Handa, B. K., Adarsh Kumar, T. N. Sondi and D. K. Goel, 1985, Pollution of surface waters by metals in Uttar Pradesh *IAWPC Tech. Annual* XII : 98-105.

Höglund, L. B., 1975, Avoidance reactions of the cod *(Gadus morhua)* to certain surfactants. *In :* Document FIR/TPLR/75 : 2/19 FAO Rome.

Hora, S. L., 1988, A preliminary note on the spawning grounds and bionomics of the so-called Indian shad, *Hilsa ilisha* (Ham) in the river Ganges. *Rec. India Mas.* **40** (2) pp. 147-148.

Hora, S. L. and Nair, 1940, Further observations on the bionomics and fishery of the Indian shad, *Hilsa ilisha* (Ham.) in Bengal waters. *Rec. Indian Mas.* **42** (1) : 35-50.

Hora, S. L., 1943, Evidence of distribution of fishes regarding rise in salinity of the river Hooghly. *Current Science* **12** *:* 189-90.

Hora, S. L., 1953, Further evidence from fish distribution of the rise in salinity of the river Hooghly. *Current Science* 22 : 89-90.

Hora, S. L. and T. V. R. Pillay, 1953, Further evidence from fish distribution of the rise in salinity of the river Hooghly. *Current Science.* 22 *:* 49-50.

Hornell, J. 1923, The Fishing methods of the Ganges. Mem. *Asiat. Soc. Beng., 8* (3) : 199-237.

Hutchinson, G. E., 1923, *A treatise on Limnology. Vol I Geology, Physics and Chemistry.* Wiley, New York. 1015 p.

ICEŜ, 1974, Report of working group for the International study of the pollution of the North Sea and its effects on Living resources and their exploitation, *International Council for the Exploration of the Seas Cooperative Research Report No. 39,* Denmark. 191 P. *In* : Heavy Metals in water, Sediments and Fish in the River Yamuna—Jhingran, Arun G. and H. C. Joshi *J. Inland Fish Soc. India.* **19** (1), 1987) : 13-23.

Jhingran, V. G., 1957, Some observations on the Hilsa fishery at Buxar (Bihar India) in the years 1952-54. *Indian. J. Fish.* **3** : 197-215.

Jhingran, V. G. and R . D. Chakraborty, 1958, Destruction of major carp fingerlings in a section of river Ganga and its probable adverse effects on fish production. *Indian J. Fish. 5* (2) : 291-99.

Jhingran, V. G. and A. V. Natarajan, 1969, An assessment of fisheries of the DVC reservoirs in relation to stocking. *Proc. of the seminar "The Ecology and Fisheries of fresh water reservoirs.* pp. 429-456. Sponsored by ICAR at CIFRI, Barrackpore.

Jhingran, V. G., 1975, *Fish and Fisheries of India,* (Ist edition), Hindustan Publishing Corporation (India) 6-UB, Jawahar Nagar, Delhi-110007.

Jhingran. A. G. and K. K. Ghosh, 1978, The Fisheries of the Ganga River System in the context of Indian Aquaculture. *Aquaculture,* **14** : 141-162.

Jhingran, A. G., and R. A. Gupta 1987, A Reappaisal of the fisheries ecology of the Ganga river System. Abstract. *Symposium on the impact of current land use pattern and water Resources Development on Riverine Fisheries.* April, 25-27, CICFRI, Barrackpore, West Bengal.

Jhingran, A. G. and H. C. Joshi, 1987, Heavy metals in water, sediments and fish in the river Jamuna. *J. Inland Fish Soc.* India *19* (1) : 13-23.

Jhingran, A. G. 1986, Factors relevant to the management and conservations of Ganga Fishery Resources. *Paper presented at the Indo-US workshop on Biological Conservation and Biomonitoring* held at Jawaharlal Nehru University, New Delhi. 10-14 November.

Jhingran, Arun G., 1989, Impact of environmental perturbations on the fisheries ecology of river Ganga. *A. Synopsis.* CICFRI. Barrackpore. W. B.

Jhingran, V. G., 1959, Studies on the age and growth of Cirrhina mrigala (Ham.) from the river Ganga. *Proc. Nat. Inst. Sci. India.* Part B, **25** : 107-137.

Jhingran, Arun, G., 1988, Aquatic pollution with special reference to the Ganga river system. *In* : *Souvenir, National Symposium on Environmental Impacts on Animals and aquaculture,* Kalyani, May, 14-16.

Jhingran, A. G. and Babulal, 1989, Primary production Energy fixation and potential fishery of the Hooghly-Matla estuarine system. (Abstract). *National seminar on management of fisheries in Inland open water system of India.* Central Inland Capture Fisheries Research Institute, Barrackpore (West Bengal).

Jhingran, A.G. and Babulal, 1989, Impact of environmental perturbations on the fisheries ecology of river Gaga (A Synopsis). CICFRI (ICAR) *Barrackpore.* West Bengal.

Job, T.J., A. David, and M. P. Motwani, 1952, Observations on the fish and fisheries of the Damodar Basin with reference to the multipurpose projects of the valley. *J. Asi. Soc. 18* (2) : 147-172.

Job, T. J., and V. R. Pantulu, 1953, Fish trapping in India *J. Asi. Soc. Beng. 19* (2) : 175-196.

Jones, S. and P.M.G. Menon, 1988, Observations on the life-history of the Indian shad *Hilsa ilisha* (Ham.) *Proc. Indian Acad. Sci. 31* (3) : pp, 101-125.

Josh, H. C., 1985, Pesticide residue monitoring in the Bhagirathi-Hooghly stretch of the Ganga river system. *Proc. National seminar on pollution control and environmental management.* NEERI, Nagpur. March 17-19, 1985, 80-86 pp.

Joshi, H. C., 1986, DDT residues in fish in river Hooghly. *Proc. National symposium on pesticide reside and environmental pollution.* Muzaffarnagar. (October, 2-4, 1985).

Joshi, H. C., 1988, Impact of man-made environmental modifications on the riverine ecology and productivity—A case study of the river Ganga. Lecture delivered at the summer Institute on Resource management and conservation on Inland Capture Fisheries of India. 4-23 July, 1988. CICFR, Barrackpore.

Kamal, M. Y., 1969, Studies on the age and growth of *Cirrhina mrigala* (Ham.) from the river Jamuna at Allahabad. *Proc. Nat. Inst. Sci. India* Part B. *35* (1) : 72-79.

Kamal, M. Y., 1991, Riverine Fisheries in India—A Retrospect. *J. Inland Fish Soc.* India *23* (2) : 1-8.

Karr, J. R. and D. R. Dudley, 1981, Ecological perspective on water quality gools. *Environ Manage.* **5** : 55-68.

Karamchandani S. J. and M. P. Motwani 1955, Early life history, bionomics and breeding of *Rita rita* (Ham.) *J. Zool. Soc. India 7* (2) : 115-126.

Karamchandani, S. J.,and M.P. Motwani, 1956, On the larval development of four species of freshwater catfishes from the river Ganga. *J. Zool. Soc. India.* *8* (1) : 19-34.

Karmakar, H.C., S. K. Mandal, R. N. Pal and G. C. Laha, 1987, Wanton destruction of early juveniles of commercial fish species in the upper Hooghly estuary. (Abstract) *In : Symposium on the impact of current land use pattern and water resources development on riverine fisheries* April 25-27; CICFRI Barrackpore.

Khan, H. A. and V. G. Jhingran 1975, Synopsis of biological data on Rohu, *Labeo rohila* (Ham.) FAO. *Fish synopsis.* No. III, 100p.

Konar., S. K., 1969, Use of control chemicals in freshwater fishery reservoir management. *In* : Aquatic pollution with special reference to the Ganga River System.– Jhingran, Arun G. Lecture delivered at Kalyani University, 1988.

Kurian, C. V. and V.O. Sabastian, 1982, *Prawn and Prawn Fisheries of India.* Hindustan Publishing Corporation, Delhi, 286 P.

Langlois, T. H. 1941, Two processes operating for the reduction in abundance or elimination of fish species form c ertain types of water areas. *Trans, North. Amer. Wild. Conf.* **6** : 189-201.

Majumder, Diptosh, 1989, Rampant corruption in pollution centrol Board. The statesman, August 5.

Malhotra, J. C.,K. K. Ghosh, and Other, 1966, Report on fish spawn prospecting investigations 1965 2. Uttar Pradesh, Bihar and Maharastra. *Bull. Cent. Inl. Fish Res. Inst. 7* pp 76 (Mimeo) Barrackpore.

Mandal, B. K., andP. K. Nayak, 1988, Utilization of running water channels for fish production (Abstract). *National Seminar on Management of Fisheries in Inland open water system of India.* December 14-16, Central Inland Capture Fisheries Research Institute, Inland Fisheries Society of India, Barrackpore.

Mazumder, S. C., 1953, Ganga Barrage and the Bhagirathi-Hooghly Problems. *Indian Journal of Power and River Valley Development* Vol. III, No. 3.

Menon, A.G.K., 1974, A check list of fishes of the Himalayan and the Indo Gangetic plains. *J. Inl. Fish. Soc. India* (special publication No. 1).

MC Dowell, D. M. 1995, The Development of the river Hooghly for navigation *Port of Calcutta, 125 years (1870-1995).* Calcutta Port Trust, October 15, Strand Road, Calcutta.

Mc Neill, R. N., Neimanis, V.P. and Dwyer, L., 1979, Water Quality Source Book, Inland Water Directorate Canada. 89 p. In : Heavy metals in water, sediments and Fish in the river Jamuna. —Arun G. Jhingran and H. C. Joshi *J. Inland Fish. Soc. India 19* (1) : 1987 : 13-23.

Motwani, M. P.,S. M. Banerjee and S. J. Karamchandani, 1956, A case of heavy fish mortality in the river Son at Dehri-on- Son, Bihar, Caused by wastes of the Rohtas Industries Ltd. Dalmianagar. *India J. Fish.* **3** (1) : 186-96.

Motwani, M. P.,V. G. Jhingran and S. J. Karamchandani, 1957, On the breeding of India Shad, *Hilsa ilisha* (Ham) in Freshwaters. Sci. & Calt. pp. 47-48.

Mukherjee M., 1993, Silting of Hooghly may inundate Calcutta. *The Statesman, June 5.*

Muncy, R. J., G. J.Atchinson, R. V.Bulkley, B. W. Menzel,L. G. Pery, and R. C Summerfelt, 1979, *Effects of suspended solids and sediment on reproduction and early life of warm water fishes—A review.* EPA-600/3-79-0. Corvallis, Environmental Research Laboratory Corvallis, 100 p.

Naidu, M. R., 1988, *Report on a survey of the fisheries of Bengal.* Calcutta No. 266 F.

Nandy, A. C., M .M. Bagchi, P. M. Mitra and S. K. Majumder, 1980, The impact of Farakka Barrage on fisheries of the Hooghly estuary. Lecture delivered at the *Summer Institute on Brackishwater Capture and Culture Fisheries,* held at the Central Inland Fisheries Research Institute. (3rd July to 2nd August) Barrackpore.

Nandy, A. C., M. M. Bagchi and S. K. Majumder 1988, Ecological changes in the Hooghly estuary due to water release from Farakka Barrage. Mahasagar—*Bulletin of the National Institute of Oceanography 16* (2) : 209-220 pp.

Natarajan, A.V. and A. G. Jhingran, 1963, On the biology of *Catla catla* (Ham.) from the river Jamuna. *Proc. nat. Inst. Sci.* India (B), 29 (3) : 326-55.

Natarajhan, A. V., B. V. Govind, S. Parameswaran Ayyer, S. K. Sarkar, M. Ramkrishnaya, M. A. Khan, B. Ray, B. K. Banerjee and S. L. Kar, 1968, Recommendations on the development of fisheries of the D. V. C. reservoirs—Konar and Tilaiya CIFRI, Barrackpore, West Bengal.

Natarajan, A. V., J. C. Malhotra and A. G. Jhingran, 1976, Possible effects of riverine spawn collection on the fishery of Ganga river system. Group discussion for laying down a National Policy on Major Carp Seed production. Pune, 28 February 1976, 7, p. (Mimeo).

Natarajan, A. V.and R. K. Saxena, 1978, Role of floating Nurseries in Aquaculture *Lecture delivered at Summer Institute on Inland Aquaculture* held at CIFRI, Barrackpore, June-July.

Natarajan, A. V., 1989, Environmental impact of Ganga Basin Development on Gene-Pool and Fisheries of the Ganga River System. *Canadian Special Publication of Fisheries and Aqualic Sciences,* 106. *In: Proceedings of the International Large River Symposium* (LARS) Edited by Douglas P. Dodge, Canada.

National Environmental Engineering Research Institute, 1978, Water quality survey of the Hooghly estuary (1974-76), Sponsored by CMDA, Calcutta.

Nuttal, P. M. and E. D. Bielby, 1973, The effects of China clay on stream invertebrates *Environ. Pollut.* 5 : 77-86.

Oag, T. M. 1939 Report on the river Hoogly and its headwaters. The commissioners for the port of Calcutta. *In : Observations on the Hydrology of the Hooghly estuary* —Benoy B. Bose (1956) *Indian J. Fish* 3, 101-118.

Pahwa, D. V. andS. N. Mehrotra, 1966, Observations and fluctuations in the abundance of plankton in relation to certain hydrological conditions of river Ganga. *Proc. Nat. Acad. Sci.* India (B) : 36 Part II, 157-189.

Pal, R. N. 1963, Observations on fluctuations in the parasitisation of the Indian shad *Hilsa ilisha* (Ham.) of the *ilisha* Hooghly estuary. *Indian J. Helminth.* 15 (2) September, 119-126.

Paul, N. P., 1969, Disposal of effluents from molasses distillation plant as scope for development of by- product industry on distillery effluents. *First IIT- K Symposium on Industrial wastes. Indian Institute of Technology, Kanpur,* February 14-19.

Pillai, M. K. K. and H. C. Agrawal,1979, Final report on the project DDT residues and its degradation in soil, water and few aquatic animals of Jamuna in Delhi. *Dept. of Sci & Technology,* New Delhi. In : Aquatic pollution with special reference to Ganga River System-Jhingran Arun G. (MS) Lecture delivered at Kalyani University, 1988.

Pillay, T. V. R., 1958, Biology of the *Hilsa ilisha* (Ham.) in the Ganga River System. *Proc. Indo-Pacific Fish Coun.* 13 (11) : 132-139 pp.

Pillay, T. V. R. and K. K. Ghosh 1962, The bagnet fishery of the Hooghly-Matla estuarine system (West Bengal) *Indian J. Fish.* 9 (1) : 71-99.

Pillay, S. R. andH. Rosa (JR.), 1963, Synopsis of biological data on *Hilsa ilisha* (Ham) 1822. *FAO Fisheries Biology synopsis* No. 25.

Pillay, T. V. R., S. R. Pillay andK. K. Ghosh, 1964, A comparative study of the populations of the Hilsa (*Hilsa ilisha* (Ham) in Indian waters. *Proc. Indo-Pacif. Fish Coun.* 10th Session, Sec - 11 : 62-64.

Qasim, S. Z. and Qayyum A, 1962, Spawning frequencies and breeding seasons of some freshwater fishes with special reference to those occurring in the plains of Northern India. *Indian J. Fish* 8 (1) : 24-43.

Rai, L. C. 1978, Ecological studies of algal communities of the Ganga river at Varanasi *Indian J. Fish* . 5 (1) : 1-6

Rai, S. P. and R. C. Singh, 1988, Potential of Pen-Culture in ox-bow lake. (Abstract) *National Seminar on Management of Fisheries in Inland Open Water System of India.* 14-16, December, CICFRI, Barrackpore.

Raja Antony., B. T. 1985, A review of the Biology and fisheries of *Hilsa ilisha* in the upper Bay of Bengal. *In : Marine Fishery Resource Management in the Bay of Bengal Colombo, Srilanka, October BOBP/WP/37 (Ras/81/051) United Nations Development Programme, FAO of the United Nations.*

Rajyalakshami, T. 1980, Comparative study of the biology of the freshwater prawn *Macrobrachium malcolmsonii* of Godavari and Hooghly river systems. *Proc. Indian Nat. Sci. Acad.* B. 45 (1) : 77-89.

Rao, K. L. 1975, *Indian's water wealth, its assessment, uses and projection.* Orient Langman, New Delhi, pp. 39-70.

Ravishchandra, 1952, Preliminary account of the distribution and abundance of fish larval in the Hooghly estuary. *Indian J. Fish.* 9 Sec. A. pp. 48-70.

Ravishchandra and P. K. Mathur, 1979, Observations on the spawning of *Hilsa ilisha* (Ham.) in the middle stretch of the Ganga river system. *Proc. Indian Sci. Congr. 66* (3) P. 84.

Ravishchandra,V. R. Desai and S. K. Das 1984, Observations on the Bathymetric distribution of hilsa larvae in the middle stretch of river Ganga near Allahabad. *J. Bombay Nat. Hist. Soc.* 80 : 2, pp. 427-429.

Ravishchandra, R. K. Saxena and R. K. Tyagi 1990, *Hilsa ilisha* (Ham.) of Ganga its glory and down fall— A Retrospect. *Recent trends in Limnology,* 365-378 pp.

Ray, P. , Some considerations on the control and abatement of distillery wastes with special reference to aquaculture. (Unpublished).

Ray, P., 1961 a, Evaluation of toxicity of some Industrial effluents to fish by bio-assays. *Indian J. Fish. 8 :* 233-40.

——, 1961 b, Relative importance of Bio-Chemical oxygen demand (BOD) and chemical oxygen demand (COD) in stream Pollution studies. *Sci & Cult.* 27 (6).

Ray, P. and A. David, 1962, A case of fish mortality caused by precipitation of ferric iron in the river Daha at Siwan. (North Bihar) *India J. Fish* **9** : 117-22.

Ray, P.,S. J. Karam Chandani, and R. D. Chakraborty, 1964, A not on the breeding of *L. gonius* (Ham.) at Jhansi (Uttar Pradesh) in 1954. *J. Bombay Nat. Hist. Soc.* **61** (1) : 195-200.

Ray, P., S. B. Singh snd K. L. Sehgal, 1966, A study of some aspects of ecology of the rivers Ganga and Jamuna at Allahabad (Uttar Pradesh) in 1958-59. *Proc. Nat. Acad. Sci.* India. See. B.vol *XXXVI,* Part III, 1966, pp. 235-272.

Ray., P. and A. David, 1966, Effects of Industrial wastes and sewage upon the Chemical and biological composition and fisheries of the river Ganga at Kanpur (Uttar pradesh) *Envir. Hlth. 8* : 307-39 *Hlth.*

Ray, P. and B. B. Ghosh, 1976, Some ecological considerations of the Hooghly estuary polluted by domestic and industrial wastes under tropical environment. ACMRP/IABO, Expert Consulatations on Bio-assays with aqualic organisms in relation to pollution problems. Dubrovnik Yugoslavia, November 16-19. FAIO publication. FI : ACMRR/BE/8, 29th September.

Ray, P.,B ., B. Ghosh and M. M. Bagchi, 1977, The effects of variations in physico-chemical parameters on plankton bio-mass around a distillery out fall in the Hooghly estuary. *Proc. of the Seminar on Industrial wastes.* pp. 141-153. Dec. 8-9, Convened by National Environmental Engineering Research Institute and Calcutta Metropolitan Development Authority.

Ray, P. B., B. Ghosh, and M. M. Bagchi, 1979, Effects of Pulp and Paper mill waste (Soda Process) around the outfall in the Hooghly estuary. *Proc. Symp. Environ. Biol.* pp. 453-464. Academy of Environmental Biology, India.

Ray, P. and E. Mitra, 1980, Effects of suspensoids on biotic life around the outfall area of a Sulphite Pulp and Paper Mill waste in the Hooghly estuary. *Symposium on Coastal aquaculture. Cochin.* India. January, 12-18.

Ray, P., 1980, A pollution survey of the Hooghly estuary as influenced by industrial and domestic wastes under tropical environment. *3rd International Congress on industrial waste water and eastes.* Stockholm, February, 6-8, 1-11.

Ray, P. and G. Laha, 1981, Effects of Industrial and Municipal wastes on the trend of Fisheries in the Hooghly Estuary. *International Symposium Wates Resources Conservations, Pollution and Abatement.* December, 11-13, 1981 Roorkee, India.

Ray, P., 1981, *Impact of man on the Hooghly estuary. In* : Water Pollution and Management Reviews. South Asian Publication Pvt. Ltd. New Delhi, Madras.

Ray, P., and S. Bandopadhyay, 1984, Utilization of Municipal wastes of Calcutta Metropolis in Freshwater bheries for aquaculture programmes and problems associated with its recycling. *Final Report 1982-84* ICAR Research Project, CIFRI Barrackpore, (West Bengal).

Ray, P., and A. G. Jhingran, 1987, Impact of irrigation projects on fishery in the Ganga River System. Paper read at the *"Symposium on "Impact of Current Land use Pattern and Water Resources Development on Riverine Fisheries* April 25-27, CICFRI, Barrackpore,West Bengal.

Ray, P., 1993, *Aquaculture in Sundarban Delta—its perspective* (An assessment) International Books & periodicals supply service Delhi-110034.

Roy, H. K., 1949, Some potamological aspects of the river Hooghly in relation to Calcutta water supply. *Sci. and Cult.* **14** (8): 318-324.

_____, 1955, Plankton ecology of the river Hooghly at Palta, West Bengal. *Ecology.* **36** (2) : 169-175.

Saha, S. B., B. B. Ghosh, and V. Gopalakrishnan, 1975, Plankton of the Hooghly Estuary with special references to Salinity and temperature. *J. Mar. Biol. Ass. India* **17** (1) : 107-120.

Saha, S. B., 1980, Studies on the Biotic characteristics in and around the outfall areas of paper mill at Tribeni, Hazinagar and Titagarh, West Bengal. *Symp. Coast, Aquacult mar. Biol. Ass.* India, Cochin 12-18, January.

Sanyal, Tapobrata, and Arun Kumar Chakraborty, 1995, Farakka Barrage Project : Promises and Achievements *Port of Calcutta, 125 years (1870—1995)* Calcutta Port Trust, October 15, Strand Road, Calcutta.

Sarkar, Birendra Nath, 1988, "GANGADHARA LUPTA HOTE CHOLECHE" (The Ganga flow is on the way of extinct) The "Aajkal" 6th November, Calcutta.

Saxena, R. K., 1966, The fishing nets and traps in a section of the middle reaches of Ganga river system of India. *Proc. Indo-Pacific Fish Coun. 11* (II), 250-77.

Saxena, K. L., R. N. Chakraborty, A. Q. Khan, S. N. Chattopadhyaya, and H. Chandra, 1966, Pollution studies of the river Ganges near Kanpur. Envir. Hlth. **8** (4) : 270-85.

Saxena, R. K. and Ravishchandra, 1968, On the introduction of phasla jal, a gill net for catching hilsa in the Ganga and Jamuna, near Allahabad *J. Bombay Nat. Hist. Soc. 66 :* 2 pp 496-498.

Saxena R. K., 1991, Drift in status of riverine gears—A case study of middle stretch of Ganga River System. *National Workshop on low energy fishing at CIFT* Cochin, 8-9 August.

Schlosser, I. J., 1982, Tropic structure, reproductive success and growth rate of fishes in a natural and modified head water stream. *Can. J. Fish. Aqut. Sci. 39* : 968-978.

Scheweiger, G., 1961, The toxic action of heavy metal salts of fish and organisers on which fish feed. *Water Pollution Abs.* 34.9, 1744.

Sen, P. R., D. K. De and D. Nath, 1990, Experiments on artificial propagation of Hilsa, *Tonualosa ilisha* (Hamilton). *Indian J. Fish,* 37 (2) : 159-162.

Seth, G. K. and T. R. Bhaskaran, 1950, Effects of Industrial Wastes disposal on the sanitary condition of the Hooghly river in and around Calcutta *Indian J. Med. Res. 38* (4) : 341.

Sharma, Kallegala, 1994, Rejuvenating Ganga *Science Reporter,* February.

Sharma, K. D., N. Lal and P. D. Pathak, 1981, Water quality of sewage drains entering Jamuna at Agra *Indian J. Environ. Hlth.* 23 (2) : 118-122. In : Heavy metals in water, sediments and Fish in the river Yamuna,—Jhingran, Arun G. and H. C. Joshi. *J. Inland Fish Soc. India* 19 (1), 1987 13-23.

Sheldon A. L., 1968, Species diversity and Longitudinal succession in stream fishes. *Ecology 50* : 558-572.

Shetty, H.P. C., S. B. Saha, and B. B. Ghosh, 1961, Observations on the distribution and fluctuations of plankton in the Hooghly Matla estuarine system, with notes on their relation to commercial fish landing. *Indian J. Fish 8* (2) : 326-363.

Shetty, H. P. C., 1967, Report on fish spawn prospecting investigations 1966 3. Bihar, Uttar Pradesh and Punjab. *Bull. Cont. Inland Fish Res. Inst. Barrackpore* 11. p. 80.

Shetty, H. P. C., and J. C. Malhotra, 1983, A report on the survey of North Bihar in relation to effects of Gandak and Kosi river valley Projects on the fisheries of the area. Survey Report No. 7 Central Inland Fisheries Research Institute (ICAR) Barrackpore, West Bengal, India.

Singer, P. C., 1977, Influence of dissolved organies on the distribution transport and fate of heavy metals in aquatic systems. *In : Fate of pollutants in the Air and Water Environments Part I.* Suffect I. H (ed). New York pp. 155-182.

Singh, R. L. (ed), 1965 b, *India a Regional Geography.* National Geographical Society of India. Varanasi-5

Singh, Balbir, R. K. Tyagi, and M.A. Khan, 1987, Management and conservations of riverine major carp fishery around Allahabad. (Abstract) Symposium on the impact of current land use pattern and water resources development on riverine fisheries. April, 25-27, CICFRI. Barrackpore, West Bengal.

Singh, S. C., 1965 b, Changes in the courses of rivers and thier effects on Urban settlements in the middle Ganga Plain Ph. D. Thesis, Deptt. of Geography, B. H. U., Varanasi.

Smith, P. W., 1971, Illinois streams a classification, based on their fishes and an analysis of factors responsible for disappearance of native species. *Ill Nat. Hist. Surv. Biol.* Notes *76* pp. 14.

Sprague, J. B., 1964 a, Avoidance of copper-Zinc solutions by young salmon in the Laboratory. *J. Water Pollut. Control Fed. 36* : 900-1004.

Sprague, J. B., 1968, Avoidance reactions of Rainbow tront to zinc Sulphate Solutions. *Water Research.* Pergamon Press. Vol. 2. pp. 367-373 (Great Britain).

Sprague, J. B. and D.E. Drury, 1969, Avoidance reaction of Salmonid fish to representative pollutants. Reprinted from Advances in water pollution Research. *In* : Proceedings of the 4th International Congress held in Prague.

Starrett., W. C., 1950, Food relationship of the minnows of the Des Moines river Iowa, *Ecology* 31 : 216-233.

Stones, T., 1976, A discourse on biochemical Oxidation of sewage, *Effluent water treatment Journal 16 :* 352-355.

Stott, B., 1975, Developments towards a new method for investigating avoidance behavior in fish. Fourth FAO/SIDA training course on Aquatic Pollution in relation to prolecting of living resources. FIR/TPLR/75: 2/26. FAO. Rome.

Sundersan B. B., P. V. R. Subrahmanyan, and A. D. Bhide 1983, An over view of toxic and hazardous waste in India. *UNEP. Industry and Environment (Spl. issue)* pp. 70-73.

Sujansingani, K. H., 1957, Growth of the Indian shad *Hilsa ilisha* in the tidal stretch of the Hooghly. *Indian J. Fish. 4* (2) : 315-335.

The Statesman, 1988, Preper dredging could prevent floods. June 13. Calcutta.

The Statesman, 1990, Furnace oil in Damodar *A report,* April, 12

The Statesman, 1992, From fly ash to floating cash. *A report.* November 3.

The Statesman, 1992 a, Ganga work hit by Scientific flows. November 4.

Tripathi, Y. R., 1976 , Qualitative and Quantitative estimate, of spawn collection from different centres of Ganga River System. Group discussion for laying down a national policy on major carp seed production. Pune. 28, February, 1976 pp 6 (Mimeo).

Vats, Virendra, 1944, In : Rejuvenating Ganga. *Science Reporter,* February.

Venugopalan, V. K., 1979, Effect of pesticides and Heavy metals on growth and primary production of a few species of marine algae. *Expanded Abstract of Marine Environment and Related Ecosystem.* National Institute of Oceanography Goa. November 13-15.

Von Brandt, A., 1958, *Modern fishing gear of the world.* 274-296 Fishing news (Book) London.

___, (1964), *Fish Catching method of the world* Fishing news (Book), London.

Water Quality Criteria, 1963, *State Water Pollution Control Board.* Publ. 3-A. U. S. A.

Wabber, J., 1972, Effects of Toxic metals in Sewage on Crops Agricultural Development and Advisary Service. Leeds 404, Wat Pollut. Control.

World Health Organisation, 1976, Report on Pesticide residues in food. *Tech. Rep. Ser.* 592, 28.

Author Index

Agrawal, HC., 137, 140.

Allen, K.R., 156.

Atchinson, G.J., 157.

Ayyer, S. Parameshwaran, 147.

Babu Lal, 59, 60, 70, 100, 119.

Bagchi, M. M., 2, 36, 52, 55, 58, 59, 60, 62, 64, 65, 66, 67, 68, 69, 70, 73, 76, 77, 79, 80, 91, 92, 96, 116, 117, 119, 154, 165, 168, 181, 183, 184.

Balon, E.K., 157.

Bandopadhyay., S., 157.

Banerjee, B.N., 15.

Banerjee, B.K., 147.

Banerjee, M.K., 15.

Banerjee, R.K., 36, 52, 59, 60, 69, 149, 152, 153, 154, 189.

Banerjee, S. M., 35, 145, 147.

Basu, A.K., 52, 55, 69, 70, 147, 152.

Bera, S., 13, 15, 24, 77, 147, 182.

Berkman, E. Hilary, 156, 157.

Bhanot, K. K., 128, 130.

Bhargava, D.S., 43, 50, 54, 69, 70, 74, 75, 76, 77, 170, 171, 173, 189, 196.

Bhaskaran, T.R., 35, 69, 145, 188.

Bhattacharya, G. P., 36, 116, 135, 183, 215.

Bhattacharya, Kapil, 11, 174, 180, 182.

Bhattacharyya, R., 25, 33.

Bhattacharya, R.K., 109.

Bhattacharya, S.K., 14, 15, 188.

Bhinde, A.D., 36, 41, 54.

Bhawmik, M.L., 132, 134.

Bielby, E. D., 157.

Bilgrami, K.S., 45, 50, 54, 73, 74, 75, 112, 115, 128, 196.

Bose, B.B., 69, 73, 74, 76, 77, 78, 79, 95, 180.

Bose, N.K., 182.

Brown, V. M., 161.

Brungs, W. A., 161.

Brusvan, M. A., 157.

Bulkley, R. V., 157.

Calamari, D., 169.

Chakraborti, A.K., 185.

Chakraborty, S., 189.

Chakrabarty, P.K., 36, 147.

Chakrabarty, R. D., 15, 70, 76, 78, 85, 119, 137, 138, 147, 157.

Chakrabarty, R. N., 35, 36, 52, 53, 145, 188.

Chandra, H., 35, 52, 53, 54, 55.

Chandra, K., 52, 53, 69, 127.

Chaphekar, S. B., 11, 34, 54, 69, 112, 178.

Chattopadhaya, Enakshi, 188.

Chatterjee, S.M., 14, 15, 188.

Chatterjee, S. P., 14.

Chattopadhyaya, S. N., 35, 52, 53.

Chaudhury, A., 116, 135, 183, 215.

Chaudhury, A.R., 116, 135, 183, 215.

Chitranshi, V. R., 112.

Chutter, F. M., 157.

Cordone, A. J., 156, 157.

Dakshni, K. M. M., 141.

Das, S.K., 127.

Das Gupta, S.P., 6, 9, 10, 18, 22, 23, 28, 30, 42, 44, 45, 46, 47, 49, 50, 54, 70, 73, 74, 75, 76, 77, 79, 178, 181, 195, 196.

David, A., 35, 36, 52, 53, 58, 77, 78, 116, 120, 124, 131, 132, 146, 147, 158, 181, 184.

Day, F., 137, 176.

De, D. K., 52, 62, 64, 65, 66, 67, 68, 70, 80, 116, 120, 127, 129, 130, 131, 135, 147, 165, 168, 183, 215, 222.

De, K. C., 135.

De, R. N., 116, 135, 215.

De, Sarkar, 25, 33.

Desai, V.R., 127.

Devassy, V. P., 69, 87.

Deoras, P. J., 29, 173.

Dhaneshwar, R.S., 36, 52, 112, 147, 153.
Dhirendra Kumar, 112.
Doudoroff, P., 162, 190, 196.
Drury, D. E., 161.
Dudley, D. R., 156.
Dutta Munshi, J. S., 45, 50, 52, 54, 73, 74, 75, 112, 115, 128, 196.
Dutta, N., 69, 76, 95, 136, 180, 183, 215.
Dutta, P., 116, 135, 183, 215.
Dutta, S. N., 135, 136, 217, 218, 222.
Dwyer, L., 142, 171.
Edwards, R. W., 161.
Ellis, M. M., 1156, 190.
Faruqui, A. J., 135, 202, 204, 205.
Forstner, U., 159.
Fujiya, M., 163, 165, 166, 167, 170, 189, 196.
Gammon, J.R., 157.
George, M. G., 35, 137, 140, 145.
Ghosh, A. N., 109, 127.
Ghosh, B. B., 34, 52, 55, 57, 58, 59, 60, 62, 64, 65, 66, 67, 68, 69, 70, 75, 77, 80, 95, 99, 154, 157, 165, 168.
Ghosh, J. J., 41.
Ghosh, K. K., 71, 74, 85, 102, 104, 108, 109, 126, 127, 131, 135, 143, 158, 175, 176, 177.
Ghosh, S. B., 95, 116, 135, 183, 215.
Goel, D. K., 141.
Gopalakrishnan, V.,34, 36, 69, 70, 95, 147.
Gorman, O. T., 157.
Govind, B. V., 147.
Gupta, S. D., 190.
Gupta, R. A., 110, 112.
Haldar, D. D., 52, 59, 69.
Hamilton, Fransis, 116, 181, 184.
Handa, B. K., 141.
Höglund, L. B., 193.
Hora, S. L., 116, 127, 128, 181, 184.
Hornel, J., 202, 203, 204, 205, 207.
Jhingran, A. G., 10, 26, 35, 71, 85, 100, 102, 104, 105, 108, 109, 110, 112, 119, 120, 127, 131, 137, 140, 143, 158, 165, 175, 176, 177, 190.
Jhingran Arun G., 106, 107, 115, 140, 141, 142, 168, 175.
Jhingran, V. G., 119, 126, 127, 129, 147, 157, 158, 190.
Job, T. J., 135, 147.
Jones, S., 127, 128, 129.
Joshi, H. C., 35, 52, 62, 63, 68, 69, 70, 79, 80, 81, 137, 140, 141, 142, 165.
Kamal, M.Y., 120, 159, 190.
Kar, S. L., 147.
Karmakar, H.C., 120.
Karmachandani, S. J., 35, 124, 127, 129, 147, 157.
Karr, J. R., 156, 157.
Kaushik, N. K., 35, 137, 140.
Kelley, D. W., 156, 157.
Khan, A. Q., 35, 52, 53.
Khan, M. A., 110, 147, 190.
Konar, S. K., 168, 169.
Kumar, Adarsh, 141.
Kutty, M. K. A., 36, 52, 147.
Kurian, C. V., 124.
Laha, G. C., 36, 48, 69, 116, 117, 118, 120, 135, 183, 215.
Lal, N., 141.
Langlois, T. H., 157.
Majumder, Diptosh 173.
Majumder, H. S., 116, 135, 183, 215.
Majumder, S. K., 58, 69, 70, 73, 74, 76, 77, 79, 91, 92, 96, 116, 117, 119, 181, 183, 184.
Malhotra, J. C., 69, 76, 95, 120, 126, 131, 180.
Mandal, B. K., 132.
Mandal, N. C., 116, 135, 183, 215.
Mandal, S. K., 120.
Mathur, P. K., 127.
Mc Dowell, 182.
Mazumdar, S. C., 179.
Mc Neill, R. L., 142, 171, 181.
Mehrotra, S. N., 52, 53, 69, 70, 76, 78, 83, 84, 85, 87, 88, 89, 100, 101, 158, 175, 176.
Menon, A. G. K., 102, 128.
Menon, P. M. G., 127, 128, 129.
Menzal, B. W., 157.
Mhatre, G. N., 11, 34, 54, 69, 112, 178.
Mitra, E., 52, 55, 56.
Mitra, P. M., 74, 116, 135, 183, 215.
Motwani, M. P., 35, 124, 127, 129, 145, 147.
Mukherjee, G., 52.
Mukherjee, M., 13, 14, 155, 156.
Mukhopadhyaya, M. K., 36, 154.
Muncy, R. J., 157.
Nagpal, T. D., 127.
Naidu, M. R. 135.
Naidu, K. K., 127, 128.
Namasudra, A. K., 116, 135, 183, 215.
Nandy, A. C., 69, 70, 73, 74, 76, 77, 79, 91, 92, 116, 117, 119, 181, 182, 183, 184.

Natarajan, A. V., 102, 103, 120, 124, 143, 144, 147, 158, 177, 179, 190.
Nath, D., 127.
Nattal, P. M., 157.
Nayak, P. K., 132.
Neimonis, V. P., 142, 171.
Oag, T. M., 179, 180.
Pahwa, D. V., 69, 70, 76, 78, 83, 84, 85, 87, 88, 89, 100, 101, 175, 176.
Pakrasi, B. B., 189.
Pal, R. N., 36, 69, 120, 147.
Pandit, P. K., 116, 135, 183.
Pantulu, V. R., 135, 207, 215.
Pathak, P. D., 147.
Paul, A. R., 116, 135, 161, 183, 215.
Paul, N. P., 188.
Perry, L. G., 157.
Pillai, M. K. K., 137, 140.
Pilley, S. R., 109, 128.
Pilley, T. V. R., 109, 127, 131, 135.
Prather, K. V., 157.
Qasim, S. Z., 35, 124, 145.
Qayyum, A., 124.
Rabeni, F. C., 156, 157.
Rai, L. C., 52, 54.
Rai, S. P., 132.
Rajyalakshmi, T., 124.
Ramakrishnaiya, M., 147.
Rao, C. S. G., 36, 147, 153.
Rao, K. L., 17, 18, 22.
Rao, K. V., 109.
Ravishchandra, 112, 113, 127, 130, 131, 135, 177.
Ray, B., 153, 147.
Ray, D., 36, 52, 147.
Ray, P., 13, 16, 25, 26, 34, 35, 36, 37, 39, 48, 51, 52, 53, 55, 56, 57, 58, 59, 60, 61, 69, 70.
Rosa, H. (Jr.), 72, 73, 74, 75, 76, 78, 85, 86, 89, 101, 116, 117, 118, 137, 138, 146, 147, 157, 162, 164, 171, 175, 183, 188, 189, 190.
Roy, H. K., 69, 73, 78, 79, 90.
Saha, B. K. 116, 135, 183, 215.
Saha, S. B., 66, 69, 77, 95, 96, 99.
Sabestian, V. O. ,124.
Sahay, R., 135, 202, 204, 205.
Sanyal, T., 188.
Sarkar, Birendranath, 9, 11, 13, 155, 174, 175,
Sarkar, N. D., 116, 135, 183, 215.
Sarkar, S. K., 147.
Saxena, K. L., 35, 52, 53.
Saxena, R. K., 112, 113, 127, 130, 135, 159, 177, 190, 198.
Schlosser, I. J., 157.
Schweiger, G., 163.
Sehgal, B. N., 70, 130, 131.
Sehgal, K. L., 69, 70, 74, 76, 78, 85, 89, 137, 138, 175.
Sen, P. N., 127, 129.
Seth, G. K., 69.
Sharma Kollegala, 194.
Sharma, K. D., 141.
Sheldon, A. L., 156.
Shetty, H. P. C., 69, 77, 95, 96, 99, 126, 131.
Shumway, D. K., 190, 196.
Siddiqui, A. Q., 35, 145.
Singh, Balbir, 110.
Singh, M. P., 112.
Singer, P. C., 164.
Singh, R. C., 132.
Singh, R. L., 19, 20, 21, 31.
Singh, S. C., 34.
Smith, P. W., 156, 157.
Singh, S. B., 52, 53, 66, 69, 70, 76, 78, 85, 89, 127, 137, 138, 175.
Sprague, J. B., 161, 193.
Sondi, T. N., 141.
Soni, J. K., 141.
Srivastava, G. N., 52, 53, 69, 127.
Srivastava, S. K., 35, 137, 140.
Starrett, W. C., 157.
Stones, T., 163.
Stott, B., 193.
Subrahmanyan, P. V. R., 36, 41, 54.
Sujansingani, K. H., 131.
Summerfelt, R. C., 157.
Sunderesan, B. B., 36, 41, 54.
Tripathi, Y. R., 127, 157, 158.
Tribedi, R. C., 35, 36, 52, 153, 145, 147.
Tyagi, R. K., 110, 112, 113, 127, 130, 177.
Vats Virendra, 196.
Vennugopalan, V. K., 162.
Von Brandh, A., 222.
Webber, J., 128.
Wittmann, G. T. W., 159.